AF894572

MAGNETIC FIELDS OF GALAXIES

ASTROPHYSICS AND SPACE SCIENCE LIBRARY

A SERIES OF BOOKS ON THE RECENT DEVELOPMENTS
OF SPACE SCIENCE AND OF GENERAL GEOPHYSICS AND ASTROPHYSICS
PUBLISHED IN CONNECTION WITH THE JOURNAL
SPACE SCIENCE REVIEWS

Editorial Board

R. L. F. BOYD, *University College, London, England*
W. B. BURTON, *Sterrewacht, Leiden, The Netherlands*
L. GOLDBERG, *Kitt Peak National Observatory, Tucson, Ariz., U.S.A.*
C. DE JAGER, *University of Utrecht, The Netherlands*
J. KLECZEK, *Czechoslovak Academy of Sciences, Ondřejov, Czechoslovakia*
Z. KOPAL, *University of Manchester, England*
R. LÜST, *European Space Agency, Paris, France*
L. I. SEDOV, *Academy of Sciences of the U.S.S.R., Moscow, U.S.S.R.*
Z. ŠVESTKA, *Laboratory for Space Research, Utrecht, The Netherlands*

VOLUME 133

A. A. RUZMAIKIN, A. M. SHUKUROV, AND D. D. SOKOLOFF

Space Research Institute, Academy of Sciences,
Moscow, U.S.S.R.

MAGNETIC FIELDS OF GALAXIES

Kluwer Academic Publishers

Dordrecht / Boston / London

Library of Congress Cataloging-in-Publication Data

CIP

Ruzmaĭkin, A. A. (Aleksandr Andreevich)
Magnetic fields of galaxies.

(Astrophysics and space science library; v. 133)
Bibliography: p.
Includes index.
1. Galaxies — Magnetic fields. I. Shukurov, A. M., 1949— II. Sokoloff, D. D. III. Title. IV. Series.
QB857.5.M34R89 1987 523.1′12 87—36917

ISBN-13: 978-94-010-7776-7 e-ISBN-13: 978-94-009-2835-0
DOI: 10.1007/978-94-009-2835-0

Published by Kluwer Academic Publishers,
P.O. Box 17, 3300 AA Dordrecht, The Netherlands.

Kluwer Academic Publishers incorporates the publishing programmes of
D. Reidel, Martinus Nijhoff, Dr. W. Junk and MTP Press

Sold and distributed in the U.S.A. and Canada
by Kluwer Academic Publishers,
101 Philip Drive, Norwell, MA 02061, U.S.A.

In all other countries, sold and distributed
by Kluwer Academic Publishers Group,
P.O. Box 322, 3300 AH Dordrecht, The Netherlands

All rights reserved.
© 1988 by Kluwer Academic Publishers
Softcover reprint of the hardcover 1st edition 1988

No part of the material protected by this copyright notice may be reproduced or
utilized in any form or by any means, electronic or mechanical
including photocopying, recording or by any information storage and
retrieval system, without written permission from the copyright owner.

TABLE OF CONTENTS

PREFACE

Magnetism, when extended beyond normal frameworks into cosmic space is characterized by an enormous spatial scale. Because of their large sizes the nature of magnets such as the Earth and the Sun is entirely different from the nature of a horseshoe magnet. The source of cosmic magnetism is associated with the hydrodynamic motions of a highly conductive medium. In this aspect, cosmic magnets resemble a dynamo. However, currents in the dynamo flow along properly ordered wires, while chaotic, turbulent motions are dominant inside stars and liquid planetary cores. This makes more intriguing and surprising the fact that these motions maintain a regular magnetic field.

Maintenance of magnetic fields is even more impressive in huge magnets, i.e. galaxies. In fact, we are living inside a giant dynamo machine, the Milky Way galaxy. Although the idea of the global magnetic field of our Galaxy was clearly proposed almost 40 years ago, firm observational evidence and definite theoretical concepts of galactic magnetism have been developed only in the last decade. This book is the first attempt at a full and consistent presentation of this problem. We discuss both theoretical views on the origin of galactic magnetism and the methods of observational study. Previous discussions were on the level of review articles or separate chapters in monographs devoted to cosmic magnetic fields (see, e.g., H. K. Moffatt, 1978, E. N. Parker, 1979 and Zeldovich *et al.*, 1983).

Some readers will expect to see a detailed discussion of the popular problems of quasars and radiogalaxies. However, one can only touch upon them, asking some questions concerning the origin and maintenance of magnetic fields in these objects. In this book we would like to unveil the new depth, fascination and novelty which have been discovered in those places where they were hardly expected: well known spiral galaxies such as the Andromeda Nebula and the Whirlpool galaxy. Indeed, recently discovered and studied global structures of magnetic fields in spiral galaxies which resemble rings and two-armed spirals are very impressive.

Our subject is not "*monumentum aere perennius*". The field of galactic magnetism is very young and is developing rapidly. Many questions are far from being answered, many ideas are disputed. However, the information already available deserves to be discussed in a monograph rather than a review paper.

The authors do not intend to hide their theoretical background. On the one hand, the principal concept of galactic magnetism as a product of a hydromagnetic dynamo belongs to theoreticians. On the other hand, after personal encounter with observational data processing and after numerous discussions with those who see a radiotelescope every morning we recognize and appreciate even more the importance of observational efforts and hope that this feeling is reflected in this book.

We would like to thank those who have helped us, directly or indirectly. Among these are Yulia Baryshnikova, Rainer Beck, V. L. Ginzburg, S. A. Molchanov, R. Wielebinski, and Ya. B. Zeldovich. Special thanks are due to Rita Bujakaitė, Natalia Furdina, Natalia Gula, Nina A. Sokolova and Karina Ter-Saakova for their help in the preparation of the manuscript.

CHAPTER I

INTRODUCTION

Magnetic fields in galaxies have a modest strength, which is conveniently measured in microgauss. Galactic magnetic fields are notable for another property: a huge spatial scale. This is expressed in kiloparsecs. These magnets, having an enormous scale but still much neglected, are the subject of our book.

Six out of every ten observed galaxies are spirals and a major proportion of recent observational and theoretical results in the investigation of galactic magnetic fields have been obtained for spiral galaxies. It is therefore not accidental that our attention mainly concerns spirals and primarily the Galaxy, our birthplace.

The first views on galactic magnetic fields began being assembled shortly after the Second World War. In those years, thorough investigation of cosmic rays began at the same time as the construction of accelerators for elementary particle research. In those years the radiotelescope became the peacetime brother of radar.

It had been noted even before the War (see, e.g., Alfvén, 1937) that electromagnetic fields within stars, near stars and in interstellar space could have strength and scale sufficient to accelerate charged particles up to energies observed in cosmic ray experiments. In 1949, E. Teller, R. Richtmeier and H. Alfvén proposed and advocated the idea that cosmic rays have a solar origin and are confined near the Sun by an interplanetary magnetic field. Directions of motions of primary cosmic rays that are presented by protons, electrons, α-particles and a small quantity of heavy nuclei, are distributed very isotropically. The radius $r(\text{cm}) \sim \varepsilon/300\,ZH$ of trajectory of a relativistic particle with the charge Ze and energy $\varepsilon(\text{eV})$ in the magnetic field $H(\text{G})$ should not exceed the scale of the planetary system. Richtmeier and Teller (1949) therefore concluded that the magnetic field $H \simeq 10^{-6}\,\text{G}$ is required for confinement of particles with $\varepsilon \simeq 10^{14}\,\text{eV}$ in the region of the size $r \simeq 3 \times 10^{17}\,\text{cm}$. Such a field would keep particles with the energy $10^{17}\,\text{eV}$ in a region comparable to the thickness of the Galactic disc, $3 \times 10^{20}\,\text{cm}$. The magnetic field can be amplified from some initial weak field by motions of a highly conductive interstellar medium. If the magnetic field is amplified until it reaches the energy equipartition with motions, $H^2/8\pi = \rho v^2/2$, it would have the strength $H \simeq 3 \times 10^{-6}\,\text{G}$ for the interstellar gas density $\rho = 10^{-24}\,\text{g cm}^{-3}$ and typical gas velocity $v = 10\,\text{km s}^{-1}$ (Alfvén, 1949).

E. Fermi (1949) did not agree with the hypothesis of the local, solar origin of cosmic rays and suggested that they fill at least the whole Galaxy. This suggestion naturally implies the existence of magnetic fields at still larger scales. Fermi noted that vast expanses of interstellar space and the high conductivity of interstellar gas imply a remarkable stability of Galactic magnetic fields. On the other hand, moving inhomogeneities of the magnetic field can effectively accelerate the particles.

In these early debates, true pictures were formed of the strength and scale of the Galactic magnetic field.

Impressive evidence of the interstellar magnetic field then came from the discovery of the polarization of starlight (Hall and Mikesell, 1949; Hiltner, 1949a, b). Electric vectors of the light coming from distant stars are orientated closely to the Galactic plane. The fairly coherent distribution of polarization planes suggests a universal origin of the orientation of dust particles which scatter and absorb the starlight (for instance, the Davis-Greenstein mechanism).

Astronomers have noticed imprints of magnetic fields in the morphological peculiarities of gas distribution in the Galaxy. The founder of the Crimean Astrophysical Observatory, G. Shajn suggested in 1955 that elongation of diffuse nebulae along the galactic plane is due to the large-scale interstellar magnetic field.

Nonthermal radioemission of the Galaxy was interpreted as a synchrotron emission, i.e. magnetobremsstrahlung of relativistic electrons gyrating in the magnetic field (Alfvén and Herlofson, 1950; Kiepenheuer, 1950; Ginzburg, 1953; Shklovsky, 1956). The detection of polarized radioemission in the Galaxy (Razin, 1958; Westerhout *et al.*, 1962; Wielebinski *et al.*, 1962) supported these earlier theoretical predictions. From those times until now, synchrotron radioemission has provided an efficient way to investigate interstellar magnetic fields in our Galaxy and in external radiosources, too. This emission originates in extended regions of the intercloud medium and thereby demonstrates the existence of the general Galactic magnetic field.

However, these were indirect items of evidence. Galactic magnetic fields have remained a hypothetical object. Of paramount importance, thus, was a direct detection of the Zeeman splitting of the λ 21 cm absorption line of atomic hydrogen in gas clouds projected onto strong radiosources. This detection was an enormously difficult problem. The splitting $\Delta\nu = eH/mc \simeq 10$ Hz should be detected on the background of Doppler line broadening which is about 10^4 Hz at $T = 10^2$ K. According to Verschuur (1979), the first five successful detections have consumed 5000 hours of telescope time! The idea behind the detection of the Zeeman effect is based on comparison of intensities of right- and left-circularly polarized wings of the absorption line (Bolton and Wild, 1957). This method has allowed the detection of magnetic fields in several neutral hydrogen clouds in the Galaxy (Verschuur, 1968). However, the clouds have much higher densities than the interstellar gas on average and the fields within them are a result of compression and distortion of the Galactic magnetic field by a cloud.

The most effective indicator of the large-scale Galactic magnetic field has proved to be the Faraday rotation of the polarization plane of radioemission from extragalactic sources (radiogalaxies and quasars) and galactic objects (pulsars). The rotation angle is proportional to the integral of the product of electron density and the magnetic field component along the direction to the source. Hence, fields at large scales, although weak, give a considerable contribution to the Faraday rotation. First determinations of Faraday rotation measures were performed by Australian astronomers (by Morris and Berge, 1964 and Berge and Seielstad, 1967 for extragalactic sources and by Manchester, 1972, 1974 for pulsars), who applied their results to estimations of the strength and configuration of the large-scale Galactic field. The strength was estimated as $(2\text{—}3) \times 10^{-6}$ G = 2—3 μG and the dominant field component turned out to be nearly parallel to the Galactic plane. Chapter IV below is partially devoted to extensions of this method and review of new results.

The magnetic field of the Galaxy is now an undoubted reality for an observer. A fascinating problem of establishing its global configuration and strength stands as a further step of investigation. In addition, a non-trivial extension has been made from our Galaxy to external spiral galaxies.

Observations carried out and interpreted by German and Japanese radio-astronomers (see Beck, 1986; Sofue *et al.*, 1986) have revealed that in the spiral galaxy M31 the large-scale magnetic field is concentrated in a nearly axisymmetric ring, as predicted by Ruzmaikin and Shukurov (1981), whereas in M33, M51, NGC 6946 and some other galaxies it has a bisymmetric structure akin to a two-armed spiral. The galaxy IC 342 has the field in the form of an extended axisymmetric configuration. Typical scales of magnetic fields are as large as several kiloparsecs along the radius and a few hundred parsecs across galactic discs. Strengths of these fields are few microgauss.

An enormous electromotive force 10^{12} Volts is required to act during the whole galactic lifetime $t_g \sim 10^{10}$ years in order to build up such a field (Hoyle, 1958). Responsibility for such a petrifying figure is due to the scale of the field. Indeed, the Faraday's induction law gives the following estimate of the e.m.f.:

$$\int \mathbf{E}\, d\mathbf{l} = -\frac{1}{c}\frac{\partial}{\partial t}\int \mathbf{B}\, d\mathbf{S} \simeq \frac{BhR}{ct_g} \approx 10^{12}\ \text{V},$$

where $h \simeq 0.4$ kpc and $R \simeq 15$ kpc are the half-thickness and the radius of the Galactic ionized gaseous disc, respectively.

What can create such a magnetic field? Thermal, chemical and other battery effects are inadequate because they cannot provide fields of the observed scales. Hoyle (1958) was thus forced to suggest that the field has the primordial origin, i.e. is given *ab initio*. However, this hypothesis, which delivers up the origin of the magnetic field to an obscure pregalactic past, cannot explain why galaxies still possess the large-scale component of the field despite the destructive action of turbulent motions of conductive interstellar gas. Magnetic fields that are created during a galaxy formation due to relative motions of protons and electrons, which are to different extents scattered by background relic photons (Harrison, 1970; Mishustin and Ruzmaikin, 1971), are too weak. Even as a seed field, subject to amplification by the dynamo process, these fields cannot compete with magnetic fields ejected from stars (Syrovatskii, 1970).

Amplification and maintenance of galactic magnetic fields are due to motions of conductive gas within a galaxy. Since the scale L of magnetic field is enormous and comparable at least with the half-thickness, h, of the ionized gas layer in a galaxy, magnetic diffusivity ν_m associated with Coulomb dissipation or ambipolar diffusion turns out to be much smaller than the effective turbulent diffusivity $\nu_T \sim hv$ (v is the characteristic turbulent gas velocity). In other words, the dimensionless magnetic Reynolds number $R_m = hv/\nu_m \simeq \nu_T/\nu_m$ is great. The inequality $R_m \gg 1$ implies that evolution of the magnetic field is governed primarily by motions of the conductive medium. The velocity that has appeared above can be associated with the velocity of random motions of gas clouds, $v \simeq 10$ km s^{-1}. In addition, interstellar gas participates in a still faster regular rotation of a galaxy with angular velocity generally falling with distance from the galactic center. A flow that provides self-excitation for a magnetic field (that is exponential growth of an initially weak field) is called the hydromagnetic dynamo. This term is also often applied to the very process of magnetic field generation.

The concept of the hydromagnetic dynamo was proposed by J. Larmor (1919) as an explanation of terrestrial and solar magnetism (see Figure I.1, where his short but impressive talk delivered in 1919 is reproduced). First models of the galactic dynamo were constructed by Parker (1971) and Vainshtein and Ruzmaikin (1971). According to their results, field generation is possible due to the joint action of non-uniform (differential) rotation and the mean helicity of turbulent gas motions, which is proportional to $\langle \mathbf{v} \operatorname{rot} \mathbf{v} \rangle$ (Steenbeck *et al.*, 1966). Mean helicity, i.e. the presence of a distinguished sense of screw motion, is a basic property of interstellar turbulence that is not typical of a common laboratory turbulence. This property is acquired under the influence of Coriolis forces in a density-stratified rotating medium (see Chapter VI).

The major part of our book is devoted to the theory of the galactic hydromagnetic dynamo. The solution of the problem of generation of the large-scale magnetic fields in spiral galaxies, taking proper account of the observed hydrodynamic properties of galaxies, can be represented in an almost completely analytical form. It is the thinness of galactic discs, whose aspect ratio is of order 0.01, which allows this. Mathematical formalism and asymptotic approaches used here are quite similar to adiabatic and quasi-classical approximations in quantum mechanics, which have acquired a modern form in the works of Maslov and his collaborators (see, e.g., Maslov, 1965; Maslov and Fedorjuk, 1981).

Successful prediction of the ring-like distribution of the generated axisymmetric magnetic field (Ruzmaikin and Shukurov, 1981), discovered shortly afterwards in some galaxies, provides the basis for belief in adequacy of our approach. In many galaxies, fulfilled are conditions for the excitation of non-axisymmetric magnetic

6. *How could a Rotating Body such as the Sun become a Magnet?* *By* Sir Joseph Larmor.

The obvious solution by convection of an electric charge, or of electric polarisation is excluded; because electric fields in and near the body would be involved, which would be too enormous. Direct magnetisation is also ruled out by the high temperature, notwithstanding the high density. But several feasible possibilities seem to be open.

(1) In the case of the sun, surface phenomena point to the existence of a residual internal circulation mainly in meridian planes. Such internal motion induces an electric field acting on the moving matter: and if any conducting path around the solar axis happens to be open, an electric current will flow round it, which may in turn increase the inducing magnetic field. In this way it is possible for the internal cyclic motion to act after the manner of the cycle of a self-exciting dynamo, and maintain a permanent magnetic field from insignificant beginnings, at the expense of some of the energy of the internal circulation. Again, if a sunspot is regarded as a superficial source or sink of radial flow of strongly ionised material, with the familiar vortical features, its strong magnetic field would, on these lines, be a natural accompaniment: and if it were an inflow at one level compensated by outflow at another level, the flatness and vertical restriction of its magnetic field would be intelligible.

(2) Theories have been advanced which depend on a hypothesis that the force of gravitation or centrifugal force can excite electric polarisation, which, by its rotation, produces a magnetic field. But, in order to obtain sensible magnetic effect, there would be a very intense internal electric field such as no kind of matter could sustain. That, however, is actually got rid of by a masking distribution of electric charge, which would accumulate on the surface, and in part in the interior where the polarisation is not uniform. The circumstance that the two compensating fields are each enormous is not an objection; for it is recognised, and is illustrated by radioactive phenomena, that molecular electric fields are, in fact, enormous. But though the electric masking would be complete, the two distributions would not compensate each other as regards the magnetic effects of rotational convection: and there would be an outstanding magnetic field comparable with that of either distribution taken separately. Only rotation would count in this way; as the effect of the actual translation, along with the solar system, is masked by relativity.

(3) A crystal possesses permanent intrinsic electric polarisation, because its polar molecules are orientated: and if this natural orientation is pronounced, the polarisation must be nearly complete, so that if the crystal were of the size of the earth it would produce an enormous electric field. But, great or small, this field will become annulled by masking electric charge as above. The explanation of pyro-electric phenomena by Lord Kelvin was that change of temperature alters the polarisation, while the masking charge has not had opportunity to adapt itself: and piezo-electric phenomena might have been anticipated on the same lines. Thus, as there is not complete compensation magnetically, an electrically neutralised crystalline body moving with high speed of rotation through the æther would be expected to produce a magnetic field: and a planet whose materials have crystallised out in some rough relation to the direction of gravity, or of its rotation, would possess a magnetic field. But relativity forbids that a crystalline body translated without rotation at astronomical speeds should exhibit any magnetic field relative to the moving system.

The very extraordinary feature of the earth's magnetic field is its great and rapid changes, comparable with its whole amount. Yet the almost absolute fixity of length of the astronomical day shows extreme stability of the earth as regards its material structure. This consideration would seem to exclude entirely theories of terrestrial magnetism of the type of (2) and (3). But the type (1), which appears to be reasonable for the case of the sun, would account for magnetic change, sudden or gradual, on the earth merely by change of internal conducting channels: though, on the other hand, it would require fluidity and residual circulation in deepseated regions. In any case, in a celestial body residual circulation would be extremely permanent, as the large size would make effects of ordinary viscosity nearly negligible.

Fig. I.1. In this report, Larmor (1919) has first suggested the idea of a hydromagnetic dynamo.

modes that can be associated with observed bisymmetric magnetic structures (Ruzmaikin *et al.*, 1986; Sawa and Fujimoto, 1986; Baryshnikova *et al.*, 1987a, b).

Along with the large-scale field, the turbulent hydromagnetic dynamo excites a small-scale, fluctuation magnetic field. The total field should have a highly inhomogeneous, intermittent distribution: regions of relatively weak intensity coexist with rare but strong concentrations of the field in layers or ropes (Molchanov *et al.*, 1985; Zeldovich *et al.*, 1985a). These theoretical results open wide possibilities for the exploration of fluctuative magnetic fields in galaxies.

Structure and strength of magnetic fields at small scales, comparable with clouds' sizes, is especially important for star formation. The collapse of rotating clouds or their fragments from low densities 10—100 cm^{-3} up to stellar densities requires the outward transport of angular momentum which is effectively accomplished by magnetic stresses (see, e.g., Mestel and Paris, 1984). This fact links together the problem of galactic magnetic fields and the exciting and rapidly developing field of star formation.

We will spend our time mainly on spiral galaxies partly because theoretical results are more advanced in this direction. One cannot pass by the importance and beauty of magnetic phenomena in quasars, astrophysical jets and radiogalaxies although there is only one chapter in this book devoted to this subject. One can only attempt to give the reader a general view of the present level of understanding of magnetic field origin in these energetic, distant astronomical objects. More details can be found in excellent reviews of Miley (1980), De Young (1984), Begelman *et al.* (1984), Bridle and Perley (1984) and Asseo and Sol (1987).

This book is written by theoreticians who specialize in the dynamo theory and its applications to astrophysics. We have attempted to discuss these problems as clearly as possible and to present the current state of affairs. Discussion of contiguous problems, especially observational ones, is rather fragmentary and considers only one aspect: their connection with galactic magnetism.

We realize that some people would consider the problem of origin of galactic magnetism a fruitless mind game. Indeed, we devote our book to the origin of things that are remote and do not promise practical profits. But, as ancient wisdom says,

> Those whose restless minds study the world's structure,
> Search for beginnings of things and wonder why the Moon wanes,
> Want to know why the Sun fades,
> And where fatal comets travel, with purple manes, . . .
> Should inquire into the origin of magnet
>
> (Claudianus, 29 (48), Magnet, 1—4)

CHAPTER II

GALAXIES

II.1. Shapes

The galactic identity of an astronomical object is established simply by looking at photographs. There were about 4000 photographs of galaxies in the collection used by Hubble. The catalogues compiled by G. de Vaucouleurs and van den Bergh included 2599 and 935 samples respectively and the modern assemblages comprise up to 32000 pieces. Figure II.1 shows a few drawings made by Vorontsov-Velyaminov (1978, p. 42) from the photographs of his catalogue.

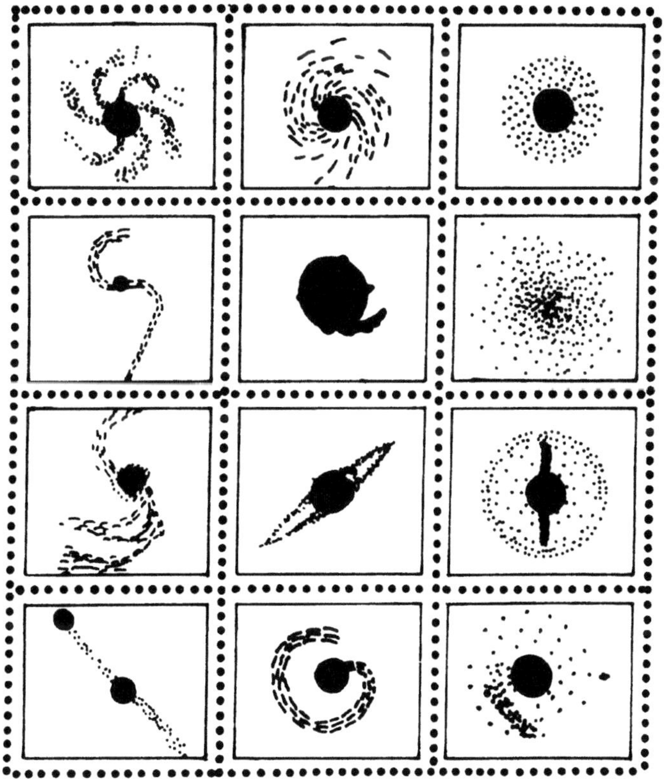

Fig. II.1. Galaxies in drawings by Vorontsov-Velyaminov (1978).

While each galaxy is unique, there are two major types — spirals and ellipticals. Some illustrative examples are given in Figure II.2.

The visible shapes of galaxies reflect mainly the distributions of stars repre-

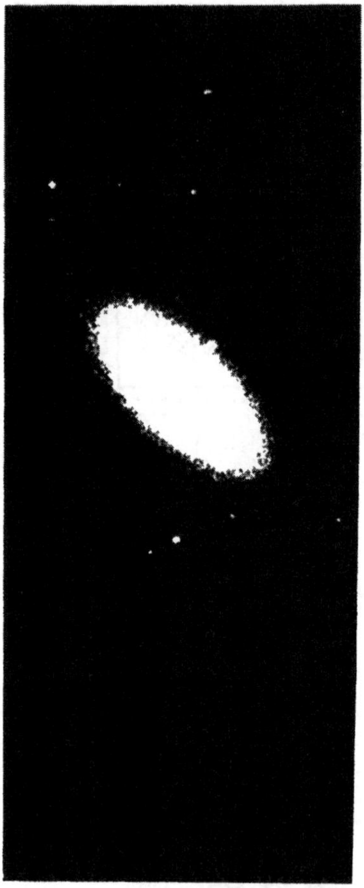

Fig. II.2. Examples of elliptical and spiral galaxies. (a). Elliptical galaxy NGC 4564.

Fig. II.2(b). A giant elliptical galaxy M87 situated in the center of the rich galaxy cluster in the Virgo constellation. To the lowest right the jet is visible. This galaxy is surrounded by strong magnetic fields discussed in Chapter VIII.

senting a major part of their bright mass. The brightness in an elliptic galaxy monotonically and smoothly decreases from the center to the periphery approximately as $\log I \propto -r^{1/4} + \text{const}$ (de Vaucouleurs, 1974). In spiral galaxies the brightness distribution has its breaks (Figure II.3). At least three components may be distinguished: a bulge, a spheroidal one and the disc. In the bulge the mass is distributed approximately in the same way as in elliptic galaxies, their dimensions differing by a factor of several. The disk brightness exponentially decreases with the radius, $I \propto \exp(-kr)$ (Freeman, 1970), where $k^{-1} \simeq 4$ kpc for our Galaxy. In some spiral galaxies, for example the Galaxy and M31, it would be more correct to describe the flat component as a broad ring, rather than a disc: disc density has a deep minimum within $r \lesssim 3$ kpc (Kormendy, 1977; Barabanov and Zasov, 1979; Einasto *et al.*, 1980; Rohlfs and Kreitschmann, 1981).

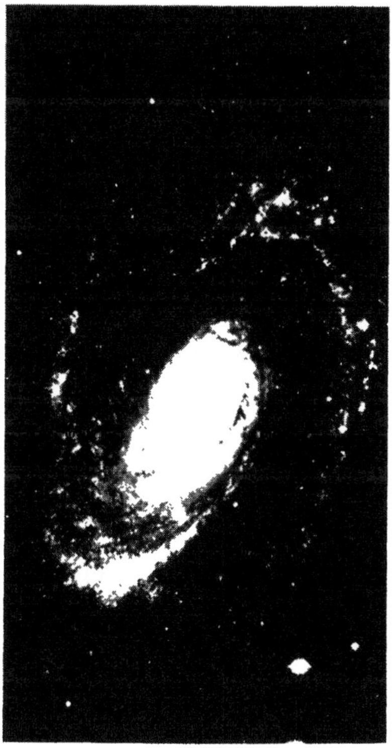

Fig. II.2(c). A nearby spiral galaxy M81 whose large-scale magnetic field is studied in considerable detail from radio observations (Chapter IV).

Fig. II.2(d). A spiral galaxy NGC4565 seen edge-on. A thin disc, dark due to absorption by dust, is clearly seen. ((a), (b) and (d) from *Cambridge Atlas of Astronomy*, Cambridge Univ. Press, 1985; (c) from Sandage, *The Hubble Atlas of Galaxies*, p. 19, 1961).

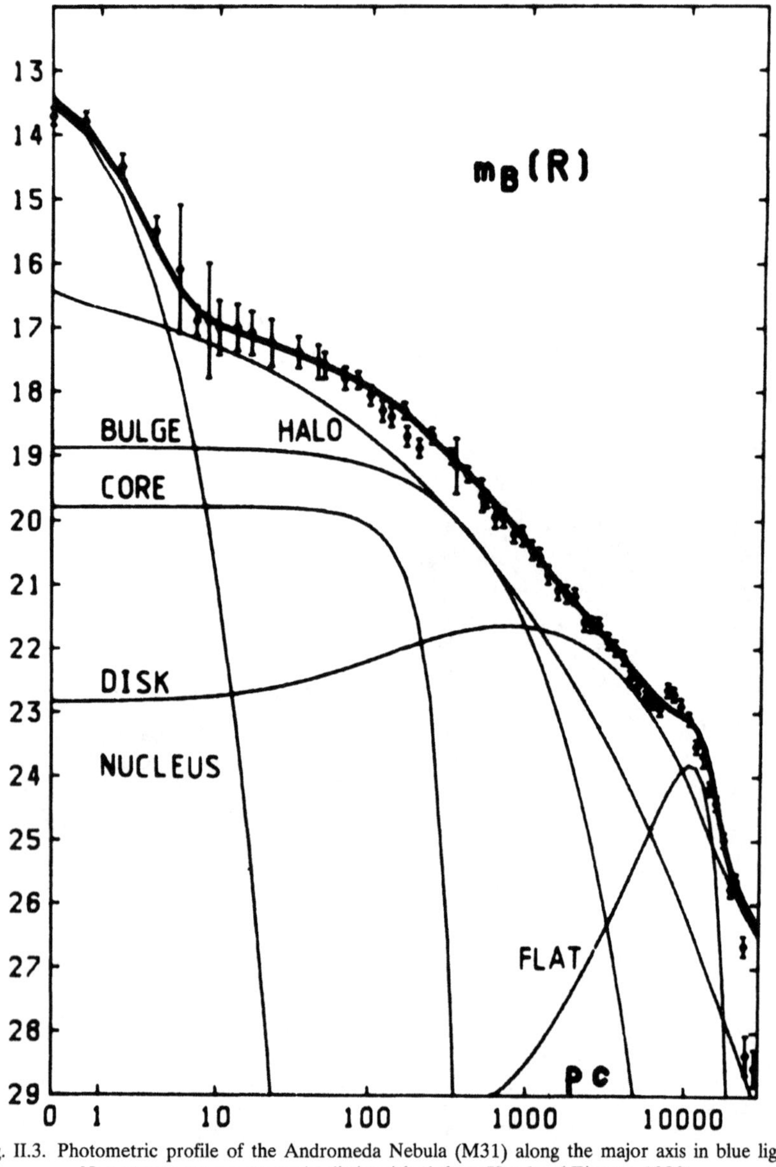

Fig. II.3. Photometric profile of the Andromeda Nebula (M31) along the major axis in blue light. Numerous components can be distinguished (from Haud and Einasto, 1983).

According to modern ideas, along with the visible, there is an invisible galactic component in the form of a massive spheroidal halo. Such a component with the density distributed as $(1/r^2)$ is invoked, for example, to explain the observed radial distributions of rotational velocities in spiral galaxies (Einasto *et al.*, 1974; Ostriker *et al.*, 1974; see also Rubin, 1983). The size of the invisible halo may be very large — up to 100—200 kpc across (Freeman, 1986). Stellar discs of spiral galaxies have the form of strongly oblate two-axis ellipsoids with an aspect ratio of 200 pc/15000 pc ~ 1 : 100 in our Galaxy.

II.2. Spiral Galaxies

A distinctive feature of spiral galaxies is a pronounced disc with a more or less distinct spiral pattern. The spiral arms may be distinguished against the disc background only because young bright stars are concentrated in them. For example, in the Andromeda Nebula (M31) 85 per cent of light is emitted by the general disc population and only 15% or less from the spiral structure (Baade, 1963, Ch. 6).

The spiral pattern is well seen in the distribution of visible light and interstellar gas, but for the mass distribution the spiral structure is only a weak disturbance. In fact, the gravitational field of the Galactic spiral arms is only 5 per cent of the average axisymmetric field while the maximal gas density in the arms is a few times larger than that between them (see, e.g., Roberts and Yuan, 1970).

Both these features of spiral galaxies, the pronounced disc and the spiral pattern, are due to their fast rotation. The Sun moves around the Galaxy's center at the velocity of 250 km s^{-1} (this value may be somewhat lower, see Chapter VI). The rotation velocity in elliptic galaxies is about half this value. In most cases the spiral pattern consists of two arms which are trailing, i.e. they are twisted clockwise when observed from the north pole of a galaxy. Rare cases of leading arms seem to be a result of external influence from nearby galaxies (Pasha, 1985). Although the galaxies with multi-armed and multi-level spiral patterns may be encountered, the two-armness is a typical feature. The spiral pattern rotates slower than the disc as a whole, therefore interstellar gas enters spiral arms at the inner edge.

In the region of developed spiral pattern the difference between the arm and disc velocities exceeds the speed of sound in the interstellar gas, so that a shock wave forms along the inner edge of a spiral arm. The standard picture of density waves (Roberts, 1969) predicts that star formation becomes more effective in the gas compressed by the shock wave, and rapidly evolving young stars give off a bright spiral pattern on photographs of galaxies. However, recent results of Elmegreen and Elmegreen (1986) imply that relation between spiral shock waves and star formation may be less straightforward. In any case, importance of spiral shock waves for interstellar magnetic fields is due to the fact that the shock waves compress and refract magnetic lines.

II.3. Gas and Dust

So far the discussion was restricted mainly to the stellar population of galaxies. An inspection of any blue plate of a spiral galaxy indicates an intense star formation and, thus, the presence of a rather large (as compared with elliptic galaxies) amount of interstellar gas. In the optical range, the interstellar gas may be seen only if it is strongly heated and ionized by a nearby young hot star — these are HII regions outlining the spiral arms*. When observed in other ranges, spiral galaxies look quite different from their optical appearance. Radiation in the λ 21 cm radio spectral line is emitted by neutral hydrogen, the λ 2.6 mm emission shows up the CO molecule distribution connected with the molecular hydrogen distribution; the infrared radiation and the optical absorption reveal the interstellar dust; the hard gamma ($\varepsilon > 100$ MeV) photons are emitted when cosmic rays interact with the interstellar gas and the radiation field; the ultraviolet radiation is emitted by multiply ionized atoms of heavy elements. The interstellar gas is a multi-component system with distinctly different properties of its components.

Figure II.4 shows the optical photograph of the Andromeda Nebula (M31) with superimposed view in the λ 21 cm neutral hydrogen line at the resolution 160×240 pc (Unwin, 1980a, b). The interstellar gas concentrates in the disc. The large-scale distribution of neutral hydrogen clearly outlines the spiral arms. Its distribution is very irregular in smaller scales.

The half-thickness of HI layer near the Sun is about 100 pc (Burton, 1976). The neutral hydrogen distribution along the radius in a spiral galaxy is usually quite featureless (Figure II.5). However, there are important exceptions as our Galaxy (see Burton, 1976; Sanders *et al.*, 1984) and M31 (see Emerson, 1974, 1976), where the interstellar gas (both atomic and molecular) is concentrated in the rings (Figures II.4 and II.6). The HI density in the central parts of many galaxies is small due to ionization by the stellar radiation field.

A large amount of molecular hydrogen has been detected in the interstellar medium of the Galaxy (Scoville and Solomon, 1975) and nearby spirals (Morris and Rickard, 1982) through the λ 2.6 mm emission typical for the CO molecules excited by collisions with the H_2 molecules. In fact, a major portion of the interstellar gas in the inner parts of galaxies is in the molecular state. For example, the surface HI density in the central regions of IC 342 is 2.4 $M_\odot$ pc^{-2} (Newton, 1980b), while the molecular hydrogen surface density is 50 $M_\odot$ pc^{-2} (Morris and Lo, 1978). In the Galaxy, the total H_2 and HI masses are nearly equal but differently distributed: 80 to 90 percent of molecular hydrogen and only 20 to 30 per cent of neutral hydrogen reside within the solar orbit (Sanders *et al.*, 1984; see Figure II.6). Just like atomic gas, in our Galaxy and M31 the molecular hydrogen

* According to the nomenclature adopted by astronomers, HI denotes the neutral hydrogen, HII is the ionized hydrogen, CIV is triply ionized carbon, etc.

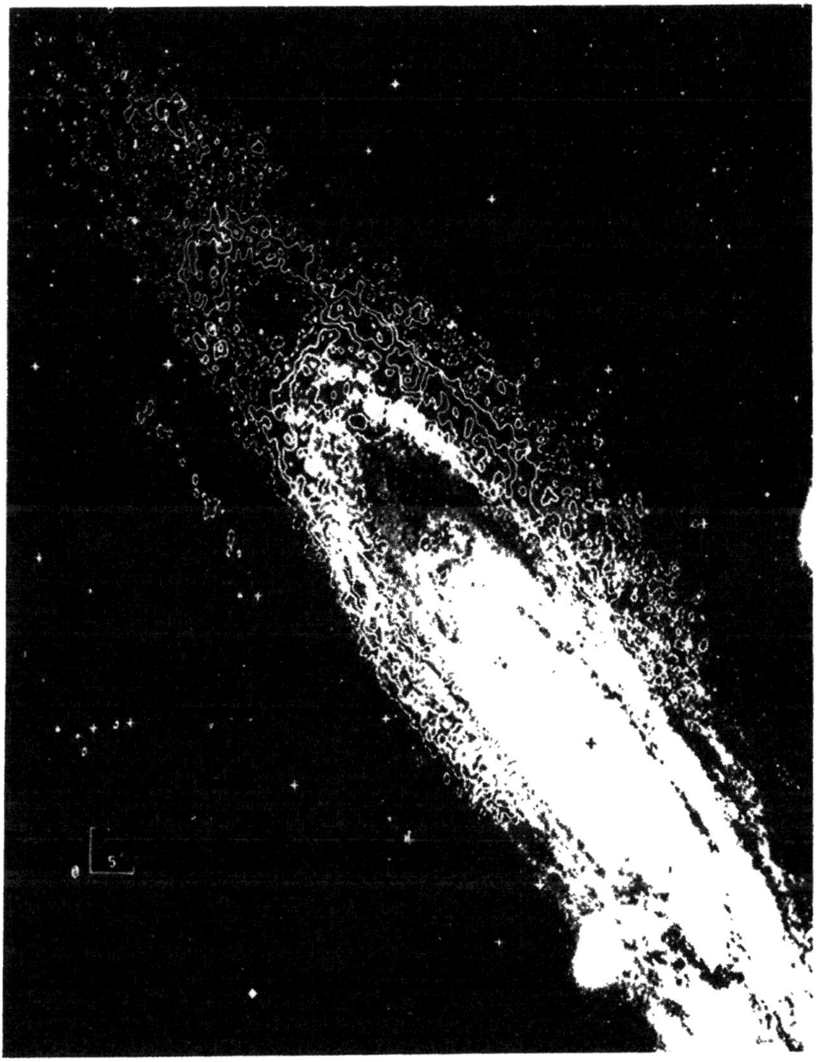

Fig. II.4(a, b). Distribution of interstellar neutral hydrogen (white contours) in the Andromeda Nebula differs markedly from the distribution of stars (from Unwin, 1980a, b).

is concentrated in rings at 6—8 kpc radius while in majority of galaxies its density monotonically decreases with distance from the center (see Morris and Rickard, 1982). The spatial distribution of molecular gas is extremely irregular, H_2 is concentrated in giant molecular clouds of 20 to 80 pc diameters and (1—30) $M_\odot$

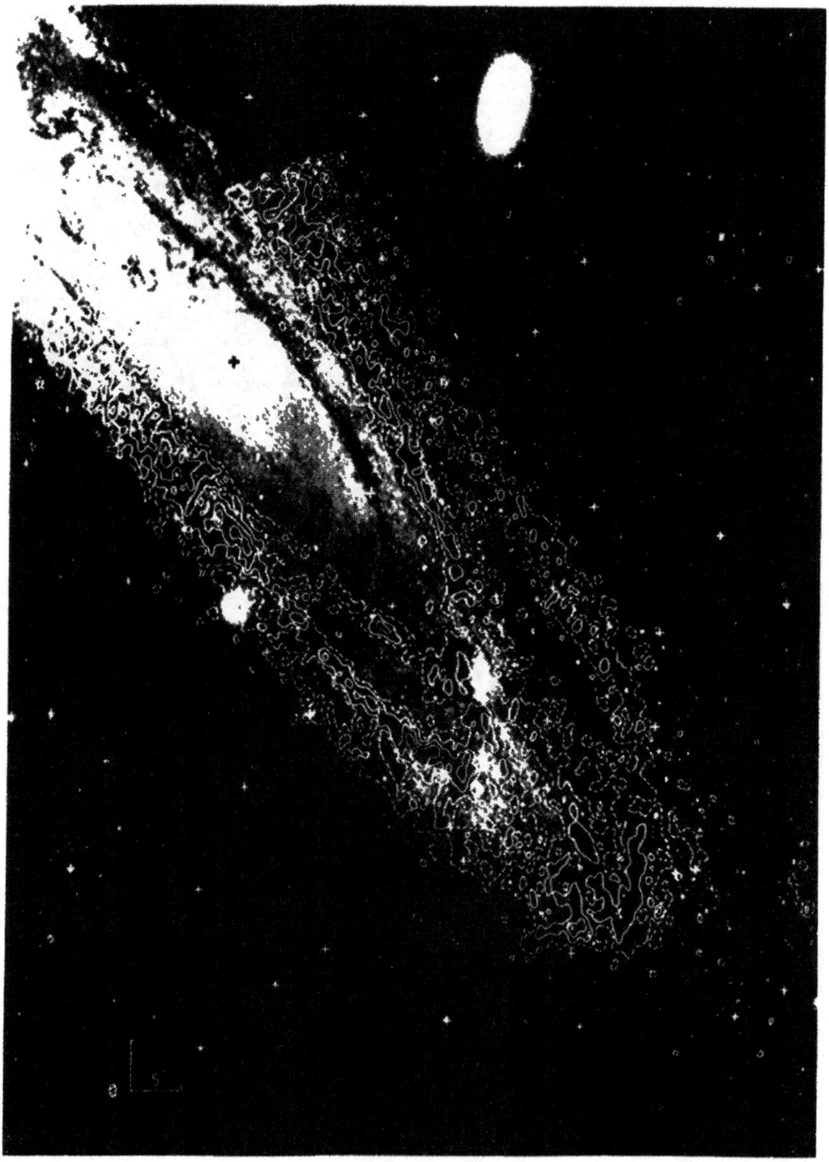

Fig. II.4(b).

masses; the half-width of their distribution around the central plane of our Galaxy is about 50 pc (Sanders *et al.*, 1984).

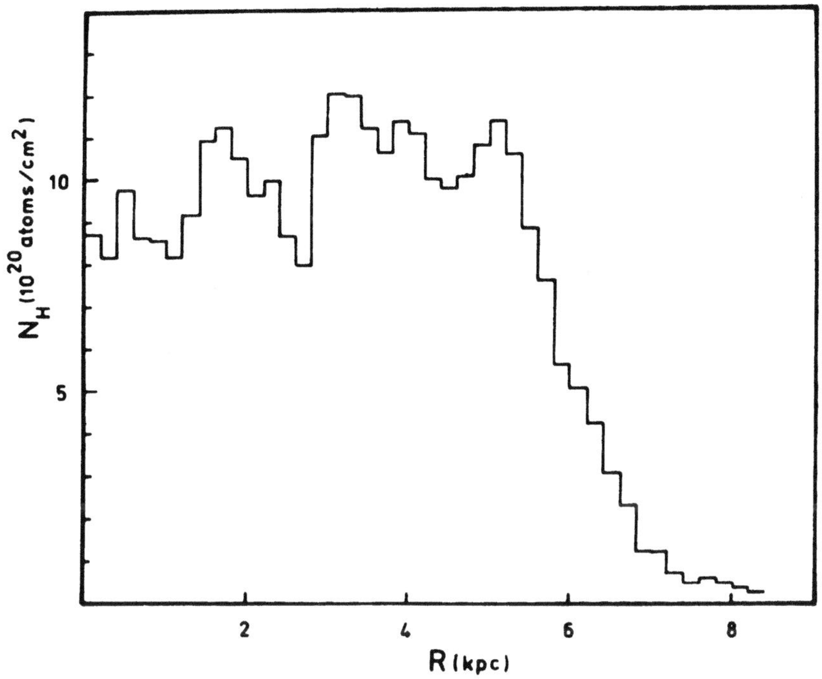

Fig. II.5. The radial distribution of interstellar neutral hydrogen in spiral galaxy M33 (Newton, 1980a).

Observations of extragalactic radio sources show that their radiation is scattered in the interstellar space by the density fluctuations of free electrons. Investigation of this phenomenon, along with other methods (see Chapter VI), has allowed to obtain a (still rather crude) estimate of the thickness of the diffuse ionized hydrogen disc in the vicinity of the Sun. According to Readhead and Duffet-Smith (1975), if the ionized hydrogen distribution is described by the exponential law $n_e \propto \exp(-z/h_0)$ then $h_0 \approx 660$ pc; for the homogeneous step distribution the disc half-thickness is estimated as 440 pc. Ruzmaikin and Sokoloff (1977b) have determined the half-thickness of the electron disc as 400 pc by comparing the Faraday rotation measures for pulsars and the extragalactic radio sources. The thermal electron density is about 0.03 cm^{-3}. Thus, near the Sun another component of the interstellar gas — a diffuse ionized gas layer — has a thickness 4—5 times as large as that of the HI layer.

In the mid-seventies the first UV telescopes were put into orbit. Observations of the Galaxy in this range yielded an unexpected result: multiply ionized nitrogen (NV) and oxygen (OVI) absorption lines were detected (Jenkins and Meloy, 1974; York, 1977), which indicates the presence of low-density gas with temperature

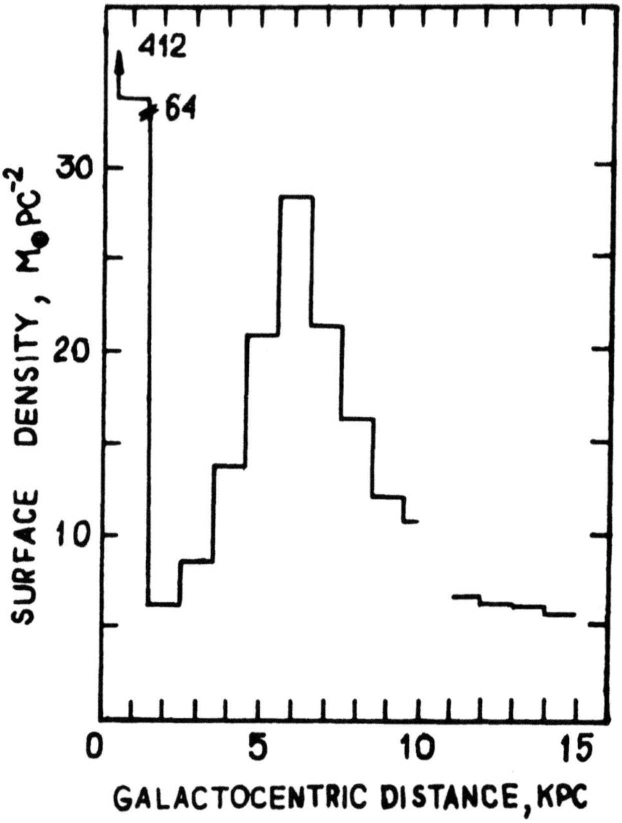

Fig. II.6. The radial distribution of interstellar gas in the Milky Way galaxy. Shown is the sum of surface densities of molecular and neutral hydrogen (after Sanders *et al.*, 1984). The gas concentrates in a ring around $r = 6$ kpc.

$T \simeq 3 \times 10^5$ K between stars. Thus an idea of a multiphase interstellar medium has been further developed (see Cox and Smith, 1974; McKee and Ostriker, 1977): the clouds of neutral hydrogen ($T \simeq 100$ K, $n \simeq 20$ cm^{-3}, $n_e \simeq 0.05$ cm^{-3}) occupying a small fraction of volume and enveloped by a warm intercloud medium ($T \simeq 10^4$ K, $n \simeq 0.2$ cm^{-3}, $n_e \simeq 0.02$ cm^{-3}) coexist with very hot rarefied tunnels ($T \simeq 10^6$ K, $n = n_e \simeq 2 \times 10^{-3}$ cm^{-3}) which are formed by merging supernova remnants. Apart from these phases, there are dense molecular gas clouds ($T \simeq 20$ K, $n \simeq 2 \times 10^3$ cm^{-3}, $n_e/n \simeq 10^{-8}$) and intermediate phases arising at transitions between basic phases. The HI clouds occupy a few per cent of the Galactic volume (see Burton, 1976) while as much as 50 per cent volume may be occupied by the tunnels (Jenkins and Meloy, 1974). These results have further developed earlier concepts of the multiphase nature of the interstellar medium, which appeared in the 'sixties (see review by Kaplan and Pikelner, 1974).

Interstellar dust is just another component of the interstellar medium which is formed as a result of chemical reactions in the interstellar gas and/or in atmospheres of cold carbon stars (see Kaplan and Pikelner, 1979).

An average size of the dust particle is less or about 5×10^{-5} cm. Several types of the dust particles with different structures and chemical compositions seem to exist. The interstellar dust concentrates into dust clouds and dense dark molecular-dust complexes, exemplified by the Orion Nebula, where the star formation proceeds. An importance of the interstellar dust for studying the magnetic fields is connected with the fact that the interstellar dust particles are oriented by magnetic fields, causing polarization of optical radiation of distant stars (Davis and Greenstein, 1951; see, e.g., Dolginov *et al.*, 1979) and thus exposing interstellar magnetic fields.

The possible spiral structure of the Galaxy, as revealed by HI observations, is schematically shown in Figure II.7 (see also Liszt, 1985).

II.4. Cosmic Rays

The charged particles with relativistic energies, which fill the interstellar space and the regions around galaxies are called the cosmic rays. They manifest catastrophic events in the interstellar environment — supernova explosions, shock waves, etc. The cosmic rays consist mainly of electrons, protons and helium nuclei, but nuclei with mass numbers up to 80 are also present (Ginzburg and Syrovatskii, 1964; Ginzburg, 1984; Ginzburg and Ptuskin, 1985). Energies of the cosmic ray protons may reach 10^{20} eV and those of electrons 10^{12} eV. However, the particles with energies from tens of MeV per nucleon to tens of GeV per nucleon make a basic contribution to the cosmic ray energy density ($w_{c.r.} \approx 10^{-12}$ erg cm^{-3}). The energy density of the electron component in the Galaxy is about one per cent of the total energy density of the cosmic rays.

The cosmic rays, or more exactly their electron component makes interstellar magnetic fields visible: relativistic electrons in the interstellar magnetic fields of 5 μG strength emit the synchrotron radiation with maximum intensity at wavelengths from 10 m to 10^{-3} cm for the electron energies from 10^9 eV to 10^{12} eV. Such nonthermal radio waves are received from all spiral galaxies; especially powerful synchrotron emission comes from quasars and active galaxies, Seyfert galaxies and radio galaxies. Unfortunately, detailed data on the cosmic ray energy and spatial distributions are rather scarce even for our Galaxy. Therefore, it is somewhat difficult to gain direct information on magnetic fields by observing the synchrotron radiation.

Independent data on the cosmic ray spatial distribution in the Galaxy may be

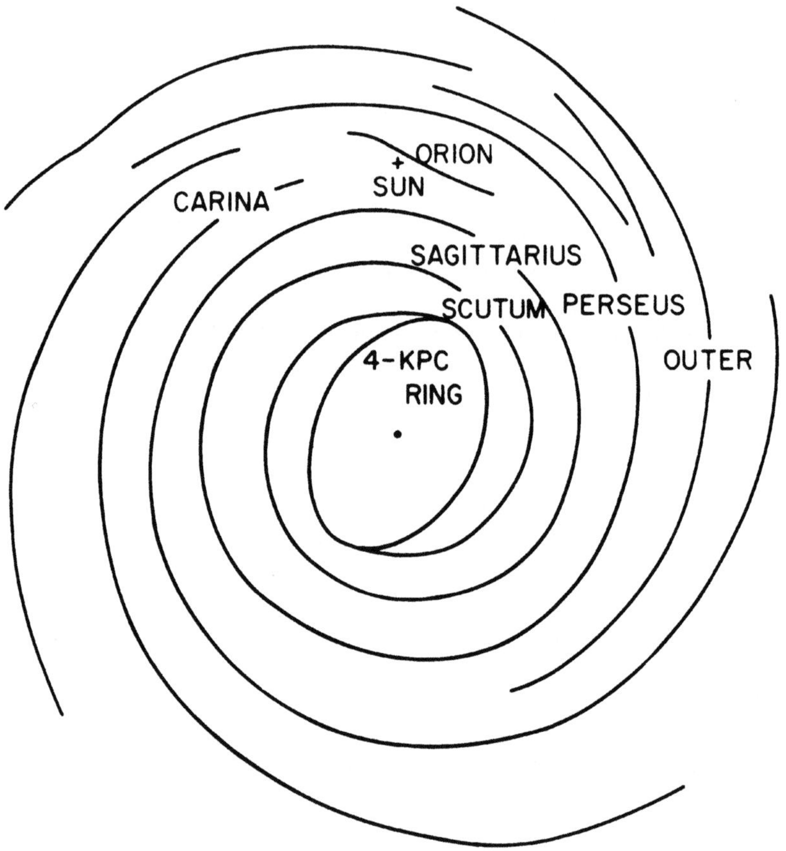

Fig. II.7. Schematic view of the spiral arms in the Galaxy (Simonson, 1976).

obtained by observing the γ radiation with the photon energy in excess of 35 MeV. The γ-quanta with energy of about 70 MeV are emitted in decay of π^0-mesons forming through interactions between the proton-nucleon component of the cosmic rays and the interstellar gas nuclei. Another source of the γ-quanta is the bremsstrahlung of relativistic electrons. An analysis of the γ-emission distribution in the 35—100 MeV range obtained by the SAS-2 satellite showed (Issa *et al.*, 1980) that the spectral density of the cosmic ray electrons decreases monotonically in the range $7 \leqslant r \leqslant 13$ kpc of galactocentric distances and correlates with the HII regions distribution. For $r < 2$ kpc the relativistic electron intensity is at least by an order of magnitude less than that near the Sun.

Cosmic rays, the charged particles, are confined within galaxies by the interstellar magnetic fields. If the pressure of relativistic particles greatly exceeds the magnetic pressure, they would inevitably escape from a galaxy. In the opposite case cosmic rays accumulate until the pressures are nearly equalized. The stationary energy density is determined by the order of magnitude relation

$$H^2/8\pi \simeq w_{\mathrm{c.r.}} \simeq 10^{-12}\,\mathrm{erg\,cm^{-3}}.$$

This equality leads to an estimate of the Galactic magnetic field strength, $H \simeq 10^{-5}$—10^{-6} G.

Thus, even general speculations on cosmic ray dynamics provide a useful estimate of interstellar magnetic field strength. In fact, similar arguments resulted in the emergence, some 40—50 years ago, of the very concept of interstellar magnetic fields. Until now, analysis of polarization and intensity properties of synchrotron emission, intrinsically associated with cosmic ray electrons, remains the most powerful tool for the exploration of cosmic magnetic fields. The universal property of Galactic and extragalactic sources of emitting radio waves largely through a synchrotron mechanism reveals how amazingly widespread are cosmic rays and cosmic magnetic fields.

CHAPTER III

OBSERVATION OF MAGNETIC FIELDS

III.1. Synchrotron Emission

A direct indicator of magnetic fields in galaxies and other astronomical objects is their synchrotron emission. It is the emission of charged particles moving with relativistic velocities in the magnetic field (see, e.g., Ginzburg and Syrovatskii, 1964; Pacholczyk, 1970; 1977).

The trajectory of a charged particle in a magnetic field has the form of a spiral winding about the magnetic line. The particle acceleration is determined by the radius of the spiral which, in particular, is proportional to the particle mass. Therefore, electrons make a basic contribution to this emission. The relativistic electron energy distribution is usually described by the isotropic power law

$$N_e \, \mathrm{d}E = KE^{-\gamma} \, \mathrm{d}E \tag{III.1}$$

where N_e is the electron number density for the electron energies in the range $(E, E + \mathrm{d}E)$. Such an approximation is justified because the synchrotron emission observed over a wide range of frequencies in a wide class of sources has a power-law spectrum. The coefficient K in formula (III.1) is determined by the total number of relativistic electrons per cm^3 and by the electron energy range over which the distribution (III.1) is valid. After corrections for atmospheric and instrumental losses as well as for the solar modulation, measurements of the relativistic electron flux at the Earth yield the following expression for the relativistic electron number density per energy interval of 1 GeV in the interstellar medium:

$$N_e = 1.54 \times 10^{-11} \left(\frac{E}{1\ \mathrm{GeV}} \right)^{-(3.15 \pm 0.2)} \quad \text{particles cm}^{-3}\ \mathrm{GeV}^{-1} \tag{III.2}$$

for high energies $4.5 < E < 63.5$ GeV (Golden *et al.*, 1984; see also Nishimura *et al.*, 1980 and references therein). Direct estimates at lower electron energies which are more interesting for our subject, are rather difficult because the conditions of the relativistic electron penetration into the heliosphere from the interstellar medium strongly vary during the 22-year cycle of the solar activity (the solar modulation). An analysis of the Galactic non-thermal background radio emission shows that at $E = 1$—3 GeV the spectral index is different, $\gamma \simeq 2.7$.

The emissivity (i.e. the energy emitted from unit volume per unit time per unit frequency interval per unit solid angle) for the electron ensemble with the distribution (III.1) in the magnetic field H is given by

$$\sigma_I \propto K\nu^{-(\gamma-1)/2} H_\perp^{(\gamma+1)/2} \qquad \text{(III.3)}$$

where ν is the emission frequency and $H_\perp$ is the magnetic field component perpendicular to the line of sight. The factor of proportionality omitted in (III.3) depends only on the spectral index γ (see Ginzburg and Syrovatskii, 1964).

For electron energy distribution (III.1) and with proper account for spectral distribution of emission by a single electron, it can be shown that the maximum of emission produced by electrons which have energy E is found at the frequency

$$\nu_{\text{eff}}(\text{MHz}) \simeq 16.0\, H_\perp(\mu\text{G}) E^2(\text{GeV}) \qquad \text{(III.4)}$$

(see Webber, 1983). In the magnetic field with the strength $H \simeq 1$—$2\ \mu$G, which is typical for spiral galaxies, electrons with energies $1 \lesssim E \lesssim 100$ GeV emit in the range $10 < \nu < 3 \times 10^5$ MHz.

By measuring the emission intensity from a source, the product $H_\perp^{(\gamma+1)/2} KL$ may be found, where L is the size of the emitting region along the line of sight, and γ, determined from the observed emission spectrum, usually lies in the range from 2 to 4; the typical galactic values of γ are between 2.4 and 2.8. The geometric factor L may, in principle, be determined from the linear size of the source, that is from its angular size and the distance.

Extraction of data on magnetic fields from measurements of the synchrotron emission intensity is hampered by the uncertain knowledge of the density of relativistic particles in the source. Until recently, the direct information available about the cosmic ray density concerned only the near-Earth space. Lately, satellite observations of the cosmic diffuse γ-radiation have made possible an independent determination of the relativistic electron density within the Galaxy at large distances from the Sun. Of course, interpretation of these observations is not free from considerable difficulties. These observations are compatible with the assumption that the cosmic rays density is rather homogeneous and, in particular, it is the same within spiral arms and between them; the relativistic electrons density decreases toward the Galaxy's boundaries (see, e.g., Issa *et al.*, 1980; Ginzburg, 1984, §6.9). As for the extragalactic synchrotron emission sources, the number of relativistic particles is most often estimated for them from the equipartition arguments (see below). In fact, relations similar to (III.3) are frequently used for evaluating the properties of cosmic rays after the magnetic field strength and configuration have been determined independently.

Measurements of the synchrotron emission intensity in normal spiral galaxies may be used for estimating the magnetic field strength by assuming, for example, the relativistic electron density in those objects to be the same as in our Galaxy. A typical example of such considerations may be found, e.g., in Beck (1982). The magnetic fields in spiral galaxies, thus determined, are several microgauss. It should be noted that the local density of cosmic rays near the Earth may prove to be lower than the average value over the disc (see, e.g., Nishimura *et al.*, 1980); then the magnetic fields calculated with local data would be overestimated.

It is a widely used method of estimating the magnetic field strength to suppose equipartition of energy between relativistic particles and magnetic fields in a synchrotron source (Burbidge, 1956; 1959). This presumption is warranted in the following way. By knowing a total energy (luminosity) J emitted by the source in a frequency range $\nu_1 < \nu < \nu_2$, one can obtain the relation between the total relativistic electron energy and the average magnetic field strength as

$$E_e = cJH^{-3/2} \tag{III.5}$$

where c is a known function of ν_1, ν_2 and γ (see, e.g., Pacholczyk, 1970, ch. 7). Now, the heavy relativistic particles are assumed to have a total energy proportional to the electron energy, $E_p = kE_e$; the value of k depends on the particle acceleration mechanism. For the cosmic rays observed near the Earth, $k \sim 100$. However, some interpretations of the extragalactic radio source observations require $k \sim 1$ (Begelman *et al.*, 1984). Then the total energy stored in the source is

$$E_{\text{tot}} = c(1+k)JH^{-3/2} + V(H^2/8\pi + \rho v^2/2),$$

where V is the source volume, and $\rho v^2/2$ is the turbulent energy density. Furthermore, the equipartition between the magnetic and kinetic energy densities is assumed, $H^2/8\pi \simeq \rho v^2/2$. It is observed that the total energy, as a function of H, reaches a minimum proportional to $J^{4/7} V^{3/7}$ when

$$H_{\text{eq}}^{7/2} = 3\pi c(1+k)J/V.$$

From this it follows that the total relativistic particles energy $(1+k)E_e$ is approximately equal to the total magnetic energy $VH^2/8\pi$. A modification of these arguments with the total pressure (instead of energy) minimization may be found in Burns *et al.* (1979).

For many extragalactic sources such an estimate seems to be the only possible one. It provides the order of magnitude estimate of the magnetic field. For example, $H_{\text{eq}} \approx 0.1$ G in compact components of jets and $H_{\text{eq}} \simeq 3 \times 10^{-7}$ G in their rarefied halo, while $H_{\text{eq}} \simeq 10^{-2}$ G in nuclei of active galaxies and in quasars (see Begelman *et al.*, 1984 and referenecs therein). Such estimates should be treated carefully because some independent estimates of magnetic fields in compact extragalactic radio sources made using the observed polarization (Pacholczyk, 1977, Chapter 7) show that relativistic particles may contain a major portion of the total energy in many sources. Thus, the synchrotron emission intensity only indicates the presence of a magnetic field in a given region (for example, its localization within a ring or nuclear region of a spiral galaxy), but not the field strength or its configuration.

Study of the frequency distribution of the synchrotron emission energy flux yields another way for estimating the magnetic fields in extragalactic sources. As the synchrotron emission propagates within the source, it is absorbed by the relativistic electrons themselves. Since the absorption coefficient of the relativistic electron ensemble with the power-law distribution (III.1) is proportional to $H_{\perp}^{(\gamma+2)/2}\nu^{-(\gamma+4)/2}$ (Pacholczyk, 1970, expression (3.52)), the spectral intensity of emission sharply drops at low frequencies. The characteristic frequency, where this effect becomes essential, is related to the magentic field strength and the angular dimension of the compact source, Ω, by

$$\nu_0^{\gamma+4} \propto H_{\perp}^{1/2}\Omega^{-2}$$

(Slysh, 1963; Pacholczyk, 1977, Ch. 2). Compact extragalactic radio sources often cannot be resolved even with the best telescopes. In this case an upper estimate of the source angular size can be obtained from the observed variability of the source. In this way an upper estimate of the magnetic field strength is obtained. For the compact radiosources whose angular dimensions can be obtained directly, the low-frequency cut-offs of their spectra give typically $H_{\perp} \simeq 10^{-4\pm 1}$ G, provided the interpretation of spectra is correct (Kellermann and Pauliny-Toth, 1969).

In some specific cases, information about the magnetic field can be gained by supplementing the data on synchrotron emission with data on relativistic electrons obtained from some additional knowledge or model calculations. An important example is the Galactic non-thermal radio background. In those regions where the Galactic radio background intensity is nearly homogeneous (i.e. $l^{\mathrm{II}} = 180°$, $b^{\mathrm{II}} = 3°$, the anticenter direction; $l^{\mathrm{II}} = 180°$, $b^{\mathrm{II}} = 60°$, a minimum of the radiohalos emission) a break in the synchrotron spectrum is observed at the frequency $\nu \approx 100$ MHz. At lower frequencies the emission spectral index is $\alpha = (\gamma - 1)/2 \approx 0.5$—$0.6$, while at higher frequencies it smoothly increases up to 0.8—1.0 (see Ginzburg, 1984, Chapter 4; Webber *et al.*, 1980; Chernyshev and Stankevich, 1975 and references therein). By associating this break (see Longair, 1981) with a change in the relativistic electrons energy spectrum at $E \simeq 2$ GeV from $\gamma \simeq 2$—2.2 to $\gamma \simeq 2.6$—3, the magnetic field strength may be estimated. Breaks in the energy spectrum of relativistic electrons are attributed to energy losses and energy dependence of electron diffusion coefficient (Ginzburg, 1984, Chapter 4, Section 10). Letting $\nu_m = 100$ MHz and $E = 2$ GeV in equation (III.4) we obtain the estimate

$$H_{\perp} = 1.6 \times 10^{-6}\ \mathrm{G}.$$

This value should be compared with the estimate 2×10^{-6} G for the large-scale magnetic field strength in the Galaxy, obtained from an analysis of the Faraday rotation of the pulsars and extragalactic radio source emission (see the next chapter). The magnetic field estimates from the observed synchrotron emissivity of the interstellar medium and the local density of relativistic electrons (III.2) lead to greater values of $H_\perp \simeq 8$—$10\ \mu$G. Sometimes this discrepancy is attributed to lower local density of cosmic rays as compared to the Galactic disc in average (see, e.g., Nishimura *et al.*, 1980). However, it seems plausible that the point is that the synchrotron emissivity depends not on the strength of the large-scale magnetic field *B*, but on the mean square of the total field, $H^2 = B^2 + b^2 \simeq 4B^2$, which gives $H \simeq 2B \simeq 4$—$6\ \mu$G (Webber *et al.*, 1980), a value which agrees better with observations. In addition, a contribution from the Galactic radiohalo should be taken into account in the analysis of the observed synchrotron intensity distribution (Badhwar and Stephens, 1977; Ginzburg, 1984, Ch. 4).

In some extragalactic objects that are bright both in radio and X-ray ranges the magnetic field strength may be determined under the presumption that the X-ray emission is a result of the inverse Compton scattering of relic microwave photons by the 1 GeV relativistic electrons which are also responsible for synchrotron emission. X-ray observations of such sources thus give the density of relativistic particles from the intensity of X-ray emission and the temperature of the relic background. After that the synchrotron emission data are used to estimate the magnetic field strength. For the nearest radiogalaxy Cen A such estimates yield the average value of $B \simeq 0.7\ \mu$G, while the energy equipartition results in $H \sim 0.8\ \mu$G (Cooke *et al.*, 1978). The main difficulty of such an approach is due to the ambiguity in interpretation of X-ray emission: it may prove to be the thermal bremsstrahlung of hot gas (Feigelson *et al.*, 1981) or the synchrotron emission of very high energy electrons (Schreier *et al.*, 1981). In the last case the minimum pressure condition in bright spots yields $H \sim 35$—$45\ \mu$G, while between the spots the field must be 5—10 times as weak. In many cases (e.g. M87) the bremsstrahlung model of X-ray emission proves to be the most suitable (see Sarazin, 1986). In these cases comparison of X-ray and synchrotron emissions yields a lower estimate of the magnetic field in the source.

III.2. Polarization of Synchrotron Emission

In a general case, the emission of a single relativistic electron moving in magnetic field is elliptically polarized. The direction of electric vector rotation depends on a sign of the angle between the magnetic field and the line of sight. When the relativistic electron revolves in a circle perpendicular to the field, and the line of sight lies in the orbit plane, the elliptic polarisation degenerates into the linear one (Landau and Lifshitz, 1962). When the trajectory of the electron is not planar but has a helical form with angle ψ between the total velocity vector and the field direction, it is known that the strength of the magnetic field H should be replaced by the effective value $H \sin \psi$. The emission of such an electron is linearly polarized when observed at an angle ψ to the magnetic field. Passing from a single particle to an electron ensemble, one should average over angles ψ, and, as a result, the polarization is always almost purely linear. The polarization plane is orthogonal to the projection of the magnetic field vector on the sky. (We do not mention here the presence of a weak circular polarization). The polarization degree p is determined as the ratio of the linearly polarized emission intensity to the total intensity. This quantity is often expressed as a percentage. If the magnetic field and the electron distribution are homogeneous and the source is optically thin with respect to synchrotron emission, the polarization degree reaches its maximum

$$p_0 = \frac{\gamma + 1}{\gamma + 7/3} \tag{III.6}$$

(see, e.g., Ginzburg and Syrovatskii, 1964) and depends only on the electron spectrum index.

We have mentioned above that the optical thickness of the source with respect to synchrotron reabsorption increases with decreasing frequency of emission. Therefore, for small frequencies, a synchrotron source becomes optically thick. The degree of linear polarization of an optically thick synchrotron source is given by

$$p_{\text{thick}} = \frac{3}{6\gamma + 13}$$

while the polarization plane is parallel to the projection of the magnetic field on the sky. Emission from optically thick regions is much less polarized as compared with the optically thin case: $p_{\text{thick}} \approx 10\%$ for $\gamma = 2.7$. All discussions below refer to the case of a small optical depth since in majority of cases synchrotron sources are optically thin at observation frequencies.

In both normal and active galaxies as well as in quasars, the magnetic field is never strictly homogeneous. One may distinguish a regular field component **B**, the characteristic scale of which is comparable with the object's size, and a fluctuation component **b**. In normal spiral galaxies these components are comparable in value.

Obviously, the random field component reduces the polarization degree of synchrotron emission. For the electron energy spectrum index $\gamma \simeq 2.7$ the polarization degree (III.6) in the strictly homogeneous magnetic field is $p_0 \simeq 73\%$. However, in normal spiral galaxies p rarely exceeds 30%, usually reaching only 10–20%. In radio galaxies and quasars the polarization degree is usually only few per cent. What is an explanation for this drastic difference between the observed polarization and maximally possible one? The polarization degree depends considerably on the orientation of the regular component, statistical properties of fluctuation fields, the ratio of strengths for these components, and a size of emission region.

When statistically isotropic random fluctuations **b** are imposed upon the homogeneous field **B**, the polarization degree from the source, whose extent greatly exceeds a correlation length of a random component, is

$$p = p_0 \begin{cases} \dfrac{(\gamma+3)(\gamma+5)}{32}\dfrac{B_\perp^2}{b^2} & \text{for } B_\perp^2 \ll b^2 \\ 1 - \dfrac{2}{3}\dfrac{b^2}{B_\perp^2} & \text{for } B_\perp^2 \gg b^2 \end{cases} \tag{III.7}$$

where $B_\perp$ is the homogeneous field component perpendicualr to the line of sight and b^2 is the mean square strength of the random component (Korchak and Syrovatskii, 1961).

These expressions allows us, in principle, to estimate the ratio of the fluctuation of the magnetic field to the transverse component of the regular field from observed polarization. The physically interesting quantity is, of course, the strength of the total regular field. Burn (1966) has proposed the following simple expression for the case when all components of the regular field are equal:

$$p \approx p_0 \frac{\text{Energy in regular field}}{\text{Total magnetic energy}} \tag{III.8}$$

This expression is suitable only for the *statistical* investigation of a sufficiently large sample of sources with randomly directed regular fields. However, when individual sources are considered, the formulae of Korchak and Syrovatskii (or other analogous results — see below) are appropriate.

Note that according to (III.7—8) the sources with a purely random field $B = 0$ emit unpolarized waves. It is true only in the limit of an infinite number of turbulent cells within the source. Meanwhile, the maximal turbulent cell size in spiral galaxies is about 100 pc, and the emission region extends only for a few hundreds of parsecs in galaxies observed face-on. The necessity to take into account the effects connected with a finite number of turbulent cells in the synchrotron emission region has become vital due to observations with high angular resolution.

Consider the effect of a finite number of turbulent cells within the observed region (Razin, 1958). We consider the source of size h, in which the magnetic field is quasi-homogeneous within cells of size l, but the field direction randomly changes from cell to cell with uniform distribution of direction angles between 0 and 2π radians. The polarization degree for each cell is equal to p_0 and at random orientations of magnetic fields within cells the polarization degree of the total emission from N cells obviously is given by

$$p = p_0/\sqrt{N}. \tag{III.9}$$

Therefore, for finite N the polarization degree differs from zero even for a purely random magnetic field. For sufficiently small N, say of about 100, $p \approx 7\%$. We emphasize that for small N expression (III.9) is inapplicable. For instance, if $N = 2$ then p is either zero (when the fields in cells are antiparallel) or p_0 (for the parallel fields).

Of course, in the limit $N \to \infty$, the polarization for the source with random field (III.9) is zero, in full accord with (III.7) and (III.8). With an increase in observation resolution the effect of the finite cell number becomes more important.

The polarization properties of synchrotron emission taking account of the finite number of turbulent cells have been studied by Kuznetsova (1983). The emission of a relativistic electron ensemble with a power-law energy spectrum was considered. The magnetic field was represented as a superposition of homogeneous and chaotic fields; for the random field components b_x, b_y and b_z, independent Gaussian distributions were adopted* with zero mean values and the dispersions $b^2/2$. The expressions for the polarization degree in various limiting cases are as follows. For $b^2/2 \gg B_\perp^2$ and $(b^2/2)/B_\perp^2 > N^2$

$$p \simeq \frac{p_0}{\sqrt{N}}\,[1 + 0.12N(B_\perp/b)^4],$$

where $B_\perp$ is the homogeneous field component perpendicular to the line of sight, $N = V/l^3$ is the number of cells with quasi-homogeneous field, V is the source volume. From this expression, for $B_\perp = 0$ we obtain the relation (III.9).

For $b^2/2 \ll B_\perp^2$ Kuznetsova obtains the relation (III.8).

In the case when $b^2/2 \simeq B_\perp^2$

$$\Phi \simeq \frac{p_0}{\sqrt{N}}\,2.3\exp(-0.02N)\Phi(\tfrac{3}{2}; 1; 0.02N),$$

where Φ is the confluent hypergeometric function. Comparison of the last equation with the expression (III.8) obtained by Burn (1966) shows that (III.8) is sufficiently accurate for $N \gtrsim 50$ (see Figure III.1); for the smaller number of turbulent cells within the source the relation (III.8) overestimates the contribution of the homogeneous magnetic field component.

The above relations allow estimations of the number of turbulent cells in the sources with purely chaotic fields and, thus of the turbulent scale. For example, the sample of active compact radiosources considered by Jones *et al.* (1985) is characterized by a polarization scatter of about 2.5%. Presuming that unpolarized emission from optically thick regions within sources gives half of the integral flux, one obtains that polarization from optically thin regions is 5%. Expression (III.9) gives then $N \approx 200$, i.e. the turbulent scale is $N^{1/3} \approx 5$ times smaller than the size of the emitting region.

* This presumption cannot be considered rigorous since it does not take into account that $\nabla\mathbf{H} = 0$ (see Zeldovich *et al.*, 1985b).

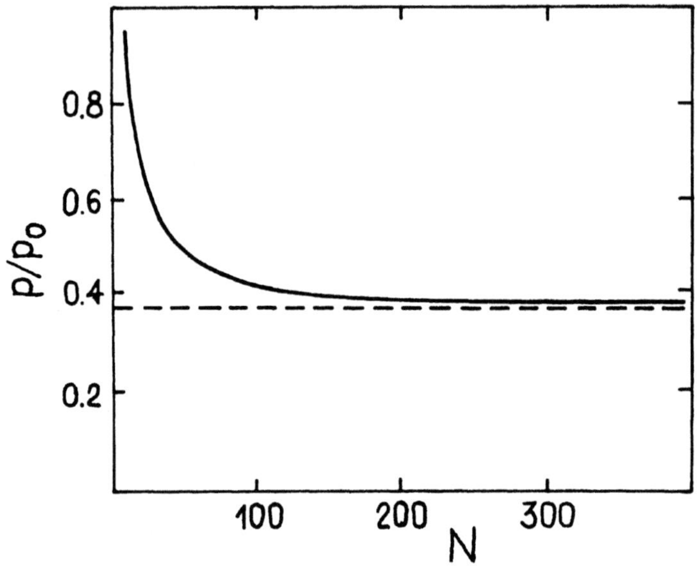

Fig. III.1. Polarization of synchrotron emission in a combination of regular and random magnetic fields as a function of the number of correlation cells within emitting region (Kuznetsova, 1983). The broken line shows the asymptotic value $N \to \infty$ according to Burn (1966).

Let us note another case when, in a purely chaotic magnetic field, the synchrotron emission is linearly polarized. This is a two-dimensional chaotic distribution of magnetic field, i.e.

$$\mathbf{B} = 0, \qquad \mathbf{b} = \{b_x, b_y, 0\},$$

which was studied by Laing (1981) in application to astrophysical jets. For the frozen-in magnetic field

$$b_x, b_y \propto r_0/r, \qquad b_z \sim (r_0/r)^2,$$

where r is the jet radius (r_0 is its initial value), and the z axis is directed along the jet axis. Therefore, in expanding jets the field tends to be two-dimensional at large distances from the nucleus, $r \gg r_0$. When an observer is located in the plane xy the emission has the polarization degree p_0. Generally,

$$p \approx p_0 \frac{\cos^2 \beta}{1 + \sin^2 \beta},$$

where β is the angle between the line of sight and the magnetic field plane. At observation along the jet, $\beta = \pi/2$ and $p = 0$. This model was generalized by Jones *et al.* (1985).

In normal spiral galaxies the regular field component usually lies in the plane of the disc. Segalovitz *et al.* (1976) have proposed a model devised for interpretation of observations in spiral galaxies. Consider the disc portion covered by the beam of a radiotelescope. Divide this region into cells so small that the field in them may be considered homogeneous and unidirectional. Combination of the regular and small scale magnetic fields is modelled by assuming that the field directions are identical in the fraction f of the cells while over the remaining cells (fraction $1 - f$) fields are distributed randomly. The field strength is the same in all cells. The polarization degree of emission from such a region with the large number of cells ($N \to \infty$) was calculated by Segalovitz *et al.* (1976) to be

$$p = p_0 \left[1 + \frac{(1-f)\pi^{1/2}\Gamma[(\gamma+5)/4]}{2(\sin\theta)^{(\gamma+1)/2} f \Gamma[(\gamma+7)/4]} \right]^{-1} \qquad \text{(III.10)}$$

where Γ is the Euler gamma function, and θ is the angle between the line of sight and the direction of the regular field. For $f \to 0$ (purely chaotic field) $p \to 0$, for $f \to 1$ (purely regular field) $p \to p_0$. For the regular field closed in the disc plane, the polarization degree averaged over the disc can be obtained by replacing the factor $(\sin\theta)^{(\gamma+1)/2}$ by its averaged value. Formula (III.10), or its averaged variant, is proposed for interpretation of polarization observations of spiral galaxies with parameter f considered as the measure of magnetic field ordering.

The reader may be surprised at the complete disagreement between expressions (III.7—8) and (III.10) obtained for apparently similar physical systems. The reason lies in different ways of defining the average (regular) field component. Korchak and Syrovatskii as well as Burn consider the averages over the ensemble of realizations of a random magnetic field. Those who wish to replace ensemble averages by volume averages must consider the field distribution in which both the regular and the fluctuation fields are present at any position. Meanwhile, in the model of Segalovitz *et al.* the average field is summed from unidirectional fields in a separate population of cells occupying a certain part of the whole source's volume. In this part of the source the field is purely regular. In the remaining part of the source the field is completely chaotic. Therefore, the quantity f cannot characterize the regularity of the magnetic field when the procedure of averaging is defined correctly.

We should note that observations of spiral galaxies never reveal a population of cells with completely aligned magnetic fields. In other words, the common and mathematically justified way of ensemble or volume averaging corresponds to situations met in actual galaxies (regular plus fluctuation fields at any given place) perfectly well. It is to be noted also that the turbulent dynamo theory yields characteristics of the mean and fluctuation fields with these components defined through the ensemble averages, or the volume averages in the sense described above.

There is one more point which deserves attention. Strictly speaking, the model of Segalovitz *et al.* is inapplicable to description of magnetic field configurations also because in this model the solenoidality of the magnetic field is violated at the boundary of every cell. However, after proper correction this model may be applied to description of *intermittent* distributions of magnetic fields, which the dynamo theory predicts to be common in turbulent media, in particular, in the interstellar gas (Molchanov *et al.*, 1985; Zeldovich *et al.*, 1987). In the case when regular magnetic fields, their Gaussian fluctuations and the intermittent component are altogether present in a source of synchrotron emission, a combination of the models of Korchak—Syrovatskii—Burn and of Segalovitz *et al.* seems to be most relevant.

We have already said in the beginning of this section that the synchrotron emission of an individual electron is elliptically polarized. When the velocity distribution of relativistic electrons is isotropic, the degree of circular polarization from an ensemble of electrons strongly decreases but does not vanish completely (Legg and Westfold, 1968; Melrose, 1971). Anisotropy of the electron velocity distribution enhances the resulting circular polarization (Sazonov, 1969b). Thus, in addition to a strong linear polarization, the synchrotron radiation possesses a weak circularly polarized component (see Pacholczyk, 1977, Sect. 3.5). The degree of intrinsic circular polarization of the synchrotron emission varies with the frequency as

$$p_V = p_1 \nu^{-1/2},$$

where the proportionality factor p_1 is directly related to the transverse magnetic field within the source. This gives the following estimate of the transverse magnetic field:

$$H_\perp = \left[\frac{p_1/100\ C}{400(\gamma + 2)} \right]^2 \times 10^9\ \mathrm{G},$$

where $C = 1$ for optically thin sources and C is a certain known function of γ for optically thick regions. A measurement of *intrinsic* circular polarization of the synchrotron source allows a direct estimate of the magnetic field strength independent of the energy density of energetic electrons!

The observed degrees of circular polarization of extragalactic radiosources are extremely low and do not exceed 0.5% (see Weiler and de Pater, 1983) which complicates observations. Moreover, the intrinsic polarization is not the single, and often not the principal source of the circular polarization of the synchrotron radiation. Influence of both relativistic and thermal plasma, both within a source and along the pathlength, anisotropy induced by magnetic field, and variations in direction of the latter — all these factors contribute to the observed circular polarizations (Sazonov, 1969a, b; Pacholczyk and Swihart, 1970, 1974; Pacholczyk,

1973; Pacholczyk, 1977, Sects. 3.7 and 3.8; Hodge, 1982). However, different components of the circular polarization have distinct frequency dependencies which allows, in principle, to single out the intrinsic contribution and thus estimate the magnetic field (Cheng *et al.*, 1985).

With an increase in angular resolution of radiotelescopes, not only integral (averaged over the whole source) intensity and polarization measurements of the synchrotron emission have become available but also finer characteristics are being measured, such as correlation functions of the fluctuations in the intensity and polarization. In principle, these data may yield an information about the magnetic field spectra over a range of scales. Klimishin and Basilevich (1963) have made first attempts in this direction (see Kaplan and Pikelner, 1970, Sect. 23) and have shown that the spatial spectra of fluctuations of the Galactic nonthermal radio background and interstellar starlight polarization are compatible with the concept of interstellar turbulent magnetic fields. A low accuracy of observational data and difficulties in their interpretation had prevented more detailed quantitative conclusions. These difficulties have not been overcome yet but some efforts can be referred to that develop this approach.

Chibisov and Ptuskin (1981) consider the brightness fluctuations of the Galactic background radio emission produced by turbulent fluctuations of the magnetic field. They assume that the spatial spectrum of chaotic magnetic fields is a power-law

$$b(l) = b_0(l/l_0)^q,$$

where $b(l)$ is the amplitude of magnetic field fluctuations at the scale l, l_0 is the energy-range scale of interstellar turbulence, i.e. a maximal size of turbulent cells, b_0 is the magnetic field strength at this scale. Such fluctuations of magnetic field result in the synchrotron emission intensity fluctuations

$$\frac{\delta I}{I} \propto \theta^{q+1/2}\left(\frac{R}{l_0}\right)^q,$$

where $\theta \simeq l/R$ is the angular separation of two investigated directions, R is the radial extent of the emission region. The magnetic field B whose scale exceeds the energy-range scale of turbulence l_0 causes anisotropy of the autocorrelation function of intensity fluctuations at large separations $\theta \gtrsim l_0/R$. Since $l_0 \simeq 100$ pc and $R \simeq 1$ kpc in the Galaxy, observations are needed with resolution higher than few angular minutes. Such observations were performed and analyzed by Dagkesamansky and Shutenkov (1987) (see Chapter IV).

Fluctuations of the magnetic field and the thermal and relativistic electron density in the source result in fluctuations of the polarization parameters of synchrotron emission. The longitudinal and transverse (with respect to the large-scale quasi-uniform magnetic field) autocorrelation functions of magnetic fluctuations, $f(r)$ and $g(r)$, are related to the respective correlations of the Stokes parameter u fluctuations* by

$$\begin{aligned} \sigma^2 f(r) &\propto \langle u(r_0)u(r_0+r)\rangle_{\perp}, \\ \sigma^2 g(r) &\propto \langle u(r_0)u(r_0+r)\rangle_{\parallel}, \end{aligned} \qquad \text{(III.11)}$$

where σ is a mean square value of the magnetic field fluctuations, and the angular brackets mean averaging over r_0. An example of such an analysis of observations may be found in Spangler (1983). Expressions analogous to (III.11) may be applied to observations of both extended extragalactic radio sources and radio background in our Galaxy.

So far we have considered mostly the synchrotron emission in vacuum. The only effect of the medium mentioned above is the synchrotron reabsorption. Now we consider the effects of thermal plasma within the synchrotron source and along the propagation path.

Both the relativistic electrons and the thermal plasma in the source itself and along the line of sight take their effect upon spectral and polarization features of emission. The degree of polarization may be decreased and the above formulae for p are only the upper estimates for it. The most important phenomenon decreasing the polarization degree is the Faraday rotation of the polarization plane in the emission sources. The Faraday effect takes a special place in investigations of interstellar magnetic fields not only because it complicates the interpretation of synchrotron emission observations but rather because it yields the most reliable independent information about the strength and configuration of galactic magnetic fields. Let us now discuss the Faraday effect, and in particular how it affects the synchrotron emission.

III.3. Faraday Rotation

The polarization plane rotation of linearly polarized emission in a plasma with magnetic fields is called the Faraday effect.

The rotation is, obviously, a consequence of asymmetry between the left and the right and may arise only in the medium with violated mirror symmetry, the

* If the intensity of a partially linearly polarized emission is I_1 along a certain direction and I_2 along the perpendicular one, and the polarization plane makes an angle θ to this direction then the quantities

$$\begin{aligned} I &= I_1 + I_2, \\ Q &= (I_1 - I_2)\cos 2\theta, \\ U &= (I_1 - I_2)\sin 2\theta \end{aligned}$$

called the Stokes parameters, completely characterize such an emission. (No admixture of circular polarization is assumed.) In formula (III.11) the value of Stokes parameter U normalized by the total emission intensity I is used.

symmetry with respect to reflections $\mathbf{r} \to -\mathbf{r}$. The cause of this violation may be the magnetic field **H** which is an axial vector (or differently speaking a pseudovector). Accordingly, the pseudoscalar quantity $\mathbf{H} \cdot \mathbf{k}$ arises where **k** is the wave propagation direction.

We consider first a plane circularly polarized electromagnetic wave which propagates in a cold plasma along the x-axis. In the absence of magnetic field, electrons in the electric field of the wave rotate in circular orbits in the plane yz so that

$$z + iy \equiv \xi = \frac{|e|E}{m\omega^2} \exp(i\omega t).$$

The effect of ambient plasma upon propagation of an electromagnetic wave is equivalent to the change of the wave phase velocity from the speed of light c to $c(1 - \omega_p^2/\omega^2)^{-1/2}$, where $\omega_p^2 = 4\pi n_e e^2/m$ is the plasma frequency.

In the presence of a magnetic field, electrons also circulate about magnetic lines at the frequency $\omega_H = |e|H/mc$. When the electromagnetic wave propagates along the magnetic field (with both directions coinciding with the x-axis) the electron motion is described by

$$m\ddot{\mathbf{r}} = \mathbf{E}\exp(i\omega t) + \frac{e}{c}\,\dot{\mathbf{r}} \times \mathbf{H},$$

which yields after integration

$$\dot{\xi} + i\omega_H \xi = i\,\frac{|e|E}{m\omega}\,\exp(i\omega t),$$

or

$$\xi = \frac{|e|E}{m\omega(\omega + \omega_H)}\,\exp(i\omega t).$$

Thus, in the right-polarized wave field ($\omega < 0$) the electron moves in a circle of smaller radius than in the left-polarized wave field ($\omega > 0$). For radio waves in the interstellar medium the difference is not large: for $H \sim 2 \times 10^{-6}$ G, $n_e \sim 1$ cm^{-3}, Larmor and plasma frequencies are $\omega_H \sim 40\ s^{-1}$ and $\omega_p \sim 5 \times 10^4\ s^{-1}$ respectively, while $\omega \sim 10^8$—$10^{11}\ s^{-1}$.

By comparing the expressions for the electron motion in the wave field in the magnetic field and without it, we see that in the presence of a magentic field ω^2 should be replaced by $\omega(\omega \pm \omega_H)$ for the left- and right-polarized waves. Accordingly, the phase velocities are different and equal to

$$u_\pm = c[1 - \omega_p^2/\omega(\omega \pm \omega_H)]^{-1/2},$$

where the upper sign corresponds to the left polarization and the lower sign to the right one.

Now we consider a linearly polarized wave (with **E** parallel to the y-axis at $x = 0$) propagating along the magnetic field. The electric field of the wave may be decomposed into a sum of two components with right and left circular polarizations and of equal amplitudes, $E = E_+ + E_-$. In the magnetic field these waves propagate with different velocities, and after the time interval t the field components are

$$(E_y, E_z)_\pm = \tfrac{1}{2} E\,[\cos\omega(t - xu_\mp^{-1}),\ \pm\sin\omega(t - xu_\pm^{-1})].$$

While at $x = 0$ the superposition of these waves yields $(E_y, E_z) = (E, 0)$, after passage of the distance $\mathrm{d}x$ their superposition $(E_{y+} + E_{y-}, E_{z+} + E_{z-})$, remaining linearly polarized, has the z-component with

$$\frac{E_z}{E_y} = \tan\left[\frac{\omega}{2}\left(\frac{1}{u_+} - \frac{1}{u_-}\right)\mathrm{d}x\right].$$

In other words, the polarization plane has been rotated by the angle

$$\mathrm{d}\varphi = \frac{1}{2}\,\omega\left(\frac{1}{u_+} - \frac{1}{u_-}\right)\mathrm{d}x.$$

Since $u_+ < u_-$, the rotation angle sign is positive when the magnetic field is directed towards an observer (i.e. along the direction of the wave propagation). By taking into account now that in the astrophysical conditions ω_p, $\omega_H \ll \omega$ (see above), we obtain

$$\mathrm{d}\varphi = \frac{\omega_H}{2c}\left(\frac{\omega_p}{\omega}\right)^2 \mathrm{d}x.$$

After the wave has travelled over the distance L, the rotation angle becomes

$$\varphi = \frac{e^3}{2\pi m^2 c^4}\,\lambda^2 \int_0^L n_e H\,\mathrm{d}x + \varphi_0,$$

where $\lambda = 2\pi c/\omega$ is the wavelength and φ_0 is the initial polarization angle within the source. In the case of arbitrary orientation of magnetic field the integrand contains a longitudinal (line-of-sight) component of the magnetic field. The above expression may be written in a more convenient form:

$$\varphi = RM\lambda^2 + \varphi_0.$$

The factor designated by two characters, *RM*, is called the rotation measure. It is an adopted definition that $RM > 0$ for the magnetic field directed towards the observer. In astrophysical applications, it is convenient to measure distances in parsecs, the magnetic field in 10^{-6} G = 1 μG, the wavelength in meters and the electron density in cm^{-3}:

$$RM = 0.81 \int n_e \mathbf{H} \cdot \mathrm{d}\mathbf{r} (\mathrm{rad\ m^{-2}}). \qquad \text{(III.12)}$$

Despite the fact that the phase velocity difference for the waves with right and left polarizations is small ($\omega_H \ll \omega$) the polarization plane rotation angle in the interstellar medium proves to be rather large because the effect accumulates over long distances (accordingly, the numerical coefficient before the integral (III.12) in RM is of order unity when units adequate for galactic conditions are used). It is quite clear also that the Faraday effect is most noticeable at radio wavelengths $\lambda \simeq 1$–30 cm.

The expression (III.12) demonstrates that the Faraday rotation of the polarization plane is fairly suitable for studying large-scale fields that make a main contribution to the integral $\int n_e H \, \mathrm{d}r$ at distances comparable with the size of a magnetized region. Analysis of the Faraday rotation of polarized emission from pulsars and extragalactic sources yields the most reliable information about large-scale magnetic fields of the Galaxy. The Faraday effect also reveals the properties of magnetic fields in nearby spirals and some other extragalactic sources. The ways in which information on magnetic fields can be extracted from Faraday rotation data are discussed in the next chapter.

Now we come back to the polarization of synchrotron emission. As we shall see, the Faraday rotation in the emission region plasma leads to a decrease of the polarization degree in comparison with the values given in the previous section. First, the rotation angle of the polarization plane is proportional to the distance travelled by waves within the source. Therefore, the polarized waves emitted at different depths in an extended source experience the polarization plane rotation by different angles. This effect decreases the polarization of the total emission. If the source may be represented as a slab where magnetic field and thermal electron distribution are both homogeneous, the polarization degree is given by

$$p = p_0 \left| \frac{\sin(RM\lambda^2)}{RM\lambda^2} \right| \qquad \text{(III.13)}$$

(Razin, 1958; Burn, 1966), where RM is the total rotation measure produced in the whole slab. Involvement of a random component of the field smoothes the dependence of polarization p on λ, specifically p becomes non-vanishing at any λ.

The second effect of Faraday rotation that depolarizes the synchrotron emission is due to the combined action of the random magnetic fields and the finite size of a telescope beam: if there are several cells within the beam with differently directed internal magnetic fields then the Faraday rotation is different within different elements of the solid angle encompassed by the beam. Clearly, this effect also diminishes the resulting polarization. At a cell size l much less than the object extension L along the line of sight, the resultant polarization degree is

$$p = p_0 \exp\left(-2RM^2_{\text{random}} - \frac{L}{l}\lambda^4\right) \tag{III.14}$$

(Burn, 1966).

A general expression for the polarization for a slab with partially regular field was obtained by Burn (1966). It has the form

$$p = p_0 \left|\frac{1 - \exp(-S)}{S}\right|, \tag{III.15}$$

where

$$S = RM^2_{\text{random}} \frac{L}{l} \lambda^4 - 2iRM_{\text{reg}}\lambda^2$$

with

$$RM_{\text{random}} = 0.81\, n_e bl,$$

$$RM_{\text{reg}} = 0.81\, n_e BL,$$

where b and B are, respectively, the random component and the regular field projection onto the line of sight ($b^2/2$ is the variance of the presumed Gaussian distribution of chaotic field), l and L are, respectively, the chaotic field scale and the source size along the line of sight in the same units as in (III.12). Here, the fluctuations in n_e are neglected in RM_{random}. For statistically independent n_e and b the term 0.81 $\sigma_{ne}B$ should be added to the given expression (where σ_{ne} is a variance of n_e). However, actually these quantities apparently are not statistically independent (Ruzmaikin and Sokoloff, 1977b). When fluctuations of the electron density and magnetic field are strongly correlated, the factor 0.81 in the expression for RM_{random} should be replaced by a proper weight factor.

Finally, the depolarization is caused by a finite transmission bandwidth of the telescope receiver. Any radio telescope actually receives emission within a certain wavelength interval $\Delta\lambda$ around the observation wavelength λ. The corresponding dispersion in the polarization plane rotation angle is given by

$$\Delta\varphi = 2RM_{\text{reg}}\lambda\,\Delta\lambda,$$

and the polarization degree reduces to

$$p = p_0 \exp(-\Delta\varphi) \tag{III.16}$$

(Razin, 1958).

Thus, the maximal polarization of synchrotron emission p_0 given by (III.6) might be reached only in an infinitely thin source with strictly homogeneous magnetic field when observed with a receiver of an infinitely narrow transmission bandwidth. Any deviation from each of these very unrealistic conditions results in a decrease of the polarization. While discussing numerous individiual effects, we have given the expressions (III.7—9, 13—16) for the polarization for each effect acting independently. In order to obtain the synchrotron emission polarization in the source where several effects take place simultaneously one should simply multiply out the corresponding contributions. For example, if the source is observed, with the field $\mathbf{H} = \mathbf{B} + \mathbf{b}$ (homogeneous plus random) of the extent L and the random component scale $l (l \ll L)$ at negligibly small bandwidth, the observed polarization of the emission would be

$$p \approx p_0 \frac{B_\perp^2}{H_\perp^2} \left| \frac{\sin RM_{\text{reg}} \lambda^2}{RM_{\text{reg}} \lambda^2} \right| \exp \left(-2 RM_{\text{random}}^2 \frac{L}{l} \lambda^4 \right) \tag{III.17}$$

where expressions (III.8), (III.13) and (III.14) have been used.

III.4. Light Polarization by Dust

Polarization studies of the light coming from distant stars in our Galaxy also give impressive information about the galactic magnetic field. Figure III.2 shows the starlight polarization planes in galactic coordinates l^{II}, b^{II} (the galactic center corresponds to the longitude $l^{\text{II}} = 0$, the disc plane is the equator $b^{\text{II}} = 0$). Does the picture remind you of iron filings tracing the field lines of a permanent magnet? It should be kept in mind, however, that the Galactic magnetic field which orientates the stellar light polarization planes is quite unlike the permanent magnet field, and properties of dust are far removed from those of iron filings. This section deals with the information which can be extracted from observations that produced the picture in Figure III.2.

The starlight polarization is caused by interstellar dust particles orientated by the galactic magnetic field. The dust particles have a non-spherical form, therefore the scattering and absorption of starlight occurs differently in different planes, which leads to its polarization. The correlation observed between the polarization degree and the degree of starlight reddening, also caused by the interstellar dust, is a serious argument in favour of such an interpretation. The connection of light scattering by small particles lying along the line of sight with reddening is familiar to everybody through the fascinating red colours of sunsets and sunrises. Light of unreddened stars is generally non-polarized. For the stars with red color excess, the polarization degree p has a fairly wide distribution, but its maximum is related

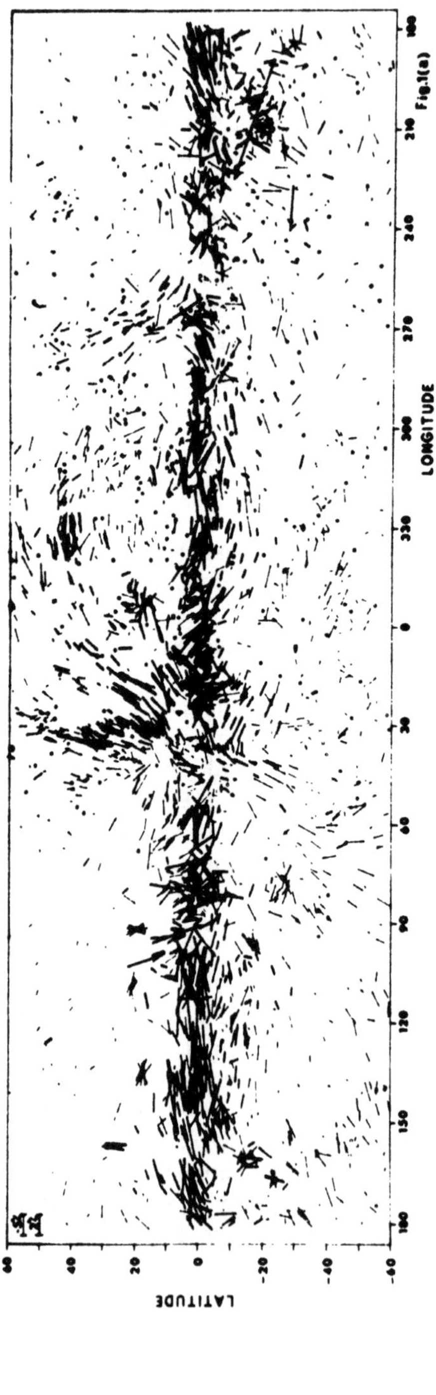

Fig. III.2. Map of the sky with orientations of electric vectors of polarized emission from distant reddened stars. These lines outline configuration of interstellar magnetic lines near the Sun (Mathewson and Ford, 1970).

to the colour excess E^{*}_{B-V}:

$$\frac{p}{E_{B-V}} \lesssim 9\% \text{ mag}^{-1} \quad \text{(III.18)}$$

where E_{B-V} is the difference between extinctions in blue and visible light. In many particular cases, however, the correlation between reddening and polarization is not very prominent (Martin, 1978, Ch. 9). According to observations, in a typical situation the polarization degree increases by 3 per cent along a path of 1 kpc in the galactic plane.

Data on the polarization of the non-thermal galactic radio background and on the Faraday rotation in the interstellar medium indicate that the galactic magnetic field is parallel, on average, to the Galactic plane. Therefore, the picture in Figure III.2 means that the dust particles absorb and scatter mainly the light with electric vector perpendicular to the magnetic field. This means that anisotropic dust particles must be aligned with their largest dimension across the field.

Maximal polarization on dust is observed in the optical range, $\lambda \simeq 4000$–6000 Å. Hence, the size of polarizing dust particles must, by the order of magnitude, be 10^3 Å $= 10^{-5}$ cm. A behaviour of interstellar absorption versus wavelength ($\propto 1/\lambda$ in the optical range, in contrast to λ^{-4} dependence for Rayleigh scattering) also indicates that the characteristic particle size must be comparable to an optical wavelength (Greenberg, 1968, Ch. 8; Martin, 1978, Ch. 3). However, enhancement of absorption in the ultraviolet range requires that several more types of dust particles of smaller size must be present. But there is no need in other types of particles to explain the observed polarization (the spectral polarization function has a single maximum which lies in the optical range), thus it is commonly assumed that particles of smaller size have a spherical form and only absorb and scatter the light without polarizing it.

In order to obtain detailed information about magnetic fields causing, eventually, the stellar light polarization, a quantitative theory of dust particle alignment by a magnetic field is needed that would be based on confident concepts of the dust particle structure. Unfortunately, neither confident concepts nor quantitative theory have been worked out yet. Nevertheless, some attempts have been made to analyze quantitatively interstellar magnetic fields through the starlight polarization observations. We discuss them below. Meanwhile, some information may be gained even without knowing specific mechanisms of alignment, coming only out of a general idea of the causal relation between starlight polarization and interstellar magnetic fields.

First of all, it can be seen from Figure III.2 that the polarization planes are maximally ordered and parallel to the Galactic plane at galactic longitudes $l^{\mathrm{II}} \simeq 140°$ and $l^{\mathrm{II}} \simeq 320°$, while in orthogonal directions the polarization planes are distributed practically randomly. This peculiarity is interpreted as an indication that in the directions $l^{\mathrm{II}} \simeq 50°$ and $l^{\mathrm{II}} \simeq 230°$ we look along the magnetic lines of the large-scale field and dispersion of polarization planes in these directions manifests a small-scale fluctuation component of the magnetic field not smoothed by the regular field. Thus, the large-scale magnetic field is directed from $l^{\mathrm{II}} \simeq 230°$ to $l^{\mathrm{II}} \simeq 50°$ in the solar vicinity. Note that starlight polarization data alone cannot determine the direction of the magnetic field, allowing also an opposite direction.

Inspection of the light polarization maps for the stars located at different distances r to the Sun shows that the loop at $-30° \lesssim l^{\mathrm{II}} \lesssim 30°$, distinctly outlined in the range $50 \lesssim r \lesssim 100$ pc, is not seen at larger distances (Mathewson and Ford, 1970). This is a local structure also seen on maps of the Galactic synchrotron radio emission (see, e.g., Berkhuijsen, 1971; Price, 1974 and references therein), known as the North Galactic Spur. The Spur is probably an envelope of an old supernova remnant (Spoelstra, 1972). The distance to it does not exceed 200 pc.

Statistical analysis of fluctuations of the starlight polarization plane orientations yields information about chaotic magnetic fields. Consider a group of stars located at the same distance from the Sun in the direction perpendicular to the large-scale field **B** (Martin, 1978, Ch. 9; Spitzer, 1978, Ch. 8). An average orientation of the starlight polarization planes is parallel to the Galactic plane and is determined by the regular magnetic field. If the polarization angle θ_p is measured from the Galactic plane

$$\langle \theta_p \rangle = 0,$$

where angular brackets denote the average over all stars observed. Consider also the dispersion of θ_p distribution around the mean value,

$$\alpha^2 = \langle \theta_p^2 \rangle.$$

We introduce the Stokes parameters (see the footnote on page 32) normalized by a total emission intensity, which characterize the polarized emission,

$$q = p \cos 2\theta_p$$
$$u = p \sin 2\theta_p$$

where p is the total polarization. When polarization is produced in randomly located clouds every one of which gives the polarization degree p_c with the polarization angle θ_{pi}, the Stokes parameter dispersions are

$$\sigma_q^2 = p_c^2 \sum_i \langle \cos^2 2\theta_{pi} \rangle \simeq p_c^2 n,$$
$$\sigma_u^2 = p_c^2 \sum_i \langle \sin^2 2\theta_{pi} \rangle \simeq p_c^2 \cdot 4\alpha^2 n, \qquad \text{(III.19)}$$

where it is assumed that $\theta_{pi} \ll 1$, and n is the number of clouds along the line of sight. Here we assume that separate terms in the sums are statistically independent, i.e. the magnetic field scales under investigation far exceed the cloud size or, in other words, $n \gg 1$.

For the open star cluster Stock 2 located at the distance 316 pc in the direction $l^{\mathrm{II}} = 133°$, $b^{\mathrm{II}} = -1.7°$ observations yield $\sigma_q \simeq 7.6 \times 10^{-3}$, $\sigma_u/\sigma_q \simeq 0.8$ and $p = np_c = 2.3 \times 10^{-2}$ (Martin, 1978, p. 156). Hence,

$$n \simeq \left(\frac{p}{\sigma_q}\right)^2 \simeq 10.$$

Now from the relation $\sigma_u/\sigma_q = 2\alpha$ we obtain

$$\alpha \simeq 0.4 \text{ rad} = 23°.$$

This quantity is a relative measure of magnetic field fluctuations:

$$\alpha\,(\text{rad}) \approx \frac{b}{B+b},$$

where a random component b of the field is assumed to be isotropic. Therefore,

$$\frac{b}{B} \simeq \frac{\alpha}{1-\alpha} \simeq 0.7.$$

Since Stock 2 is located at about 300 pc and the cloud number is about 10, the obtained value of chaotic field refers to the scale $\lesssim 10$ pc (cf. Spoelstra, 1972; Voshchinnikov and Marchenko, 1982) and characterizes magnetic fields in the dust clouds rather than typical (correlation-scale) magnetic fluctuations. On the other hand, another possible interpretation is that the fluctuation magnetic field has a multi-scale structure (Kleeorin *et al.*, 1986) and this method, being restricted to rather small scales, gives one of intermediate scales of magnetic autocorrelation function.

Information about the correlation scale of the chaotic field may be obtained by analyzing the dependence of dispersion of polarization fluctuations on the distance to a stars sample (Jokipii *et al.*, 1969). Let $p(r)$ be the polarization for a star located at distance r from the Sun. If the polarization $q(r)$ is produced per unit length, then

$$p(r) = \int_0^r q(x)\,\mathrm{d}x,$$

where

$$q(x) = q_0 + \delta q(x), \quad q_0 = \langle q(x) \rangle, \quad \langle \delta q(x) \rangle = 0$$

and the angular brackets denote quantities averaged over the whole sample of stars under investigation. Consider again the dispersion of polarization fluctuations around the average value $\langle p \rangle = q_0 r$:

$$\sigma^2(r) = \int_0^r \mathrm{d}x_1 \int_0^r \mathrm{d}x_2 \langle \delta q(x_1)\, \delta q(x_2) \rangle$$
$$= 2 \int_0^r (r - x)\, C(x)\, \mathrm{d}x, \tag{III.20}$$

where

$$C(x) = \langle \delta q(r)\, \delta q(r + x) \rangle,$$

is the autocorrelation function, and the random function $\delta q(r)$ is assumed to be homogeneous and isotropic. The function $C(x)$ changes weakly at small x but quickly decreases for $x > l$ where l is the correlation length of the polarization fluctuations. From this it is easy to estimate the integrand in (20) and obtain:

$$\sigma^2(r) \simeq \begin{cases} r^2 C(0) & \text{for} \quad r \ll l \\ 2rlC(0) & \text{for} \quad r \gg l \end{cases} \tag{III.21}$$

The dispersion obtained as a function of distance has different forms for nearby and distant star samples. The second formula in (III.21) of course agrees with (III.19). Observations are well described by expressions (III.21) allowing (Jokipii *et al.*, 1969) an estimation of the correlation length as

$$l \simeq 150 \text{ pc},$$

which may be naturally interpreted as the correlation length of magnetic field fluctuations. Somewhat larger estimate $l \simeq 225$ pc was obtained by Nee (1980) (see also Nee and Jokipii, 1979; Osborne *et al.*, 1973).

It should be noted that the statistical results concerning the interstellar polarization and absorption should be treated rather carefully because they yield only crude estimates although they give a true order of magnitude. Among embarassing though inevitable assumptions are homogeneity and isotropy of statistical properties of dust clouds spatial distribution (the clouds are known to be strongly concentrated in the disc and spiral arms, and the correlation length obtained is comparable with the dust clouds disk thickness) and the presence of a single population of clouds. Some other complications are discussed by Kaplan and Pikelner (1970, Sections 12 and 14) and Heiles (1974, Section IIIa).

In order to obtain more comprehensive quantitative information about interstellar magnetic fields one needs to know how the magnetic field aligns the dust particles. The fact that the particles must be oriented by their larger dimensions perpendicularly to the field excludes a concept of ferromagnetic particles: they would, just like iron filings, be arranged along the magnetic lines. The most popular alignment mechanism is paramagnetic relaxation proposed by Davis and Greenstein (1951) (see also Jones and Spitzer (1967) and review by Aannestad and Purcell (1973) and references therein). In a simplified form, it is as follows.

It is assumed that the cosmic dust particles responsible for the light polarization have paramagnetic properties, i.e. they are magnetized by the external static magnetic field **H** so that the induced magnetic moment per unit volume is

$$\mathbf{M} = \chi_0 \mathbf{H},$$

where $\chi_0 > 0$ is the magnetic susceptibility connected to a more widespread quantity, magnetic permeability μ, through $\mu = 1 + 4\pi\chi_0$; the magnetic induction within the dust particle is $\mu\mathbf{H}$. Paramagnetism of interstellar particles may be due, for instance, to the iron ions (an odd number of electrons in the outer orbital produces a non-zero intrinsic magnetic moment in an iron ion) or to the nuclear magnetic moments of hydrogen atoms in the ice shell of a dust particle.

A feature important for the orientation mechanism is the rotation of particles. Even if the particle motion is determined only by random collisions with atoms of ambient gas, the equidistribution of energy over the degrees of freedom in the thermal equilibrium yields

$$\tfrac{1}{2} I\omega^2 \approx \tfrac{1}{2} kT_g,$$

where I is the particle's moment of inertia, ω is the angular velocity of its rotation, T_g is the kinetic gas temperature and k is the Boltzmann constant. To give an idea of the orders of magnitude we estimate ω for spherical particles of radius $a = 10^{-5}$ cm and density 1 g cm^{-3} at $T_g = 100$ K. Then $I = 2Ma^2/5$ and

$$\omega \approx 3 \times 10^5 \text{ s}^{-1}.$$

Rather small (about 0.1%) variations of surface properties of dust particles may lead to the hydrogen atoms combining into H_2 molecules preferably in some 'active' regions over the particle surface. When escaping the particle, molecules carry away some of the kinetic energy, and if the active regions are sufficiently immobile on the surface, a 'rocket' effect arises that speeds up particle rotation (Purcell, 1975). Such a mechanism could probably spin a particle up to $\omega \approx 10^9$ s^{-1}.

The particles then must necessarily rotate, and the preferable axis of rotation seems to be the one about which the moment of inertia is maximal (see Martin, 1978, Section 8.4).

Let the rotation axis be inclined at an angle φ to the external magnetic field **H**. In the body reference frame, the particle feels the static field $H \cos \varphi$ and variable field $H \sin \varphi \, e^{i\omega t}$. Reversals of particle magnetization with frequency ω lead to dissipation of rotational moment component perpendicular to **H**, so that the rotation axis tends to align itself with the field. The dissipation is described by an imaginary part of magnetic susceptibility

$$\chi = \chi_0 + i\chi_1.$$

For sufficiently low frequencies of rotation ($\omega \lesssim 10^8\ \mathrm{s}^{-1}$ for iron ions and $\omega \lesssim 10^5\ \mathrm{s}^{-1}$ for nuclear paramagnetism of ice) χ_1 remarkably weakly depends on the particle composition being (Aannestad and Purcell, 1973)

$$\chi_1 \simeq 2.5 \times 10^{-12} \frac{\omega}{T_d},$$

where T_d is the particle temperature in K, and ω is measured in s^{-1}.

A complex value of the magnetic susceptibility means that the magnetization component arises which is perpendicular to the plane defined by the vectors **H** and $\boldsymbol{\omega}$. The magnetic moment is given by

$$\mathbf{M} = \chi_0\mathbf{H} + i\chi_1\mathbf{H} = \chi_0\mathbf{H} + \chi_1 e^{i\pi/2}\mathbf{H}.$$

The multiplier $e^{i\pi/2}$ corresponds to rotation by the angle $\pi/2$. Decomposing the particle magnetization into the components parallel and perpendicular to the rotation axis:

$$\mathbf{M} = \chi_0(\mathbf{H} \cdot \boldsymbol{\omega}) \frac{\boldsymbol{\omega}}{\omega^2} + \chi_1 \frac{\boldsymbol{\omega} \times \mathbf{H}}{\omega}. \tag{III.22}$$

The particle with magnetic moment **M** in the external field **H** is subject to the rotational torque

$$\mathbf{L} = V\mathbf{M} \times \mathbf{H},$$

where V is the particle volume. The rotational energy $K = I\omega^2/2$ obeys the equation

$$\frac{\mathrm{d}K}{\mathrm{d}t} = \boldsymbol{\omega}\mathbf{L},$$

from which we obtain, by using (22),

$$\frac{\mathrm{d}K}{\mathrm{d}t} = -V \frac{\chi_1}{\omega} [\omega^2 H^2 - (\boldsymbol{\omega} \cdot \mathbf{H})^2]$$

It is clear now that $\mathrm{d}K/\mathrm{d}t < 0$. In the stationary state $\mathrm{d}K/\mathrm{d}t = 0$ and the rotation axis is parallel to the magnetic field: $\omega H = (\boldsymbol{\omega} \cdot \mathbf{H})$. The characteristic alignment time is estimated from the last equation as

$$\tau_{\mathrm{al}} = \frac{2I}{V} \frac{\omega}{\chi_1} \frac{1}{H^2}.$$

Certainly, the particle alignment cannot be perfect due to, for example, random collisions with atoms of ambient gas. Due to collisions with atoms, particle alignment would be destroyed in a time equal, by an order of magnitude, to the period in which the mass of collided atoms becomes equal to the particle mass:

$$\tau_{\rm rel} \simeq \frac{M}{\pi a^2 n v m},$$

where M and a are the particle mass and size and v, n and m are, respectively, the thermal velocity, density and mass of a hydrogen atom. In order that the alignment mechanism be effective the following inequality should be satisfied:

$$\delta = \frac{\tau_{\rm rel}}{\tau_{\rm al}} > 1,$$

where δ for the Davis—Greenstein mechanism is given by

$$\delta \approx 3 \times 10^8 \frac{H^2}{T_d T_g^{1/2} n a}, \qquad \text{(III.23)}$$

where T_g is the gas temperature. A substantial degree of alignment means that $\delta \approx 1$. Under typical conditions, $H = 2 \times 10^{-6}$ G, $T_d = 10$ K, $T_g = 100$ K, $n = 10\ \text{cm}^{-3}$, $a = 10^{-5}$ cm and $\delta \approx 0.1$.

Note that for effective orientation of particles, aside from condition $\delta \simeq 1$, it is also necessary that the particles and gas have different temperatures (Jones and Spitzer, 1967).

As it is seen from the above estimates the mechanism proposed by Davis and Greenstein (1951) provides too weak for orientation of particles. In order to explain observations it is required that $\delta \gtrsim 0.4$ (Spitzer, 1978, Section 8.3.). The value (III.23) of δ exceeds unity only when $H \gtrsim 10^{-5}$ G. However, some modifications of the Davis—Greenstein mechanism, e.g. concentration of paramagnetic ions into clusters or spinning up the particles to very large suprathermal angular velocities by the 'rocket' effect (Purcell, 1975, 1979) seem to help in overcoming the difficulty (Martin, 1978, Section 8.4; Hong and Greenberg, 1980; Aanestad and Greenberg, 1983; Cugnon, 1983, 1985).

By knowing the orientation mechanism (for example, the one described above and subjected to necessary modifications) one may try to gain quantitative information about magnetic fields from observations. The ratio $p_{\max}/E_{B-V}$ is most sensitive to the field strength (Voshchinnikov *et al.*, 1986). For example, within the framework of the orientation mechanism of Davis and Greenstein the observed upper limit of the polarization (III.18) implies $\delta \leqslant 0.8$ for the dust grain model of Hong and Greenberg (1980). By virtue of (III.23), this gives

$$H < 1.6\, n^{1/2}\ \mu\text{G} \qquad \text{(III.24)}$$

(Voshchinnikov *et al.*, 1986). This estimate refers to the dust clouds with number densities $10 < n < 100\ \text{cm}^{-3}$.

Alignment of dust grains by magnetic fields renders interstellar medium optically anisotropic. The resulting ability of preferential extinction of the light whose electric vector is parallel to the major axes of oriented grains (known as dichroism) manifests itself in the linear polarization of the light from distant stars. This phenomenon is accompanied by linear birefringence: a phase difference between the components of the electric vector parallel and orthogonal to the magnetic field is accumulated along the pathlength through dusty interstellar space; the linear polarization is thereby supplemented by the circular one. Both theoretical arguments and observations indicate that the degree of interstellar circular polarization is very low, of order of few hundredths of percent (Martin, 1974, 1978; Martin and Campbell, 1976; Codina-Landaberry and Magalhães, 1976; Voshchinnikov, 1986; Il'in, 1987). Studies of interstellar circular polarization are especially interesting in relation to the interstellar chaotic magnetic fields because it arises only when the direction of the grain alignment varies along the pathlength. However, weakness of the circular polarization and its low sensibility to the magnetic field strength (Il'in, 1987) hampers its application to investigation of interstellar magnetic fields.

Despite the abundance of observational data and considerable theoretical developments, there are wide opportunities still unused in quantitative analysis of interstellar polarization.

The interstellar light polarization by the grains aligned by magnetic fields has been observed also in a number of external galaxies (see Elvius, 1978a, b; Martin, 1978, Ch. 12). Photoelectric polarization observations of the Andromeda Nebula show that the magnetic field there is in average parallel to the disc plane (Hiltner, 1958), just like as in our Galaxy. The polarization degree in M31 is about the same as in the Galaxy (Martin and Shawl, 1979, 1982). Detailed references about the dust observations in M31 can be found in Sharov (1982). In general, observations of optical polarization in spiral galaxies show that magnetic fields are directed along the spiral arms (Elvius, 1978a, b; Sofue *et al.*, 1986 and references therein). However, in some cases the observed polarization arises not due to transmission of starlight through clouds with aligned dust particles but rather due to scattering of light by randomly oriented dust grains. Thus, observation of optical polarization does not unambiguously indicate the presence of large-scale magnetic fields (cf. Jura, 1982). Optical polarization obviously independent of magnetic field is observed in such galaxies as M82, NGC 1068, 2685, 3623, and 4216 (Elvius, 1978b). In some cases (e.g., NGC 7331 – King, 1986) a mixture of scattered-light polarization and aligned-dust polarization is observed. A beautiful example of optical polarization pattern is given by Scarrott *et al.* (1987) for M51. These results show polarization orientations that form a spiral pattern at distances from ≈ 200 pc to 4–5 kpc from the galactic center (Figure III.3). Unfortunately, optical polarization measurements give the field orientation only leaving undetermined its direction, and thus they cannot help to distinguish axisymmetric fields

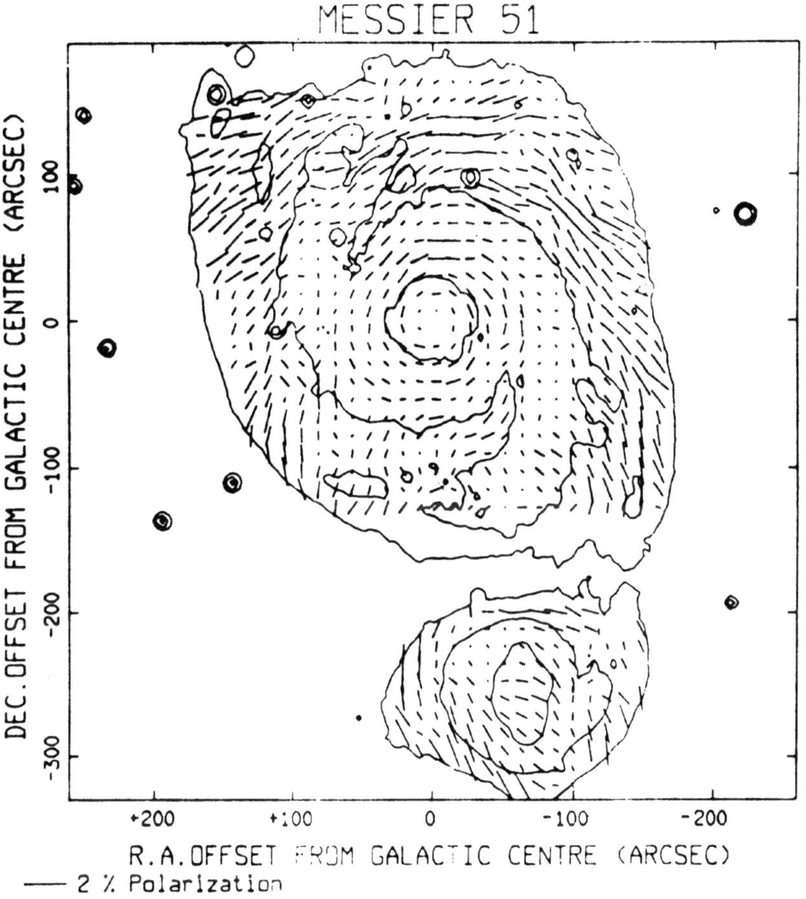

Fig. III.3. Orientations of magnetic fields in a nearby spiral galaxy M51 from optical observations of Scarrott *et al.* (1987).

from bisymmetric ones (see Chapter VII below). Note, however, that in bisymmetric structures, in contrast to axisymmetric fields, there are considerable regions with nearly zero large-scale fields and, hence, with low polarization. We also emphasize that observations in interarm regions are important, where the field direction is not affected by spiral shock waves.

III.5. Zeeman Splitting

Measuring the splitting of spectral lines in magnetic field is a classical and historically the first method of observation of cosmic magnetic fields. Only in twelve years after the discovery of Zeeman, Hale (1908) has succeeded in using it for measuring the magnetic field of sunspots. Comparison of intensities of a spectral line wings has allowed Babcock (1947) to detect the Zeeman splitting in the emission of distant peculiar magnetic stars. Ten years later Bolton and Wild (1957) proposed to use the Zeeman splitting of the 21 cm neutral hydrogen absorption line to measure magnetic fields in the interstellar medium. Such measurements were achieved a further ten years later (Verschuur, 1968).

Detection of the Zeeman effect in the interstellar space requires very long observation times. Verschuur (1979) estimates the observation time consumed by the first five confident detections as 5000 hours!

The difficulty of detection is due to the weakness of interstellar magnetic fields. The interstellar Zeeman splitting proves to be far less than the Doppler width of spectral lines. Hence the effect can be registered more easily in those sites where the field is stronger and the gas is colder: in dense clouds ($n \gtrsim 10^3$ cm^{-3}), where the gas is mainly in a molecular form, or in still more dense regions ($n \simeq 10^4$–10^6 cm^{-3}) with maser emission. The scales of such objects are very small in comparison with the galactic ones, and it is a difficult problem to establish the relation between the magnetic fields registered and the Galaxy's general field. In the present context, the most valuable measurements of the Zeeman effect are those for relatively rarefied, gravitationally unbound clouds of neutral hydrogen which may be considered as hierarchical elements of turbulent motions of interstellar gas. In this section we briefly describe the Zeeman effect and basic ideas used for its interpretation by astrophysicists.

In the absence of external fields, atomic energy levels do not depend on the direction of the total angular momentum (orbital **L** plus spin **S**) of electrons. In other words, the energy levels are degenerate with respect to the momentum directions. In the magnetic field **H** that distinguishes a certain direction, an atom acquires the additional energy $-\mu(\mathbf{L} + 2\mathbf{S}) \cdot \mathbf{H}$ which depends on the orientation of the angular momentum with respect to the magnetic field (here $\mu = e\hbar/2mc = 9.3 \times 10^{-21}$ erg G^{-1} is the Bohr magneton). Now the atomic state depends on the angular momentum projection onto the direction of the magnetic field, i.e. the degeneracy is removed. The energy levels split into $2j + 1$ equidistant levels, where j is the quantum number of the total angular momentum $\mathbf{J} = \mathbf{L} + \mathbf{S}$. The energy levels are given by

$$E_H = E_0 \pm \mu g M H, \qquad M = 0, 1, \ldots, j.$$

(Landau and Lifshitz, 1958). The factor g is called the Lande factor. In particular, it appears due to the fact that the mechanical and magnetic momenta are related differently for the electron's orbital motion (M_e, l) and spin (M_s, s): $M_l = -(e/2m)l$ and $M_s = -(e/m)s$. At $s = 0$, $g = 1$, and at $l = 0$, $g = 2$.

The selection rules dictate that transitions are allowed between the levels for which $\Delta M = 0, \pm 1$. If the Lande factor is the same for the upper and lower levels, the spectral line is split into a triplet (normal Zeeman effect):

$$\nu_\pi = \nu_0, \nu_\sigma = \nu_0 \pm g(e/4\pi mc)H$$
$$= \nu_0 \pm 1.4g(H/10^{-6}\,\mathrm{G})\mathrm{Hz},$$

where ν_0 is the basic, unshifted frequency. In a general case, when the upper and lower levels are characterized by different Lande factors the number of components may be larger (anomalous Zeeman effect). In the astrophysical conditions the normal Zeeman effect is observed in the λ 21 cm neutral hydrogen absorption line, and the anomalous Zeeman effect, in the λ 18 cm OH molecule line.

The unshifted component of the triplet, ν_π, is linearly polarized along the field, and emission at this frequency is concentrated around the direction perpendicular to the magnetic field. The σ-components, ν_σ, are generally elliptically polarized. In observing along the field, when the π-component cannot be seen, their polarization degenerates into the circular one with the line of higher frequency having the left polarization and the low-frequency line, the right polarization. When the observation is made across the field, the π-componet is visible and the complete triplet is observed. In this case the σ-components are polarized perpendicularly to the field direction. The total (integrated along all directions) intensities of the Zeeman components are equal to each other. From the Zeeman splitting observations one can, in principle, determine both a total strength and direction of the magnetic field by knowing how intensities and polarizations of the spectral components depend on an angle between the magnetic field and the line of sight. However, the observational restrictions allow, as a rule, only information about the longitudinal field component. As was noted above, the smallness of level splitting (i.e. magnetic field weakness) and a large Doppler broadening make it difficult to use the Zeeman effect for measuring interstellar magnetic fields. For the fine structure λ 21 cm line (ν_0 = 1420 MHz) of neutral hydrogen the Zeeman splitting in the field 10^{-5} G is about 30 Hz, while the line half-width due to the thermal Doppler broadening is $\Delta\nu = \nu_0 v_T/c \approx 10^4$ Hz at T = 100 K (where v_T is the thermal velocity). Therefore, in order to detect a weak magnetic field by means of the Zeeman effect the absorption lines are usually chosen in spectra of radio sources projected on the neutral hydrogen clouds: absorption lines arise in colder gas than do emission lines.

Weakness of the Zeeman splitting in the interstellar medium (in fact, a correct description is abnormal broadening rather than splitting of a spectral line) permits the Zeeman effect registration only in the regions where interstellar fields are stronger and the gas temperature is lower than the average, i.e. in sufficiently dense clouds. An important advantage of this method of field registration is that no additional information is necessary that would require observational determination, such as a density of thermal electrons and an extension of active region for the Faraday effect, a density of relativistic electrons for synchrotron emission or physical properties of dust particles for optical polarization observations. However, this merit of the Zeeman effect is counterbalanced by the narrow-range and extremal conditions in which it is useful. In fact, the method may be applied for measuring magnetic fields in objects of lesser than galactic scales. The Zeeman effect is indispensable for studying very interesting regions of the interstellar medium, centers of star formation. However, the information about the general interstellar magnetic field obtained with its aid cannot be considered direct and reliable (cf. Heiles, 1976).

In order to estimate an average field in the interstellar medium by basing it on data about the field in the clouds, one should solve an inverse evolution problem and learn how to determine initial parameters of the cloud through its observed characteristics. Attempts were made to estimate a strength of the average magnetic field in the Galaxy by observations of magnetic fields in the clouds through adoption of a certain relationship between the magnetic field strength and the gas density. The simplest argument is as follows. At the isotropic compression of the interstellar gas cloud, the gas density increases as $n \propto R^{-3}$, where R is the cloud radius, while the magnetic field strength grows as R^{-2} if the field is strictly frozen into the isotropically compressing gas and the flux HR^2 is conserved. Under these conditions the field grows with the gas density according to the law $H \propto n^{2/3}$. Such a relation was expected in early observations of magnetic field in clouds, which leaded Verschuur (1970) to a reasonable estimate of the average galactic field as 1—3 μG from data for 9 clouds. Further investigations showed (Mestel and Paris, 1984; Mouschovias, 1976a, b; Ruzmaikina, 1983) that the magnetic field considerably affects the cloud formation process making it anisotropic in the most dense parts. Therefore, the magnetic field itself grows slower than by the law $n^{2/3}$. As a result, the exponent q in the law $H \propto n^q$ lies somewhere in the range $\frac{1}{3} \lesssim q \lesssim \frac{1}{2}$ for dense interstellar clouds (Mouschovias, 1976a, b). In addition, one cannot be sure that the magnetic field may be considered frozen-in at all stages of the cloud compression. If the field dissipation is due to the ambipolar diffusion then the condition for which the field may be considered frozen-in is as

$$H(G) < 10^{-16} x^{1/2} n R \tau_c^{-1/2} \qquad \text{(III.25)}$$

(Ruzmaikina, 1983) where x is the gas ionization degree, n is the gas density, R is the cloud radius and τ_c is the characteristic evolution time (for example, the free fall time). At the initial stage of compression we may assume that it is one-dimensional and $x \propto n^{-1}$. Then (III.25) takes the form

$$H \lesssim 3 \times 10^{-24} x_0^{1/2} n_0^{3/2} R_0 n^{-1/4}.$$

For typical initial values $x_0 = 3 \times 10^{-2}$, $n_0 = 1\ \text{cm}^{-3}$, $R_0 = 10$ pc,

$$H < 5 \times 10^{-3} n^{-1/4} \text{G}. \qquad \text{(III.26)}$$

Parameters of the most dense clouds investigated by Verschuur (1970) are $H \sim 10^{-4}$ Gauss and $n \sim 10^3\ \text{cm}^{-3}$. These values obviously satisfy inequality

(III.26), which allows to consider the magnetic fields as frozen into the clouds. By comparing (III.26) with inequality (III.24) we may easily confirm for ourselves that the field may be considered frozen-in for the dust clouds with $10 \lesssim n \lesssim 100\ cm^{-3}$ which produce polarization of starlight. Thus, deviations from the law $B \propto n^{2/3}$ in clouds of rather low density ($n \lesssim 10^3$—$10^4\ cm^{-3}$) seem to be associated not with magnetic dissipation but rather with the anisotropic character of cloud compression. The field effect upon dust clouds, which is manifest in their extension along the magnetic lines, is discussed in the next section.

In most dense interstellar clouds ($n > 100\ cm^{-3}$) hydrogen is present mainly in molecular form. In this case the λ 21 cm line is weak, and the Zeeman splitting may be observed in the λ 18 cm OH molecule emission and absorption lines (Turner and Verschuur, 1970; Crutcher *et al.*, 1975; Lo *et al.*, 1975; Crutcher and Kazès, 1983; Heiles and Stevens, 1986; Kazès and Crutcher, 1986; see also Chaisson and Vrba, 1978). Magnetic fields observed in strong hydroxyl maser sources by the λ 18 cm line splitting do not exceed 10^{-2}—10^{-3} Gauss at densities 10^6—$10^9\ cm^{-3}$.

In the HII regions where hydrogen is present mainly in the ionized state, the Zeeman splitting of hydrogen recombination lines may, in principle, be measured. However, the observations allowed Troland and Heiles (1977) to obtain only the upper limits of the magnetic field strength. Observations of the Zeeman splitting in ionized carbon recombination lines arising in dense molecular clouds allowed Silvergleit (1984) to derive the upper estimates of magnetic fields in these regions as few milligauss and to draw a conclusion that the exponent q in the law $H \propto n^q$ is smaller than 2/3.

A new step in the application of the Zeeman effect measurements to the investigation of interstellar magnetic fields is made by Schwarz *et al.* (1986) (see also Bregman *et al.*, 1983) who have produced maps of magnetic field distribution in the direction of the supernova remnant Cas A. The gas that produces the observed HI absorption lines belongs to the Perseus arm. The angular size of the investigated region is $\sim 5'$ which corresponds to the linear size ~ 4.5 pc at distance ~ 3 kpc; the linear resolution of magnetic field mapping is 1 pc. The magnetic fields are very inhomogeneous: they reach values as high as 40 μG although the median value is 10—15 μG. In the whole observed region (galactic coordinates $(l^{\mathrm{II}}, b^{\mathrm{II}}) = (111°.7, -2°.1)$) the field is directed away from the observer, i.e. it has the same direction as the large-scale field in the local Orion arm. Note that Faraday rotation data indicate that the large-scale fields in the Perseus and Orion arms are oppositely directed (Agafonov *et al.*, 1987; see, however, Vallée, 1983b).

A curious situation is revealed by comparison of Zeeman measurements in HI and OH lines at the same direction of Cas A (Heiles and Stevens, 1986). Hydroxyl lines arise in dense molecular clumps while absorbing neutral hydrogen occupies less dense shells of these clumps. Thus, the result is very unexpected: the field in less dense shells, 50 μG, is 5 times stronger than in dense molecular cores!

Accumulation of field measurements in interstellar clouds allows us to infer observationally a relation between the field strength and gas density in the clouds. Figure III.4, taken from Troland and Heiles (1986), compiles the data for clouds whose density spans a very wide range of five orders of magnitude. Observational data fit well the theoretical predictions of Mouschovias (1976a, b), Scott and Black (1980), Mestel and Paris (1984) and Mouschovias (1985). According to the theory, at initial stages of cloud formation with rather strong magnetic field, clouds contract along magnetic lines so that the magnetic field is almost invariable within clouds. After the density is reached at which the effects of self-gravitation become considerable ($\sim 10^2$ cm^{-3} for a 10^3 $M_\odot$ cloud), contraction across magnetic lines begins and the internal field begins to grow. The magnetic field plays an essential role in the cloud evolution making contraction anisotropic and the exponent k in the relation $B \propto n^k$ is less than $\frac{2}{3}$, which value is typical of isotropic contraction. Probably the value of k lies between $\frac{1}{3}$ and $\frac{1}{2}$. In order to see how the magnetic field affects the cloud contraction, consider the following simplified model (Mestel and Paris, 1984). Consider a cloud which is contracting into an oblate shape (initially, along magnetic lines) with self-gravity balanced by internal pressure:

$$2\pi G\rho h^2 = a^2$$

where ρ and h are the cloud density and half-thickness, respectively, and a is the isothermal sound speed. Conservation of mass gives

$$M = \frac{4\pi}{3}\rho h R^2 = \text{const},$$

where R is the cloud radius, while conservation of the magnetic flux reads as

$$\Phi = \pi B R^2 = \text{const}.$$

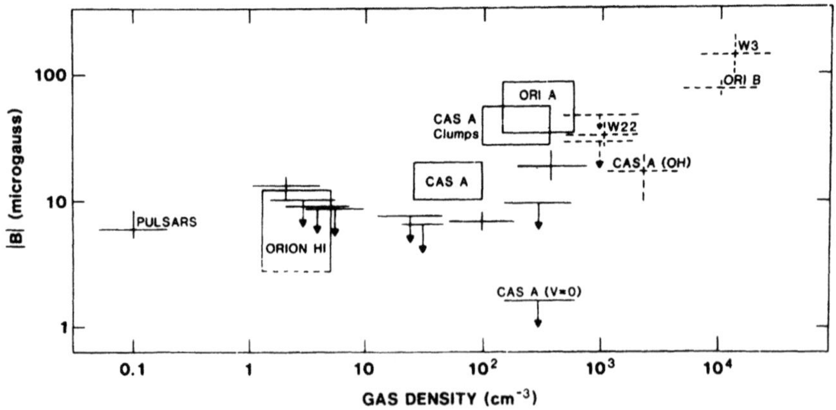

Fig. III.4. Strength of magnetic fields in interstellar gas clouds versus cloud density according to Zeeman measurements (Troland and Heiles, 1986).

Combination of these three relations gives

$$B/\rho^{1/2} = \frac{\Phi}{M}\,(8a^2/9\pi G)^{1/2},$$

For a cloud with fixed Φ and M this predicts $B \propto \rho^{1/2}$.

According to detailed calculations, magnetic fields are frozen into the cloud gas up to densities 10^5 cm^{-3}. At higher densities the magnetic flux leaks out of clouds due to ambipolar diffusion.

III.6. Other Methods

Aside from the methods for observational investigation of the magnetic fields which have been described above, there are some other methods, more restricted in their applications.

In the mid-fifties, when the only evidence of the magnetic field in the interstellar medium was non-thermal radio emission and starlight polarization, an important role was played by the observation of Shajn (1955) that the elongated shape and orientation of diffuse interstellar nebulae could be explained by an effect of magnetic fields that permeate the interstellar gas as a whole. It is essential that an idea of a large scale of the field, comparable with the Galaxy's size, was employed.

Further investigations showed that shapes and orientations of interstellar gas clouds are not always associated with the influence of a magnetic field. Many examples are known where the magnetic field within the cloud is directed along its smaller dimension or is perpendicular to the Galactic plane. Therefore, shapes of gas clouds cannot serve as a reliable indicator of the orientation nor of the strength of the Galactic large-scale magnetic field. However, Shajn's ideas cannot be overestimated in the context of the historical development of views on galactic magnetic fields. Vertical dust lanes, apparently associated with poloidal large-scale magnetic fields, were recently noticed in some galaxies by Sofue (1987).

Recently it was proposed to use observations of maser emission and polarized molecular radio lines for studying the interstellar magnetic fields (Litvak, 1971; Goldreich *et al.*, 1973; Varshalovich and Burdjuzha, 1975). Polarization of the maser emission coming from the most dense regions of the interstellar medium, where star formation takes place, allows a successful estimation of the field strength (see Western and Watson, 1984 and references therein). One more way to detect interstellar magnetic fields is associated with the polarization of molecular radio lines, for example CO, which arises when the optical properties of the source become anisotropic due to a magnetic field. This polarization is related to Zeeman splitting and, in principle, may be used for measurement of the magnetic field in an emitting region. Both emission (Goldreich and Kylafis, 1981, 1982) and absorption (Kylafis and Shapiro, 1983; Kylafis, 1983) radio lines can be used for these observations. However, the effect has not yet been detected observationally (Wannier *et al.*, 1983; Deguchi and Watson, 1984; Barvainis and Wootten, 1986).

Many methods have been developed for measuring the magnetic fields of astronomical objects (e.g., neutron stars and pulsars, stellar coronae, etc.) but they are not applicable to the investigation of relatively weak interstellar fields, so we do not discuss them here. Details about these methods may be found in Zeldovich *et al.* (1983).

III.7. Discussion

We have described the basic ideas underlying various methods of detection and measurement of magnetic fields in the interstellar medium of our Galaxy and external galaxies. The basic method of measuring magnetic fields in external galaxies is registration of the Faraday rotation. As for observations within our Galaxy, all the methods described may be used here but it is easy to see that they yield somewhat differing values of the field strength.

The typical values of fields detected by means of the Zeeman effect are tens of microgauss and even fields of a few milligauss are encountered. At the same time, the average field in the whole bulk of the interstellar medium is only 2 μG. The Zeeman effect observations reveal relatively strong fields in dense interstellar clouds. Such gravitationally bound objects live their own lives, and the internal magnetic fields often cannot be considered as a part of a general galactic field. Typical dimensions of the regions that produce the observable Zeeman effect do not generally exceed a few parsecs, and often are measured in astronomical units (for example, the maser emission regions).

Other manifestations of magnetic fields — synchrotron emission, Faraday rotation and optical polarization — bring information from more vast regions. Therefore, a necessity arises to estimate typical scales which are sampled by some method, especially as different methods yield somewhat different estimates of the field parameters (see also Spoelstra, 1977). For example, according to observations of starlight polarization, the field near the Sun is directed to $l^{\mathrm{II}} = 50°$. Data on the polarization and intensity of synchrotron emission give about the same result (Spoelstra and Brouw, 1976), while an analysis of the Faraday rotations for pulsars and extragalactic radiosources yields the large-scale magnetic field direction as $l^{\mathrm{II}} = 90°$ (Manchester, 1974; Ruzmaikin and Sokoloff, 1977b, c). Slight differences between the values obtained by various methods are interpreted as a result of the difference in sizes of regions sampled by each method. What are these sizes?

Let us begin with the starlight polarization. The polarization observations are available for stars located at distances as large as 4 kpc. However, basing their study on data about the polarization of stars from several clusters, Verschuur and other authors (see Verschuur 1970 with references therein) agree that the polarization observed is produced in the close vicinity of the Sun, no farther than 500 pc. Note that for the strictly homogeneous magnetic field the polarization degree of the starlight would grow linearly with distance. In a completely chaotic field the root mean square value of polarization grows linearly at distances less than the correlation field length and as the square root of the distance for more distant stars (see (III.21)). Obviously, observations must give a result somewhere between these two extremes. More extended regions of the size up to 10 kpc probe the observations of interstellar polarization in the near infrared range (Okuda, 1985; Kobayashi *et al.*, 1986); only preliminary results of such observations have been published yet.

Polarization of the Galactic non-thermal radio background reflects properties of a smaller region. Investigation of the linear polarization at five wavelengths performed by Spoelstra and Brouw (1976) has shown that the typical value of the Faraday rotation measure is 8 rad m^{-2}. With the average electron density $n_e \simeq 0.03$ cm^{-3} and the magnetic field $B \simeq 2\,\mu$G, the path length over which such *RM*s are produced can be easily estimated as $L \simeq 150$ pc. Note that for the wavelength 49 cm the synchrotron emission of interstellar gas is depolarized by the Faraday depolarization over the path length $L \simeq 250$ pc which follows from the equality $RM\lambda^2 = \pi$ for $B = 2\,\mu$G.

In observing the pulsar rotation measures, a scale of the registered field is determined, certainly, by a distance to the respective pulsar (in practice, by an average distance to some sample), i.e. a few hundred to a few thousand parsecs. It is more difficult to determine over what distance the Faraday rotation of extragalactic sources is created. A typical value of the rotation measure for them is 50 rad m^{-2} (Spoelstra, 1977). At the above values of n_e and B it corresponds to the distance $L \simeq 1$ kpc comparable with Galactic dimensions. Agreement between the field directions determined from pulsars, Galactic sources, and extragalactic sources as well as very stringent limits on the possible intergalactic magnetic field ($\lesssim 10^{-11}$ Gauss, see Chapter V) indicate that the Faraday rotation occurs primarily within the Galaxy and reveals the Galactic magnetic field at its largest scale.

Thus, magnetic fields of the largest scale are revealed by the Faraday rotation. Various methods of magnetic field detection sample the scales rather regularly distributed in the range from several kiloparsecs to several astronomical units.

Heiles (1976) expresses the opinion that the Faraday rotation occurs mainly in a rarefied, hot inter-arm gas, while relatively dense, cold regions within the local spiral arm make a dominant contribution to the optical polarization. In this case the difference between optical and radio data would be due to the difference between directions of magnetic field in the spiral arms and between them. However, a detailed analysis of the Faraday rotation in the Galaxy shows (see Chapter IV) that the observed rotation is produced, almost completely, within the local spiral arm. The contribution of the inter-arm gas seems to be negligible, possibly due to both lower electron density and magnetic field. In either event, the difference between the directions of the Galactic magnetic field given by different methods is interpreted as an actual difference between the average field directions in the regions sampled by different methods. A beautiful model of the field configuration that embraces the available data is a local loop of magnetic lines with dimensions 100—200 pc (Gardner *et al.*, 1969; Vallée and Kronberg, 1973).

As we have already noted, the interstellar medium is extremely inhomogeneous with at least three different phases being present. The question (see also Heiles, 1976) naturally arises: which phase the magnetic field registered by one or another method belongs to?

Let us begin with estimation of the contributions of various phases of the interstellar medium to the observed Faraday rotation measures. The neutral hydrogen clouds are characterized by the electron density $n_e \simeq 0.05$ cm^{-3}, the magnetic field $H \simeq 10$ μG and the size 10 pc; there are 3—10 clouds along the line of sight of 1 kpc. Thus, on the path of 1 kpc the HI clouds contribute to the rotation measure $RM_{\text{cloud}} \simeq 10$—40 rad m^{-2}. In the intercloud medium $n_e \simeq$ 0.02 cm^{-3} and $B \simeq 2$ μG. Since the clouds occupy negligibly small volume of the galactic disc while the hot phase may possibly occupy up to 50 per cent of the volume, the segments of 500 pc fall on intercloud medium and hot tunnels from the total path length of 1 kpc. Therefore, in the intercloud medium the rotation measure $RM_{\text{warm}} \simeq 15$ rad m^{-2} is produced. In the hot tunnels the electron density is only 0.002 cm^{-3}. Meanwhile, magnetic field in tunnels can be probably as strong as in the intercloud medium — this follows from equipartition arguments since the pressures in both phases are comparable. Therefore, the hot tunnels produce only ten per cent of the rotation measure produced in the warm intercloud medium. Hence, the basic and comparable contributions to the observed rotation measures, 50 rad m^{-2}, of pulsars and extragalactic radiosources come from the neutral hydrogen clouds and the intercloud medium.

The intensity of synchrotron emission is proportional to $KH^{\alpha}l$, where K is related to the relativistic electron density, $\alpha \approx 1.8$ and l is the size of an emitting region. If the density of cosmic rays changes only weakly with position (Ginzburg, 1984, Sect. 6.9) then the largest contribution to the observed non-thermal background of the Galaxy is made by the intercloud medium and the hot tunnels occupying a major fraction of the interstellar volume and possessing comparable magnetic fields. The 'thick disc' radioemission (e.g., Beuermann *et al.*, 1985) can be probably attributed to the hot phase of the interstellar gas.

Contributions to the starlight polarization from the clouds and intercloud medium are about the same (Heiles, 1976). In the hot ambient gas of the tunnels dust grains are destroyed rather rapidly. However, some grains may survive even there. The lifetime of graphite grains in the ambient gas with $T = 10^6$ K and $n = 2 \times 10^{-3}$ cm^{-3} is as large as 10^9 years (Burke and Silk, 1974). However, ice mantles of silicate grains, that produce the observed polarization, are evaporated in much shorter times. Barlow (1978a, b, c) and Draine and Salpeter (1979) predict rather low lifetimes not exceeding 10^7—10^8 years for the ice mantles of dust particles. Barlow concludes that the ice mantles can survive only in dense regions of interstellar space. Therefore, it seems that the hot tunnels hardly can contribute significantly to the starlight polarization.

CHAPTER IV

INTERPRETATION OF OBSERVATIONAL DATA

IV.1. The Magic of Data Processing

Astrophysics deals with objects remote in space and time. Active experiments are impossible (except for within the solar system), and observational data are usually incomplete. So, there is a vast field of interpretations, stretching between observations and a final picture of a phenomenon. This field often begins with data processing.

Here we show some of the subtleties of the primary interpretation on an example of the determination of Faraday rotation measures. The object is to obtain the RM coefficient in the following expression for the positional angle determining a polarizational plane of radio emission of a distant source (a pulsar, radio galaxy or quasar):

$$\varphi = \varphi_0 + RM\lambda^2 \quad \text{(IV.1)}$$

(see Section III.3), where φ_0 is an intrinsic positional angle formed within the source, and λ is a wavelength of the radiation detected by a radio telescope.

At first glance, the problem looks simple: one should draw the straight line (IV.1) through the observational points (λ^2, φ). Then to obtain the unknown parameters RM and φ_0, one should have measurements at least at two points. Due to inevitable random errors $\Delta\varphi_i \equiv \sigma_i$, some statistical data processing method is required. It seems that the simplest least-square method would suffice. In fact the situation is more complicated because the positional angle is determined with an ambiguity of the addition of an integer number of half-revolutions $k\pi$, $k = \pm 1, \pm 2, \ldots$ which leads to an uncertainty in RM and φ determined by the least-square method (see Figure IV.1). In practice the problem may sometimes be solved by choosing a minimal possible value of the rotation measure (Morris and Tabara, 1973) if the other obtained values of RM are great enough to be implausible purely by astrophysical considerations. It is often the case, however, that a few rotation measures prove comparable in value and plausible from the astrophysical point of view, and the exact statistical criteria are required for an unambiguous choice of RM and φ_0. Two similar statistical methods of such a choice were proposed by Vallée and Kronberg (1975) and Haves (1975). Both are based in fact on a minimization procedure of the least-square method and *a priori* restriction of the rotation measure value $|RM| < RM_{max}$. Unfortunately, in a

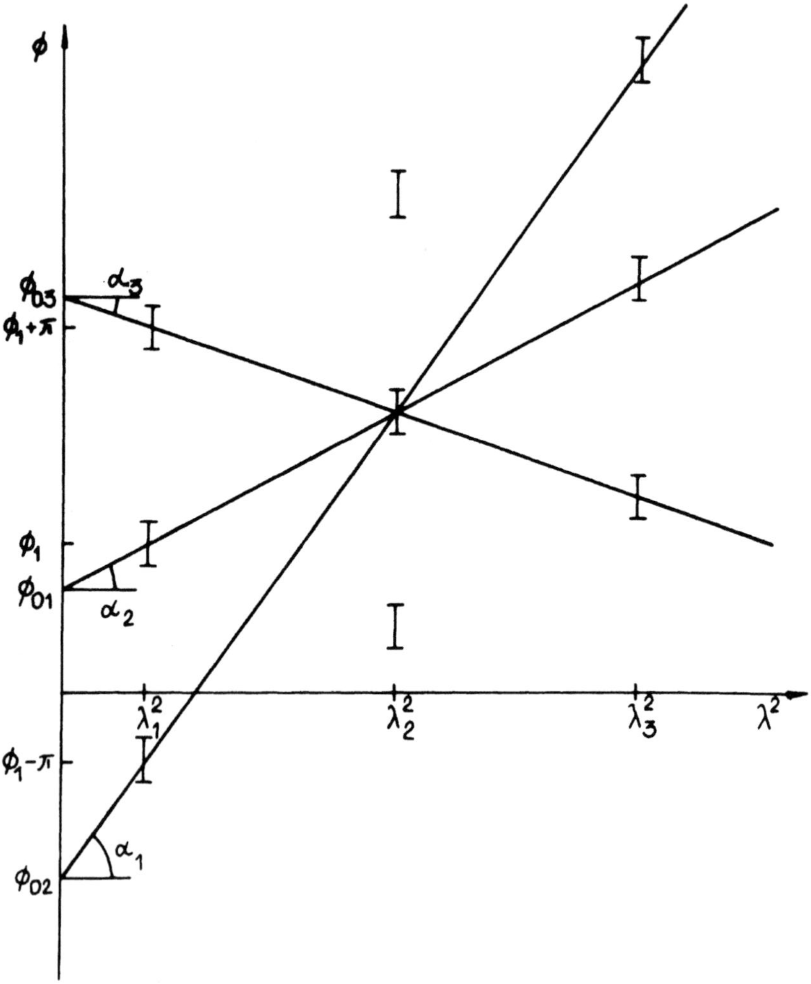

Fig. IV.1. A simple illustration of ambiguity in determination of the Faraday rotation measure and intrinsic position angle. Since position angles of polarized emission at three wavelengths, λ_1, λ_2, and λ_3, are determined with ambiguity of addition of an integer number of half-rotations, π, three straight lines can be drawn on the plane (φ, λ^2) which gives three different pairs: φ_{01} and $RM_1 = \tan \alpha_1$, φ_{02} and $RM_2 = \tan \alpha_2$, φ_{03} and $RM_3 = \tan \alpha_3$.

number of cases (Ruzmaikin and Sokoloff, 1977c) these methods yield contradictory results that, in addition, depend (Haves, 1975) on a choice of $RM_{\max}$, varying considerably when $RM_{\max}$ is changed only by a factor of a few times. It is worth noting that a reasonable estimate for $RM_{\max}$ is difficult to obtain from astrophysical considerations.

Now we describe a method for determination of the rotation measures and intrinsic positional angles from the observational data which does not involve arbitrary restrictions on RM (Ruzmaikin and Sokoloff, 1979). An important feature of the method is due to the fact (Manchester, 1972) that an ambiguity in RM and φ_0 connected with the addition of the integer number of π, vanishes in the limit of closely spaced wavelengths. This fact allows us to drop the *a priori* assumption about restriction of the rotation measures by a certain value. On the other hand, an adequate realization of the method requires the specially planned observations:

(1) Among the data obtained, the measurements at two sufficiently close wavelengths must be present (a criterion for closeness, see below).

(2) Other wavelengths at which the observations are performed must be distributed according to a certain law (see formula (IV.6)).

The statement (1) implies, in particular, that to determine very large rotation measures one needs observations at two sufficiently close wavelengths. The required restrictions are not severe, and the method works well for observations not planned beforehand. There are a number of sources, however, for which RM and φ_0 cannot be reliably determined without special observations.

An algorithm for calculation of the rotation measure and the internal positional angle is as follows. Let N measurements of angle φ_i, $i = 1, 2, \ldots N$ be known at the wavelengths λ_i, with standard errors σ_i. We arrange the measurements in an order of increasing $|\lambda_i^2 - \lambda_1^2|$. We assume as usual that the λ_i are determinate, i.e. do not contain random errors. At the end of this section we discuss the peculiarities associated with errors in the determination of the wavelengths.

We shall call the true positional angle ψ_i a continuous function of λ which possibly differs from the observed φ_i by an integer number of half-rotations,

$$\psi_i = \varphi_i + k_i \pi.$$

Without loss of generality we may put $\psi_1 = \varphi_1$. Our basic assumption is as follows. Let the wavelengths λ_1 and λ_2 be so close that $|\varphi_2 - \varphi_1|$ is less than twice the measurement error of the position angles. Then it is natural to assume that the values of true position angles are also close, i.e. the same number of half-rotations $k\pi$ should be added to φ_1 and φ_2. In practice position angles are not determined very accurately (often within tens of degrees), and we use a crude criterion given by

$$|\psi_2 - \psi_1| < \pi/2. \tag{IV.2}$$

Obviously, ψ_2 coincides either with φ_2 or $\varphi_2 \pm \pi$. If the condition (IV.2) is satisfied the straight line (IV.1) can uniquely be drawn in the plane (φ, λ^2) through the first two points (ψ_1, λ_1^2) and (ψ_2, λ_2^2). More exactly, due to the errors σ_1 and σ_2 we can draw a set of close straight lines, which give

$$\begin{aligned} RM &= (\psi_2 - \psi_1)/\Lambda_{21} \\ \varphi_0 &= (\psi_1 \lambda_2^2 - \psi_2 \lambda_1^2)/\Lambda_{21} \end{aligned} \tag{IV.3}$$

with standard errors

$$\tfrac{1}{2}\mathscr{D}\,RM = (\sigma_1 + \sigma_2)/\Lambda_{21}$$
$$\tfrac{1}{2}\mathscr{D}\,\varphi_0 = (\sigma_1\lambda_2^2 + \sigma_2\lambda_1^2)\Lambda_{21},$$

where $\Lambda_{21} = \lambda_2^2 - \lambda_1^2$.

The rotation measure thus determined cannot exceed $\pi/2\Lambda_{21}$. For example, for $\lambda_1 = 1$ cm and $\lambda_2 = 2$ cm this gives $|RM| \leqslant 5 \times 10^3$ rad m^{-2}. Thus within the observational possibilities available the method allows, in principle, the calculation of the rotation measures even for sources with very large *RM*.

A bundle of straight lines going through λ_1 and λ_2 forms a cone (see Figure IV.2 where for greater confidence the 4σ intervals for φ_1 and φ_2 are chosen). A cross-section of the cone by the straight line $\lambda = \lambda_3$ gives the range of possible values of the position angle ψ_3. As permissible we should take those values of $\psi_3 = \varphi_3 + k\pi$

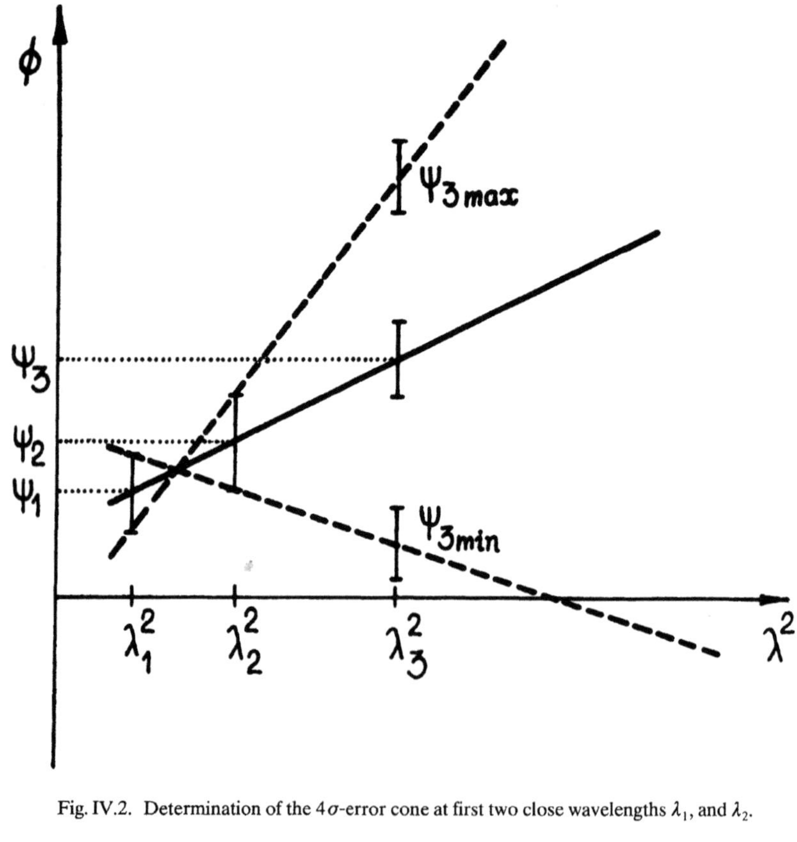

Fig. IV.2. Determination of the 4σ-error cone at first two close wavelengths λ_1, and λ_2.

which are found within the cone. Then we draw the straight lines through newly determined points and two previously established ones by the least-square method. Each of these straight lines gives rise to its own cone due to the errors σ_1, σ_2 and σ_3, the cross-section of which by $\lambda = \lambda_4$ allows us to obtain a choice of permissible values of the fourth position angle and to draw the straight line (or a few lines) through the four points altogether, etc. Two possible problems are evident. Firstly, the next position angle $\varphi_i + k_i\pi$ may not get into the cone. Secondly, the situation may arise when straight line cannot be drawn through all points within the cone with a desired degree of confidence. Hence, the numerous branchings are unlikely and the procedure may terminate before all observational points have been used. Practically it means that either the linear relation (IV.1) is invalid or some specific measurement is wrong. The latter possibility can be easily verified by eliminating the fall-out point from an analysis and repeating the procedure again. If the process goes as far as to the very end, i.e. all N observational points are located at the single straight line, then the unique values of RM and φ_0 are obtained which can be accepted as the true values of the rotation measure and position angle. If several values of RM and φ_0 are still admissible then a unique determination requires additional specialized measurements. Let us give a recommended set of observations.

The simplest way of obtaining a unique result is to take a current wavelength λ_i close to the previous one so that only a single ψ_i value fits within the error cone. For a practical crude estimate of the cone opening at the ith wavelength we may use the expression

$$\Delta\psi_i \equiv \psi_{i\max} - \psi_{i\min} \simeq 2p(\sigma_1\Lambda_{i,\,i-1} + \sigma_i\Lambda_{i,\,1})\Lambda^{-1}_{i-1,\,1}, \tag{IV.4}$$

where $\Lambda_{i,\,i-1} = \lambda_i^2 - \lambda_{i-1}^2$, the value $p\sigma$ (taken equal to 4σ in Figure IV.2) determines the standard confidence level for which the cone is given. The expression (IV.4) is exact for $i = 3$. In order to get no more than one value of ψ_i within the cone opening thus obtained we must obviously require

$$\Delta\psi_i \leqslant \pi - 2\sigma_i. \tag{IV.5}$$

We then obtain the following sequence of desired wavelengths putting for simplicity $\sigma_i = \sigma_1 \equiv \sigma$:

$$\lambda_i^2 \lesssim \frac{1}{2}\left[\frac{(\pi - 2\sigma)\Lambda_{i-1,\,1}}{2p\sigma} + \lambda_{i-1}^2 + \lambda_1^2\right]. \tag{IV.6}$$

For illustrative purposes we consider a simple example when $\lambda_1 = 1$ cm and $\lambda_2 = 4$ cm. According to relation (IV.2) these values allow calculation of $|RM|$ up to $\sim 10^3$ rad m^{-2}. Let $\sigma = 10°$, then from (IV.6) we obtain a recommended sequence of wavelengths at which the observations should be made: $\lambda_{3,\ldots} = 5.6$; 7.7; 10.1; 14.6; 20.2; 27.8; 38.3; 52.7; and 72.5 cm. In practice we surely have to take into account that different spectral ranges may originate in different parts of the source volume and thus have different initial position angles φ_0 (and, generally speaking, different Faraday rotation measures RM). In such a case observations in different spectral ranges should be treated separately (Vallée, 1980).

An expected accuracy of the rotation measure and the position angle (2σ-interval) is

$$\mathscr{D} RM \simeq 4\sigma/\Lambda_{\text{max},\,1} \approx 1.4 \text{ rad m}^{-2},$$

$$\mathscr{D} \varphi_0 \simeq 2\sigma(\lambda^2_{\text{max}} + \lambda^2_{\text{min}})/\Lambda_{\text{max},\,1} \simeq 20°.$$

A large amount of observational data now available was obtained as a result of unspecialized observations. Therefore, despite the abundance of data, for some sources the measurements may be absent for two close wavelengths or they could be distributed with large gaps over the wavelength scale. Accuracy of wavelength and position angle determination is very important. Obviously, a variation in the position angle associated with an inaccurate determination of λ_i should not exceed the error σ_i, i.e.

$$2RM\lambda_i\Delta\lambda_i < \sigma_i. \tag{IV.7}$$

This criterion should be applied to the calculated rotation measures and the results which do not satisfy it must be rejected. Note that the large values of RM are the most dangerous from this point of view. In addition, when σ_i are small, accuracy of wavelength determination becomes especially important. The latter factor is usually rather more essential for long waves. We should emphasize that the criterion (IV.7) must be applied in any method of evaluation of RM and φ_0. Consider a specific example of measurements for the source PKS 0052 + 68 at $\lambda = 31$ cm with $\sigma = 2.5°$ (Haves *et al.*, 1974). Here $\Delta\lambda = 1$ cm and the estimate is made from the published accuracy of determination of λ. The calculated rotation measure is $RM = -90.5 \pm 9.9$ rad m^{-2}. It can be easily verified that the criterion (IV.7) is not satisfied in this case. Hence, this measurement cannot be used for calculation of the rotation measure and the intrinsic position angle.

Note that usually only rounded values of wavelengths are published, which makes estimations difficult even for the known receiver bandwidth. The data with large errors σ_i ($> \pi/4$ for 2σ confidence interval) are also useless since the position angle must be within the interval $(0, \pi)$. The algorithm described above was realized by Ruzmaikin and Sokoloff (1979) for a set of measurements of position angles of Galactic and extragalactic radiosources (see Tables 1 and 2 from their paper).

For a number of sources, straight lines cannot be drawn through the observational points on the plane (φ, λ^2).

Two main factors can be pointed out that may be a reason for this situation.

Firstly, position angle measurements which do not fit the λ^2 dependence at long wavelengths where the angular resolution is poor may indicate the influence of extended nature or spatial inhomogeneity of a source. For example, the radio source Fornax A has two resolved components, Fornax A(a) 0319—37 and Fornax A(b) 0322—37. The observational results at $\lambda = 74.3$ cm, $\varphi_a = 101° \pm 10°$ and $\varphi_b = 108° \pm 10°$, cannot be fitted to λ^2 dependences for either component. However when these data are excluded, unique straight lines can be drawn for each component through the remaining points. V. A. Razin has pointed out that this effect can also be encounted for the sources which are not resolved into separate components.

Secondly, a source may be variable and the published observational results might be obtained at different phases* (Morris and Whiteoak, 1968). In this case the points violating the λ^2 dependence do not refer to a certain wavelength range but are distributed randomly. It is typical for the data we used. For example, only one observation from the existing four at 18 cm for the radiosource 0531 + 21 fits the straight line drawn through the measurements at other wavelengths.

In some cases our results coincide or are close to those of Morris and Tabara (1973), Vallée and Kronberg (1975), and Haves (1975). The main difference is an almost complete absence of large (> 200 rad m^{-2}) rotation measures in our results. For example, Vallée and Kronberg give $RM = 351 \pm 16$ rad m^{-2} for the source 1345 + 12. Our method yields 42.4 $\pm$ 10.0 rad m^{-2}.

Investigation of the sources with two resolved components, like Fornax Aa and b and 0055—01a and b shows (see Table 2 of Ruzmaikin and Sokoloff, 1979) that an intrinsic contribution of the sources to the observed rotation measures in this case is small since both components show close rotation measures. Increasing the number of such double sources as well as investigating the polarization variability of radio sources would reveal most valuable information about the intrinsic rotation measures of sources thus separating them from the Galactic contribution. In combination with the intrinsic position angles it would yield direct data on values and directions of the magnetic fields within extragalactic sources.

Among all the sources investigated by us (300), in only 147 could we manage with confidence to determine their rotation measures and intrinsic position angles. The rest do not satisfy simple criteria of reliability, i.e. either more than one-third of available data points have to be rejected because they do not fit the λ^2 dependence, or else an artificial character of implied dependence cannot be ruled out. A large portion of the sources that do not satisfy the first of these require-

* At present the long-term polarization investigations of some radiosources (so-called monitoring) are available (see, e.g., Barvainis and Predmore, 1984), which reveal the position angle variability. These investigations, however, are usually performed at one or two wavelengths, so it is difficult to say whether they manifest the variability of φ_0 or RM.

ments seem to show variability of the rotation measure or of the intrinsic position angle.

As an example let us consider the well known double radiosource Cygnus A. For the western component the position angle measurements at three wavelengths 0.95 cm (137° ± 21°), 3.02 cm (42° ± 5°), and 3.95 cm (135° ± 17.5°) are available and we have obtained

$$RM = -32 \pm 2215 \text{ rad m}^{-2}, \qquad \varphi_0 = 143^\circ \pm 121^\circ.$$

However, an occasional straight line could cross these three points with a great probability and the criterion (IV.7) is not satisfied. For the eastern component we have observational data at four wavelengths satisfying the criterion (IV.7); however, the accuracy of the result is poor:

$$RM = -116 \pm 2893 \text{ rad m}^{-2}, \qquad \varphi_0 = 38^\circ \pm 157^\circ.$$

Thus, the data available are insufficient for accurate determination of the rotation measure and the intrinsic position angle for both components of Cygnus A. The data from 25 observations of Cygnus A are known, for which the components were not resolved. In this case

$$RM = -1099 \pm 21 \text{ rad m}^{-2}, \qquad \varphi_0 = 18^\circ \pm 4^\circ.$$

However, 13 points (i.e. more than one third of all available) were rejected when deriving this result. Therefore, this result is also unreliable.

IV.2. How the Magnetic Field is Derived from Faraday Rotation Data

The next step in interpretation is the reconstruction of the magnetic field from the Faraday rotations. To do this it is certainly insufficient to know the rotation measure for only one source since the expression for the rotation measure

$$RM = 0.81 \int_0^d H n_e \, \mathrm{d}l, \tag{IV.8}$$

where d is the distance to the source in pc, contains only the integral over the line-of-sight field component. However, if we have a great enough number of sources lying in various directions we may try to reconstruct the field configuration, and even strength provided that the electron density n_e is also known.

The problem can be solved in the simplest way by using pulsars as a probe. An intrinsic contribution of these Galactic radiosources to the rotation measure is small, the distances to them are known, and n_e can be determined independently through the dispersion measures,

$$DM = \int_0^d n_e \, \mathrm{d}l \qquad \text{(IV.9)}$$

(Manchester, 1972, 1974). The simplest estimate of a weighted mean value of the line-of-sight component, which was first obtained by Manchester, can be obtained simply by dividing the integral (IV.8) by the integral (IV.9). A better estimate can be made if at least two pulsars are available with known *RM*, *d* and *DM* along nearly the same line of sight. If $\Delta d = d_2 - d_1$ is relatively small, the problem can be reduced to, crudely speaking, differentiating the integrals (IV.8) and (IV.9). However, the distances (as well as *RM* and *DM*) are determined with a considerable error δd. Hence, the differentiation operation is extremely dangerous because, due to the error, the denominator in the expression $\Delta RM/\Delta d$ and $\Delta DM/\Delta d$ may vary by a large factor.

From the mathematical point of view, the above difficulty means that the problem of reconstructing the field from the integrals (IV.8) and (IV.9) is improperly posed in the sense of Hadamard. The reason for this improperness lies deeper than in a mere inability to determine distances to the sources with sufficient accuracy. In fact, it is difficult to separate contributions made to the rotation measures and the dispersion measures by compact formations along the line of sight, for example nebulae around pulsars, from that made by a regular field at the galactic scale. A significance of such a spurious contribution may be easily illustrated by the example of the Crab pulsar. For this pulsar $RM = -42.3$ rad m^{-2} (Manchester, 1972). On the other hand, for the Crab Nebula as a whole the rotation measure is -21.6 rad m^{-2}. Thus, the intrinsic contribution from the Crab Nebula itself is about $+20$ rad m^{-2} *.

Methods of solution of improperly posed problems of such a type are associated with the solution smoothing (Tikhonov and Arsenin, 1979). It is not that the contribution from the compact object is merely rejected, but the most possibly smooth distributions of the field and electron density are sought for to satisfy the observational data. Note that without smoothing even a contribution of a purely regular field may be interpreted as a sum of contributions from compact formations.

As a first step we assume that the galactic field consists of a homogeneous component **B** directed along $(l_0^{\mathrm{II}}, b_0^{\mathrm{II}})$ in galactic coordinates and a fluctuation

* The intrinsic position angles prove to be close for the Crab pulsar and the Nebula, which indicates that both have parallel fields. Taking that the Nebula extends along the line of sight for 1.5 pc and the internal field is $B_{\|} = 3 \times 10^{-4}$ G (a lower estimate from independent data, Woltjer, 1972; Shklovsky, 1976) we obtain an estimate for the electron density within the Crab Nebula as $n_e \simeq 6 \times 10^{-2}$ cm^{-3}.

component. Let us first assume that all the radio sources (pulsars) are situated within a local spiral arm*. Then the total rotation measure is a sum of regular and fluctuation components

$$RM = RM_g + RM_f. \tag{IV.10}$$

The regular contribution of a single source is described by

$$\begin{aligned} RM_g &= 0.81\ Bn_e d \cos(\widehat{\mathbf{Bd}}) \\ &= c_0 d\left[\cos b_0^{\rm II} \cos b^{\rm II} \cos(l^{\rm II} - l_0^{\rm II}) + \sin b_0^{\rm II} \sin b^{\rm II}\right], \end{aligned} \tag{IV.11}$$

where $c_0 = \mu \langle n_e \rangle B$, $\mu = 0.81$, $l^{\rm II}$ and $b^{\rm II}$ are the galactic coordinates of the source**, $\langle n_e \rangle$ is the mean electron density.

We consider in more detail how the averaged regular component RM_g and the fluctuation component RM_f can be extracted from the observed RM (Jokipii and Lerche, 1969; Ruzmaikin and Sokoloff, 1977b). Introducing the quantity

$$\mathscr{P}(s) = n_e(\mathbf{B}\mathbf{n}_r),$$

where $\mathbf{n}_r$ is a unit vector along the line of sight we obtain

$$RM(d) = \mu \int_0^d \mathscr{P}(s)\,\mathrm{d}s,$$

where the integral is taken along the line of sight. The mean value $RM = RM_g$ is a regular contribution of the Galaxy, an expression for which is given above. The mean value of the fluctuation term equals zero.

For the purpose of the statistical analysis we should know the variance of RM_f,

$$\begin{aligned} \sigma_{RM}^2(d) &\equiv \overline{RM^2} - RM_g^2 \\ &= \mu^2 \int_0^d \mathrm{d}s_1 \int_0^d \mathrm{d}s_2 [\overline{\mathscr{P}(s_1)\mathscr{P}(s_2)} - \overline{\mathscr{P}}(s_1) \cdot \overline{\mathscr{P}}(s_2)], \end{aligned} \tag{IV.12}$$

where the bar denotes the statistical average. The quantity in the square brackets is the two-point correlation function $C(s_1, s_2)$. Assuming the galactic turbulence to be homogeneous, the correlation function depends only on $s = |s_1 - s_2|$. Simple algebra reduces (IV.12) to

$$\sigma_{RM}^2(d) = 2\mu^2 \int_0^d (d - s) C(s)\,\mathrm{d}s.$$

* Therefore, we do not use the pulsars with $d > 2000$ pc (see Table IV.1). Moreover, we reject the sources that are projected onto a strong local anomaly of the field — the North Polar Spur. This reduces the number of pulsars used down to 27.

** In order to avoid confusion in notations, we use an old-fashioned notation for the Galactic coordinates. We denote by l the correlation length and by b the chaotic magnetic field.

TABLE IV.1

Parameters of the pulsars with known rotation measures *RM* (rad m^{-2}), dispersion measures *DM* (pc cm^{-3}), and distance estimates *d* (pc).

PSR	l^{II}	b^{II}	*RM*	ξ	*d*	*DM*	η	*N*	*IZ*
1. 1749−28	1	−1	95.0	36.1	1000	51	7.0	5.67	0
2. 1706−16	6	14	−2.5	−4.8	160	25	30.4	0.41	0
3. 1604−00	11	35	6.5	5.3	400	11	−2.1	1.74	0
4. 1642−03	14	26	16.5	22.4	200	36	37.7	0.60	0
5. 1822−09	24	1	69.0	62.0	350	19	5.8	1.43	0
6. 1818−04	26	5	70.5	31.8	1500	84	10.8	9.00	0
7. 2045−16	31	−33	−10.8	−1.9	450	12	−2.4	2.05	2
8. 1929 10	47	−4	−8.6	−24.3	60	3	3.5	0.07	2
9. 1933 16	52	−2	−1.9	38.2	6000	159	−7.6	39.00	0
10. 1919 21	56	3	−18.2	−8.1	250	12	3.7	0.86	1
11. 2016 28	68	−4	−34.6	6.7	1000	14	−8.6	5.67	2
12. 2020 28	69	−5	−74.7	−10.0	1000	25	−3.9	5.67	2
13. 2021 51	88	8	−6.5	15.4	550	23	2.5	2.69	1
14. 2111 46	89	−1	−223.7	2.0	4000	141	0.7	25.67	0
15. 2154 40	90	−11	−44.0	23.4	2300	71	−2.1	14.33	0
16. 1508 55	91	52	0.8	13.0	600	20	−0.4	3.02	1
17. 2217 47	98	−8	−35.3	17.7	1500	44	−2.5	9.00	2
18. 0031−07	110	−70	10.0	13.0	500	10	−4.7	2.37	2
19. 2319 60	112	−1	−224.0	−19.8	2500	96	2.5	15.67	0
20. 0105 65	125	3	−29.0	1.9	600	30	5.4	3.02	1
21. 0138 59	129	−2	−50.0	24.3	3000	35	−15.6	19.00	0
22. 0809 74	140	38	−11.7	−5.9	200	6	−1.1	0.60	1
23. 0329 54	145	−1	−63.7	−3.3	1300	27	−6.4	7.67	2
24. 1112 50	155	61	3.2	7.4	500	9	−5.3	2.37	1
25. 0301 19	161	−33	−8.3	5.6	850	16	−6.1	4.67	2
26. 0525 21	184	−7	−39.6	−7.9	2000	51	−5.1	12.33	2
27. 0531 21	185	−6	−42.3	−9.3	2000	57	−3.4	12.33	2
28. 0611 22	188	2	69.0	23.9	1500	97	15.1	9.00	1
29. 0323 26	197	32	5.9	0.1	900	19	−5.3	5.00	1
30. 0450−18	217	−34	15.0	−1.5	800	40	6.0	4.34	2
31. 0950 08	220	44	1.8	0.7	60	3	3.5	0.07	1
32. 0834 06	229	26	24.5	8.6	450	13	−1.7	2.05	1
33. 0818−13	236	13	−2.8	−25.2	2100	41	−8.7	13.00	0
34. 0628−28	237	−17	44.2	−7.3	1600	34	−6.8	9.67	2
35. 1133 16	242	69	3.9	3.2	130	5	1.0	0.29	1
36. 0740−28	244	−2	150.0	20.0	1750	74	4.2	10.67	2
37. 1237 25	252	87	−0.6	−1.3	450	93	4.5	2.05	1
38. 0833−45	264	−3	36.6	5.2	500	69	33.7	2.37	2

Remarks to Table IV.1:

Columns 5 and 8 give the calculated fluctuations of rotation measures ξ (rad m^{-2}) and dispersion measures η (pc cm^{-3}), respectively. Column 9 gives the values of $N = [d/\mathscr{L} - 1 + \exp(-d/\mathscr{L})]$ (see IV.13)), while the value of *IZ* (Column 10) indicates the position of the pulsar in the northern (*IZ* = 1) or the southern (*IZ* = 2) Galactic hemisphere; *IZ* = 0 indicates the pulsars rejected from the statistical analysis.

It is natural to assume that there exists a characteristic scale $\mathscr{L}$ such that for $s \ll \mathscr{L}$ the value $\mathscr{P}(s_1)$ and $\mathscr{P}(s_2)$ are strongly correlated, i.e. $C(s)$ is great, but for $s \gg \mathscr{L}$ the value $\mathscr{P}(s_1)$ and $\mathscr{P}(s_2)$ may be considered independent, i.e. $C(s) \to 0$.* A simple model is

$$C(s) = C_1 \exp(-s/\mathscr{L})$$

then

$$\sigma_{RM}^2(d) = 2\mu^2 \mathscr{L}^2 C_1 [d/\mathscr{L} - 1 + \exp(-d/\mathscr{L})]$$

where C_1 is the variance of $n_e B$. Now the expression (11) can be written in the form

$$RM = c_0 d [\cos b_0^{\mathrm{II}} \cos b^{\mathrm{II}} \cos(l^{\mathrm{II}} - l_0^{\mathrm{II}}) + \sin b_0^{\mathrm{II}} \sin b^{\mathrm{II}}] + + [d/\mathscr{L} - 1 + \exp(-d/\mathscr{L})]^{1/2} \xi, \tag{IV.13}$$

where ξ is a random variable with the zero mean value. Its variance does not depend on d and is equal to (symbol δ indicates fluctuation of a corresponding quantity)

$$\sigma_\xi^2 [\sqrt{2} \mu \mathscr{L} \delta(n_e B)]^2.$$

Thus, we have parametrized the observed rotation measure by the four values c_0, l_0^{II}, b_0^{II}, and $\mathscr{L}$. The parameters c_0 and l_0^{II} will be determined by a regression method. The latter allows determination of b_0^{II} as well. However, nearly all pulsars are low-latitude sources which means that the vertical component of the magnetic field (due to which b_0^{II} appears in (IV.13)) makes a small contribution to the pulsar rotation measures. Therefore, b_0^{II} can be only determined with a large error. Taking this into account we simply put $b_0^{\mathrm{II}} = 0°$.

The regression method used is a natural generalization of the least-square method. As compared with other methods its advantage is in its simplicity. Also, the method is almost insensible to a specific form of the distribution of ξ and variations of the adopted model. The latter circumstance is especially important — as we have noted above, the accepted model is rather crude. Modification of formula (IV.13) hardly changes the parameters c_0 and l_0^{II} within the 2σ-error. The parameter $\mathscr{L}$ is crudely estimated in the following way. We divide the pulsars into two zones according to the value of d: the pulsars with $d < 900$ pc are 16 in number. Those with $d > 900$ pc are 11. Let us calculate the empirical variance of σ for each zone. According to our hypothesis the variance σ_ξ^2 does not depend on d. Therefore confidence intervals for the empirical variances in each zone, calculated for a reasonably chosen probability, must overlap. In practice it is more convenient to compare a ratio of empirical variances against the Fisher criterion,

* This scale, the turbulent correlation scale, is denoted by l_0 in the rest of the book. For understandable reasons, we denote it by $\mathscr{L}$ in this chapter.

i.e., for the representation (IV.13) to agree with observational data we must have

$$\sigma_{1\xi}^2/\sigma_{2\xi}^2 \leqslant \mathscr{F}(N_1, N_2, P), \tag{IV.14}$$

where $\sigma_{1\xi}^2$ is the larger of the two variances, N_1 and N_2 are the pulsar numbers in respective zones and P is the mentioned probability. For large d, however, formula (IV.11) is obviously inaccurate, and its use leads to an effective increase of σ_ξ in the far zone. The latter, in turn, leads to an effective rise in $\mathscr{L}$: the greater $\mathscr{L}$, the greater the mean ratio of the values $d/[\mathscr{L} - 1 + \exp(-d/\mathscr{L})]$ in the far and near zones. Hence, the method must yield an overestimated value of $\mathscr{L}$. Therefore, we take a sufficiently wide confidence interval for the empirical variances, namely, we put $P = 99.9\%$ which corresponds to three standard errors for a normally distributed random value.

In order to determine the parameters c_0 and l_0^{II} we minimize the sum (see next section for more detail):

$$S(c_0, l_0^{\mathrm{II}}) = \frac{1}{N-2} \sum_{i=1}^{N} \xi_i^2$$

$$= \frac{1}{N-2} \sum_{i=1}^{N} \frac{(RM - RM_g)^2}{d/\mathscr{L} - 1 + \exp(-d/\mathscr{L})} \tag{IV.15}$$

for different values of $\mathscr{L}$ (10, 30, 50, 100, 150, 200, 250, and 300 pc). The values of c_0 and l_0^{II}, which minimize S, are taken as the true values. The 95% confidence intervals corresponding to two standard errors are determined from the conditions

$$S(c_0, l_0^{\mathrm{II}}) \leqslant S_{\min} \left[1 + \frac{2}{N-2} \mathscr{F}(2, N-2, 95\%) \right], \tag{IV.16}$$

where $N = 27$ is the total number of pulsars in our sample. These confidence manifolds are rather complicated in the parameter space. As errors we give simpler quantities, the projections of the confidence manifolds on the coordinate axes c_0 and l_0^{II}.

The calculated values of the parameters c_0 and l_0^{II} proved to be practically independent of a choice of the correlation length. These are given by

$$\begin{aligned} c_0 &\equiv 0.81 \langle n_e \rangle B = -0.06 \pm 0.02 \text{ rad m}^{-2}, \\ l_0^{\mathrm{II}} &= 99° \pm 12°. \end{aligned} \tag{IV.17}$$

The 1σ errors are cited here. The minus sign in c_0 means that the large-scale magnetic field is directed from $l_0^{\mathrm{II}} + 180°$ to l_0^{II}. To determine the strength of the field **B** we must know the average density of galactic electrons $\langle n_e \rangle$. An estimate $\langle n_e \rangle = 0.03 \text{ cm}^{-3}$ is widely used (e.g., see Terzian and Davidson, 1976). However, it is natural, for the sake of consistency, to calculate $\langle n_e \rangle$ independently for our pulsar sample by using an expression similar to (IV.13)

$$DM = \langle n_e \rangle d + [d/\mathscr{L} - 1 + \exp(-d/\mathscr{L})]^{1/2}\eta. \tag{IV.18}$$

The value of $\langle n_e \rangle$ calculated from (IV.18) with the regression method does not depend on $\mathscr{L}$ and is equal to

$$\langle n_e \rangle = 0.035 \pm 0.01 \text{ cm}^{-3}. \tag{IV.19}$$

Now from (IV.17) and (IV.19) we obtain the field strength

$$B = -2.1 \pm 0.5\ \mu\text{G} \tag{IV.20}$$

where the confidence intervals are calculated without an error in $\langle n_e \rangle$ being taken into account.

Now we shall turn to estimate of the amplitude and the characteristic scale of the magnetic field fluctuation component. The second and the third columns of Table IV.2 list ratios of empirical variances of ξ and η in the far and near zones

TABLE IV.2

$\mathscr{L}$ (pc)	$\sigma_{1\xi}^2/\sigma_{2\xi}^2$	$\sigma_{1\eta}^2/\sigma_{2\eta}^2$	$\sigma_{1\zeta}^2/\sigma_{2\zeta}^2$	$\frac{\delta(n_e B)}{N_e B}$	$\frac{\delta n_e}{N_e}$	$\frac{\delta B}{B}$
10	5.1	1.3	2.8	3.65	1.94	3.79
30	4.6	1.4	2.6	2.14	2.34	2.22
50	4.3	1.4	2.2	1.68	1.84	1.74
100	4.0	1.6	2.2	1.24	1.35	1.27
150	3.0	1.7	2.0	1.05	1.07	1.07
200	2.7	1.8	1.9	0.95	1.04	0.96
250	2.3	1.9	1.7	0.88	0.97	0.89
300	1.9	2.0	1.6	0.84	0.92	0.83

($\sigma_{1\xi}^2$, $\sigma_{2\eta}^2$ and $\sigma_{2\xi}^2$, $\sigma_{1\eta}^2$ respectively). The ratios should be compared with $\mathscr{F}$ (10, 15, 99.9%) = 3.8 (Bolshev and Smirnov, 1965; Dreiper and Smith, 1966). $\sigma_{1\zeta}$ and $\sigma_{2\zeta}$ are respective variances of δB calculated with distances to pulsars estimated from DM in (IV.15). Use of the latter estimates allows to determine $\delta B/B$ separately. It is clear that $\sigma_{1\xi}^2/\sigma_{2\xi}^2 < \mathscr{F}$ when $\mathscr{L} > 100$ pc. The ratio $\sigma_{1\eta}^2/\sigma_{2\eta}^2$ weakly grows with the correlation length, and for all $\mathscr{L}$ under consideration it does not contradict the hypothesis (IV.18). The mean square deviations of ξ and η are, respectively,

$$\sigma_\xi = \sqrt{2}\, \mu \mathscr{L}\, \delta(n_e B) = \sqrt{\frac{S_{\min}}{N-2}},$$

$$\sigma_\eta = \sqrt{2}\, \mu \mathscr{L}\, \delta n_e = \sqrt{\frac{S_{n_e \min}}{N-1}},$$

where $S_{n_e \min}$ is the quantity similar to $S_{\min}$ but for the model (IV.18). From this we obtain the relative amplitudes of fluctuations $\delta(n_e B)/\langle n_e \rangle B$ and $\delta n_e/\langle n_e \rangle$ given in the fifth and the sixth columns of Table IV.2. If n_e and B are strongly correlated,* the value $\delta(n_e B)/\langle n_e \rangle B$ should be close to $\delta B/B$. According to the analysis of polarization of continuous galactic radio emission, $3 > \delta B/B > 1$ (Spoelstra and Brouw, 1976). From Table IV.2 we see that this range corresponds to $\mathscr{L} \lesssim 200$ pc. Hence, comparing the result obtained with the estimate of $\mathscr{L}$ obtained with the use of (IV.14) we have

$$\mathscr{L} \simeq 100\text{–}150 \text{ pc}. \tag{IV.21}$$

IV.3. The Large-scale Magnetic Field of the Galaxy According to Faraday Rotations of Extragalactic Sources

It is an attractive option to use abundant data on the rotation measures of extragalactic radio sources (radiogalaxies and quasars) instead of or in addition to the relatively small number of pulsars. Modern catalogues count several hundreds of such objects with known rotation measures. However interpretation poses considerable problems connected with the fact that the observed rotation measures are generally sums of contributions from the source itself, from the intergalactic medium between the source and the Galaxy and from the regular and fluctuation magnetic fields of the Galaxy.

The rotation measures observed for some radio sources are large (> 50 rad m^{-2}) as compared with an expected Galactic contribution (~ 20 rad m^{-2}), which agrees with an idea that the radiogalaxies and the quasars are the objects with strong internal magnetic fields (see Chapter IX). A contribution from the hypothetical intergalactic field appears to be small (see Section V.2). Therefore, it seems to be quite natural to make the following *a priori* assumption

$$RM = RM_g + \frac{RM_s}{(1 + z)^2} \tag{IV.22}$$

(Ruzmaikin *et al.*, 1978), i.e. to assume that the Galaxy and the source itself make the dominant contributions to the observed rotation measures. The multiplier $(1 + z)^{-2}$, where z is the redshift of the source, reflects the fact that the polarization plane rotation angle is proportional to the squared wavelength which scales with the redshift z as $(1 + z)^{-1}$. Hence, RM_s is the rotation measure referred to the source itself.

The Galactic contribution to the observed rotation measure can be approxi-

* We may note that our results do not agree with the assumption that n_e and B are statistically independent, since in that case $\delta B/B$ proves to be extremely underestimated and even physically meaningless (imaginary) for large $\mathscr{L}$.

mated by the following relation that is similar to (11) and follows from a simple geometry of Figure IV.3:

$$RM_g = -K \frac{\cos l_0^{\mathrm{II}} \cos(l^{\mathrm{II}} - l_0^{\mathrm{II}}) \cos b^{\mathrm{II}} + \sin b_0^{\mathrm{II}} \sin b^{\mathrm{II}}}{|\sin b^{\mathrm{II}}|}. \tag{IV.23}$$

We adopt l_0^{II} and b_0^{II} as the Galactic coordinates of the point to which the vector **B** points (see Fig. IV.3). Then the coefficient K is always positive. It is expressed in terms of the field as

$$K = 0.81 \langle n_e \rangle Bh \text{ rad m}^{-2}$$

where $\langle n_e \rangle$ is the average electron density in cm^{-3}, h is the half-thickness of the

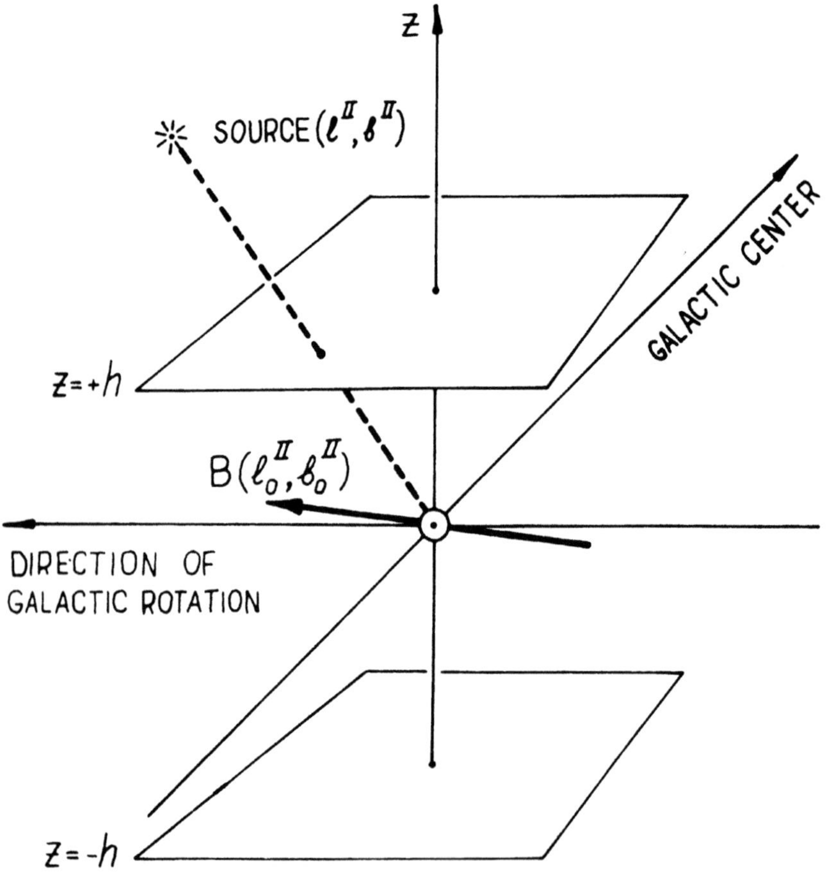

Fig. IV.3. Diagram showing directions of the magnetic field and the line of sight.

galactic disc in parsecs, B is the strength of the large-scale magnetic field in microgauss. When deriving expression (IV.23), we have considered the local spiral arm as a plane layer, i.e. the effects of the side surfaces of the arm have been ignored together with the effects of the inter-arm space and of the other arms. Therefore, we restrict our consideration to the sources with $|b^{\mathrm{II}}| > 20°$ for which the approximation (IV.23) is sufficiently accurate (the cross-section of the arm is typically 250 pc × 400 pc). Moreover, we take into account the local disturbances in the Galaxy's magnetic field (e.g. the North Polar spur) making appropriate omissions in observational data.

A most natural and accurate way to generalize the plane-layer model is to represent the arm as a cylinder with the elliptical cross section $x^2/a^2 + y^2/h^2 = 1$, where a and h are the semi-major and semi-minor axes. To calculate RM we should know the cosine of the angle between the line of sight of a given source (l^{II}, b^{II}) and the magnetic field (l_0^{II}, b_0^{II}). The expression for the angle has been already used in formula (IV.23) and we further express the distance d, passed by the radiation within the Galaxy, through the parameters h, a, l^{II} and b^{II}. It is easy to see that

$$d = \frac{(x^2 + y^2)^{1/2}}{\sin\beta} = \frac{[h^2 + (1 - h^2/a^2)x^2]^{1/2}}{\sin\beta}$$

$$x = r\cos b^{\mathrm{II}}\cos l^{\mathrm{II}}, \qquad \cos\beta = \cos b^{\mathrm{II}}\sin l^{\mathrm{II}}$$

where β is the angle between the direction to a source and the cylinder axis. The above relations allow us to express d and then to obtain the desired result,

$$RM_g = -K[\cos b_0^{\mathrm{II}}\cos(l^{\mathrm{II}} - l_0^{\mathrm{II}})\cos b^{\mathrm{II}} + \sin b_0^{\mathrm{II}}\sin b^{\mathrm{II}}] \times \times \left(\frac{h^2}{a^2}\cos^2 b^{\mathrm{II}}\cos^2 l^{\mathrm{II}} + \sin^2 b^{\mathrm{II}}\right)^{-1/2} \tag{IV.24}$$

which reduces to (IV.23) in the limit $h/a \ll 1$.

Let us consider now the second term on the right-hand side of (IV.22). We shall suppose an intrinsic rotation measure to be a random variable. It is natural to assume that the distribution function of this random variable does not depend on galactic coordinates of the sources. This assumption is equivalent to the assumption of the isotropy of the Universe. It is natural also to assume random variables RM_s to be mutually independent.

Things are more complicated when we consider RM_s as the function of z, the redshift. Even after the factor $(1 + z)^{-2}$ has been isolated the RM_s still remains dependent on z due to the evolution and selection effects: at large z brighter and more compact sources with larger RM_s can be observed (B. V. Komberg, 1977, private communication). Since an exact form of dependence of the intrinsic rotation measure of the redshift is unknown, we divide the observational data available into zones with certain Δz, and assume variance of RM_s constant within each zone. The variances obtained within each zone yield together a piecewise behaviour of $\overline{RM_s^2}(z)$ describing the evolution and selection effects.

We assume that the average value of the intrinsic rotation measure is zero over the whole sample ($\overline{RM_s} = 0$). Physically, this assumption means that the sources do not have a prevalent direction of the magnetic field, which, in turn, is connected with the isotropy of the Universe (see Section V.3).

The catalogues used below contain numerous sources with unknown redshifts. We can use these sources by isolating them into a separate zone and assigning them $z_{\text{eff}} = 0$.

In order to calculate the parameters K, l_0^{II} and b_0^{II} which characterize the Galactic magnetic field, we use again the regression method mentioned above. Now we describe it in more detail. We introduce the function residual as

$$\begin{aligned} S(K, l_0^{\text{II}}, b_0^{\text{II}}) &= \sum_n (1 + z^{(n)})^4 (RM^{(n)} - RM_g^{(n)})^2 \\ &= \sum_n RM_s^{(n)2}, \end{aligned} \qquad \text{(IV.25)}$$

where $RM^{(n)}$ and $z^{(n)}$ are the Faraday rotation measure and the redshift of the nth source, respectively, while $RM_g^{(n)}$ is the function of the desired parameters, and the source coordinates, and the redshift, that is given by (IV.23) or (IV.24). Summation is carried out over all sources of the chosen sample.

The essence of this method is that RM_s coincides with the true values of the intrinsic rotation measures when the sum of squared deviations (IV.25) is minimal, since the average value of RM_s is zero. Strictly speaking, this statement is true when RM_s are independent identically distributed random numbers and their number tends to infinity. From the minimum condition for S we find the desired parameter values.

In order to take into account the effects of selection and evolution which may violate the assumption of equal variances we divide the whole sample of sources into four groups, each one corresponding to a certain range of z variation. The first group includes the sources with $z \leqslant z_1$ the second one covers $z_1 < z \leqslant z_2$, sources from the third subsample $z > z_2$, and the sources with unknown redshifts are collected into a fourth group. Specifically, we choose $z_1 = 0.3$, $z_2 = 0.8$. It is also assumed that the evolution and the selection effects are small within each group and we can apply to each group the above method, in which the variances of RM_s are supposed to be constant. However, the variances are allowed to be different for different groups. Then we make calculations for the whole sample with contributions from each zone weighted by $\alpha_i = S_i^*/(N_i - 3)$, where S_i^* and N_i are residual and the number of sources in the ith group respectively; the number 3 in this expression is the number of parameters (K, l_0^{II}, b_0^{II}). Specifically, we minimize the residual

$$s(K, l_0^{II}, b_0^{II}) = \sum_{n, i} \frac{(1 + z^{(n)})^4}{\alpha_i} (RM^{(n)} - RM_g^{(n)})^2. \tag{IV.26}$$

It is a minimum of S that determines the desired values of the magnetic field parameters. The values of S_i^* for separate zones also give some information about the selection and evolution effects.

In the linear approximation, this method reduces to the well known least-square method. However, the residual function S non-linearly depends on the parameters l_0^{II} and b_0^{II} through the trigonometric functions. Thus, to obtain a minimum of the residual the iteration procedure should be employed. It is also possible, as Andreassian (1982) notes, to reduce the problem to a linear one through transformation of variables. The possibility of such a transformation follows from the fact that the problem is linear in the physical components of the magnetic field.

The confidence manifolds for the desired parameters are determined as follows. For S the interval

$$S \leqslant \bar{S} = S^* \left[1 + \frac{p}{N - p} \mathscr{F}(p, N - p, 95\%) \right] \tag{IV.27}$$

is the confidence set for the 95% probability level (Dreiper and Smith, 1966). Here S^* is the minimum value of S, $\mathscr{F}$ is the Fisher distribution function, N is the total number of sources, and $p = 3$ corresponds to the number of the magnetic field parameters. The criterion of 95% probability approximately corresponds to a criterion of two standard errors for the case of the normal distribution. As a matter of fact the relation (IV.27) has a simple meaning: $[(\bar{S} - S^*)/S^*]^{1/2} \approx (p\mathscr{F})^{1/2} \times N^{-1/2}$, where the factor $N^{-1/2}$ corresponds to a common law for statistical fluctuations. Those parameter values are the confident ones for which $S \leqslant \bar{S}$.

To determine this confidence manifold we approximate the residual function near the minimum by the second-order polynominal whose coefficients are determined by a matrix X. The matrix is obtained by calculating respective partial derivatives of S at the minimum point. The confidence set is an ellipsoid in space of the parameters that is given by the matrix inequality

$$(K - K^*, l_0^{II} - l_0^*, b^{II} - b_0^*) \cdot X \cdot \begin{bmatrix} K & -K^* \\ l_0^{II} & -l_0^* \\ b_0^{II} & -b_0^* \end{bmatrix} < \bar{S} - S^*. \tag{IV.28}$$

Here the asterisk marks the parameter values minimizing the residual.

In order to determine the confidence intervals from the above inequality, for example for K, we substitute the calculated values of parameters K^*, l_0^*, b_0^* and the matrix X into (IV.28) and obtain the second-order inequality with respect to $K - K^*$ and l_0^{II}, b_0^{II}. The values of $K - K^*$ thus obtained represent the desired limits of the confidence intervals. The same is valid for other parameters as well.

To solve the inequality (IV.28) we use the following approximation. Let x_{ij} be the elements of the matrix X. Then

$$\gamma_{ij} = \frac{x_{ij}}{(x_{ii} \cdot x_{jj})^{1/2}}$$

is the correlation matrix and, for example, the quantity γ_{12} is called the correlation coefficient between K and l_0^{II}. It is easy to verify that the moduli of the correlation coefficients do not exceed unity. Then the confidence interval, for example for K, can be approximately expressed as

$$\Delta K = \left(\frac{pS^* \mathscr{F}(p, N-p, 95\%)}{(N-p)x_{11}} \right)^{1/2} \times$$
$$\times \left(\frac{\gamma_{12}^2}{1-\gamma_{12}^2} + (1-\gamma_{12}^2)^{1/2} \right) \left(\frac{\gamma_{13}}{1-\gamma_{13}^2} + (1-\gamma_{13}^2)^{1/2} \right). \tag{IV.29}$$

The obtained ΔK represents a half-length projection of the ellipsoid (IV.28) onto the axis K. The similar formula obtained for the case of two parameters, for example $u_1 = K/((\overline{S} - S^*)/x_{11})^{1/2}$ and $u_2 = l_0^{\mathrm{II}}/((\overline{S} - S^*)/x_{22})^{1/2}$, is exact and can easily be derived by solving the error ellipse equation

$$u_1^2 + 2\gamma_{12}u_1u_2 + u_2^2 = 1$$

combined with the condition for the projection boundary $\mathrm{d}u_1/\mathrm{d}u_2 = 0$. When deriving formula (IV.29), we neglected corrections connected with the correlation between l_0^{II} and b_0^{II} which are proportional to $\gamma_{23}^2(\gamma_{12}^2 + \gamma_{13}^2)$ while the effects taken into account are proportional to either γ_{12}^2 or γ_{13}^2. Hence, the above formula yields sufficiently exact values of the confidence intervals when the correlation coefficients are not very large, approximately, up to 0.5. Otherwise, it rather overestimates the confidence intervals.

Data on the observed rotation measures of extragalactic radiosources are gathered in the catalogues of Mitton (1972), Morris and Tabara (1973), Vallée and Kronberg (1975) and Haves (1975). They are summarized and supplemented in the catalogues of Eichendorf and Reinhardt (1980) and Tabara and Inoue (1980).

In order to enhance the reliability (see Section IV.1), we preserve only the sources with moderate rotation measures $|RM| < 100$ rad m^{-2}, and omit low-latitude sources with $|b^{\mathrm{II}}| < 20°$, since near the equator expression (IV.23) is not valid. Moreover, we omit sources projected onto the North Polar spur region ($b^{\mathrm{II}} > 0°$, $0° < l^{\mathrm{II}} < 40°$, $270° < l^{\mathrm{II}} < 360°$) which, as an analysis shows, does not introduce qualitative changes.

The results of statistical data processing prove to be more unexpected than one would assume. For the sources of the southern hemisphere, $b^{\mathrm{II}} < 0$, the large scale component of the magnetic field is confidently revealed. It is directed approximately along the Orion Arm in which the Sun is located. As an illustration we give an example of the parameter values obtained from the catalogue by Vallée and Kronberg (1975) for the high-latitude sources with $b^{\mathrm{II}} < -20°$:

$$\begin{aligned} &K = 27.7 \pm 4.7 \text{ rad m}^{-2}, \\ &l_0^{\mathrm{II}} = 99° \pm 8.5°; \qquad b_0^{\mathrm{II}} = 5° \pm 7° \end{aligned} \tag{IV.30}$$

(the number of the used sources $N = 88$), and those obtained from the combination of the catalogues of Morris and Tabara, and of Vallée and Kronberg in the model of an elliptical arm:

$$\begin{aligned} &K = 24.2 \pm 4.7 \text{ rad m}^{-2}, \\ &l_0^{\mathrm{II}} = 87° \pm 8.5°; \qquad b_0^{\mathrm{II}} = 1° \pm 9° \end{aligned} \tag{IV.31}$$

($N = 163$). Here the 1σ errors are given. These results allow one to obtain a strength of the large-scale field from the known n_e and h. The electron density has been estimated in the previous section as $n_e \simeq 3 \times 10^{-2}$ cm^{-3}. An estimate of the half-thickness can be obtained through comparison between the results for $c_0 \equiv 0.81\ Bn_e$ in the previous section and for $K = 0.81\ Bn_e h$ obtained here. By assuming B to be the same in both quantities we have $h = K/c_0 \simeq 400$ pc, which agrees with independent estimates described in Section VI.2. Hence,

$$B(\mu\mathrm{G}) \approx 0.1\, K \left(\frac{\langle n_e \rangle}{0.03 \text{ cm}^{-3}} \right)^{-1} \left(\frac{h}{400 \text{ pc}} \right)^{-1}. \tag{IV.32}$$

The agreement between the results (IV.30) and (IV.31)* implies that the regular component of the Faraday rotation observed on the Earth is produced predominantly within the local Orion Arm. This fact may be explained by a low electron density and weaker fields in the inter-arm regions. Contributions of the other Galactic spiral arms to the regular component are not taken into account in this model because those arms occupy a rather small part of the celestial sphere. To determine the large-scale magnetic field in other arms, special windows must be chosen. In particular, Simard—Normandin and Kronberg (1979) have thus derived the data for the Sagittarius arm and Agafonov *et al.* (1987) and Vallée (1983a, b) have investigated the field in the Perseus Arm.

On the other hand, the statistical results for the Northern hemisphere prove to

* Note that the results are distinctly different if the low-latitude sources are taken into account.

be insignificant or unstable against the catalogue modifications, data processing method and selection principles. For example, for the catalogues used above we obtain

$$K = 19 \pm 7 \text{ rad m}^{-2};$$

$$l_0^{\text{II}} = 35^\circ \pm 26^\circ; \qquad b_0^{\text{II}} = -6^\circ \pm 23^\circ, \qquad |b^{\text{II}}| > 20^\circ,$$

$$N = 28$$

instead of (IV.30), and

$$K = 8 \pm 6 \text{ rad m}^{-2}$$

instead of (IV.31). Some procedures (Andreassian, 1980, 1982) even yield $K < 0$, i.e. the field directed antiparallel to that of the Southern hemisphere.

This instability is an evidence of strong distortions of the field in the Northern hemisphere. The cause of this asymmetry may lie in the predominance of loops or spurs of the North Polar spur type, or bubbles remaining from supernova explosions in the Northern hemisphere (Spoelstra, 1972, 1977; Vallée, 1984). These explosions seem to be a basic source of interstellar turbulence (see Section VI.3). It is well known that the radio emission of the loops is strongly polarized. Nevertheless the loops do not show abnormally high Faraday rotations (Spoelstra, 1977). It is also possible that the local magnetic asymmetry between Northern and Southern hemispheres is connected with rare but significant magnetic deviations of the local field from the average value (intermittency, see Molchanov *et al.*, 1985), which is typical of the hydromagnetic dynamo. In both cases the north—south asymmetry is temporal; during other epochs or in other parts of the Galaxy the Southern hemisphere may prove to be more irregular.

Another unexpected result is obtained in analysis of the second component of the total rotation measure (IV.22) which we have *a priori* interpreted as a contribution of the intrinsic rotation measure of the sources. An analysis of the difference $RM - RM_g$ for the source groups with different redshift ranges shows that actually it does not depend on z while we have expected to see such a dependence as $(1+z)^{-2}$.

This may be explained, in principle, by the evolution and selection effects of the radio sources. However, the mean square value $RM_f = RM - RM_g$ is observed to be distinctly dependent on galactic latitude. For example, for the ellipsoidal arm model $RM_f = 39$; 32 and 14 rad m^{-2} for $|b^{\text{II}}| > 0^\circ$; $> 20^\circ$ and $> 35^\circ$ respectively. Therefore we are forced to interpret it as a contribution from a chaotic Galactic magnetic field. An estimate of the mean square deviation RM_f obtained from pulsar data through extrapolation of RM_f to high latitudes (18 rad m^{-2}, Ruzmaikin and Sokoloff, 1977b) agrees with that obtained above. Let us emphasize that the estimate of RM_f obtained from pulsar data obviously characterizes the fluctuation magnetic field within the Galaxy. At low latitudes the observed radioemission passes through many correlation cells. Therefore the fluctuation component RM_f growing in proportion with a square root distance to the boundary of the Galactic disk can be considerable.

Thus, we have come to an important conclusion. A dispersion of the rotation measures of the extragalactic radiosources is small, or, more exactly, does not exceed a contribution of the Galactic magnetic field. In order to avoid misunderstanding let us note that we consider a statistical characteristic of an entire sample of the sources given in a catalogue rather than a single radio source.

IV.4. The Fluctuation Magnetic Field in the Galaxy

Thus, instead of (IV.22) it is more appropriate to adopt the following statistical hypothesis

$$RM \overset{\mathrm{s}}{=} RM_g + RM_f \tag{IV.33}$$

The symbol 's' means that the equality should be understood only in the statistical sense; it may be violated for some sources (with great RM_g). Here the regular rotation measure RM_g has the same meaning as in (IV.23), or rather as in (IV.24). The fluctuation component can be determined in the same way as was done for pulsars (see (IV.13)), by replacing the distance to pulsar d by the distance to the boundary of the spiral arm in the direction to a given source. Since $d/\mathscr{L} \gg 1$ we have approximately

$$RM_f = \left[\frac{h}{\mathscr{L}} f(l^{\mathrm{II}}, b^{\mathrm{II}})\right]^{1/2} \xi \tag{IV.34}$$

where ξ is the random variable with a vanishing mean value and unit variance. For an elliptical arm $f = (e^2 \cos^2 b^{\mathrm{II}} \cos^2 l^{\mathrm{II}} + \sin^2 b^{\mathrm{II}})^{-1/2}$ with $e = 0.5$ while for the slab model $f = (\sin b^{\mathrm{II}})^{-1}$. It is clear that the disk half-thickness is $h = d/f$.

The new hypothesis (IV.33) does not reject completely the results of the previous section for K, l_0^{II} and b_0^{II}; it refines them and determines the mean square value of the fluctuation field. Parameters of the average magnetic field agree with those obtained above within two standard errors, therefore we do not write them down here. An estimate of the fluctuation field, δB, can be found by minimizing the sum

$$S(K, l_0^{\mathrm{II}}, b_0^{\mathrm{II}}) = \sum \xi^2$$

$$\equiv \sum_n \frac{(RM^{(n)} - RM_g)^2}{h/\mathscr{L} f}$$

and equating it to $S_{\min} = (0.81\, N\mathscr{L} \langle n_e \rangle \delta B)^2$. Using the Eichendorf and Reinhardt combined catalogue we obtain

$$\delta B/B \simeq 1.7. \tag{IV.35}$$

Thus, the fluctuation component prevails over the regular one. The estimate (IV.35) agrees with the result by Spoelstra (1977) $1 < \delta B/B < 3$ obtained from an analysis of fluctuations of the continuous Galactic radioemission polarization. Agreeable results are also obtained from fluctuations of the emission intensity (see Section IV.6).

These estimates yield both the mean-square value of the fluctuation at the basic scale of interstellar turbulence and the value of this scale. More refined information on spectral properties of the fluctuation field can be obtained from structure functions of the rotation measure fluctuations for extragalactic radiosource. The first attempt at such an investigation (Simonetti *et al.*, 1984; Simonetti and Cordes, 1986) has given an upper estimate for the basic scale, $\leqslant 5°$ (i.e. $\leqslant 90$ pc if the active region has a size of 1 kpc). Note that construction of the structure functions at very small scales of a few angular minutes (tenths of a parsec at the distance 1 kpc) can be made using the rotation measures determined for different components of those double radiosources which have small intrinsic rotation measures (Simonetti *et al.*, 1984).

When formulating the hypothesis (IV.33) we assumed the contributions of the fluctuation field to the rotation measures to be independent in different directions. Obviously, it is true when generally the lines of sight toward any two or more separate sources do not pass through the same turbulent cell. Hence, the number of the sources in the investigated sample should not be excessive! A crude estimate for an elliptic arm with the length about 10 h and cross-section $h \times 2h$ gives

$$N \lesssim \frac{3\pi h\ 10\ h}{\pi \mathscr{L}^2} = 30 \left(\frac{h}{\mathscr{L}} \right)^2 \approx 500.$$

The modern catalogues include even more radiosources. Therefore it is more expedient to improve (see Section IV.1) and correctly use the existing data than to add new sources.

IV.5. The Structure of the Large-Scale Field

So far we have assumed the large-scale field to be homogeneous. Now we discuss the applicability limits of this approximation. Certainly in the transverse direction with respect to the field direction the field scale is determined by the galactic disk thickness. Along the arm the investigated region size is determined by a maximal distance passed within the Galaxy by radio emission from low latitude sources. For the catalogues used above $b_{\min} \approx 10°$; then the scale of the measured field is equal to

$$\lambda = \frac{2h}{\tan b^{\mathrm{II}}_{\min}} \approx 10\,h \approx 4\,\mathrm{kpc},$$

fairly agreeing with the Orion arm length at both directions from the Sun.

In the next approximation after homogeneity we may formulate a problem of determination of the spatial structure of the large-scale field. First of all it is interesting to find out the behaviour of the field across the disk. An important question is the parity of magnetic field with respect to the equatorial plane $z = 0$. If the basic mode of the poloidal field is dipole the vertical field component B_z must have the same sign above and below the galactic plane. The symmetry is opposite for the quadrupolar field: $B_z(-z) = -B_z(z)$. According to the mean field dynamo theory the azimuthal field component is antisymmetric in the case of dipole poloidal field, or symmetric in the case of quadrupole field, under the transformation $z \to -z$ i.e. under the reflection in the central plane of the Galaxy.

The results of §3 show that even this simplest symmetry cannot be established with confidence due to local distortions of the field in the Northern hemisphere, $z > 0$. The only certain conclusion is that the field is directed along the local spiral arm (i.e. nearly azimutally). An attempt to determine independently the symmetry of the weaker radial field component has yielded only upper estimates of the strength of this component (Ruzmaikin *et al.*, 1977, 1978; Zeldovich *et al.*, 1983, p. 251).

It could be hoped that distortions of the field in the Northern hemisphere are local and we may determine the field parity by turning to more distant regions of the Galaxy.

Study of the field within other arms requires knowledge of the Faraday rotations for a sufficient number of low-latitude sources.* Simard-Normandin and Kronberg (1979), Vallée (1983a, b) and Agafonov *et al.* (1987) are the first to attempt investigation of the field in the Perseus and Sagittarius Arms using such sources. At the same time this method may be used for determining radial variations of the large-scale field. A summary of the data on the structure of the regular magnetic field of the Galaxy is represented in Figure IV.4.

IV.6. Intensity Variations of the Galactic Non-thermal Radio Background

Determination of the Faraday rotation measures requires difficult multi-frequency polarization observations of radiosources. Meanwhile, certain information on the magnetic field can sometimes be gained in simpler ways. Let us discuss a method and the first observational results of a determination of the characteristics of

* Another way is to observe nearby galaxies visible edge-on at high resolutions.

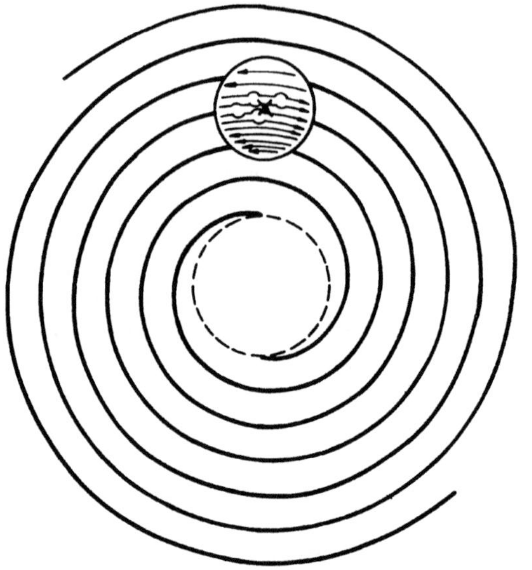

Fig. IV.4(a). The probable structure of the magnetic field in the Galaxy in the solar vicinity is shown on a schematic drawing of the Galactic spiral pattern. Near the Sun, whose position is marked by the cross, local magnetic loops are shown. The large-scale magnetic field direction in the Perseus arm is given according to Agafonov *et al.* (1987). The full circle contains the whole investigated region!

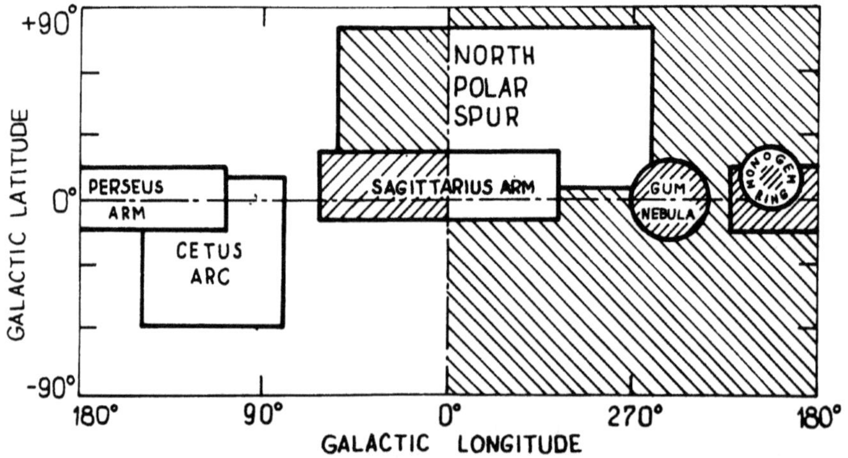

Fig. IV.4(b). The map of Faraday rotation measures of radiosources. Given are positions on the sky of the Galactic spiral arms and magnetic bubbles and spurs. The field is directed from the observer in shaded regions and toward us in unshaded ones. After Vallée (1984) and Broten *et al.* (1985).

average and fluctuation fields from intensity variations of the Galactic synchrotron emission (Chibisov and Ptuskin, 1981; Dagkesamansky and Shutenkov, 1987).

The intensity of synchrotron emission is determined by the density of relativistic electrons and by the component of the magnetic field transverse to the line of sight (see Section III.1). If relativistic electrons are distributed uniformly, which is favoured by the observed isotropy of cosmic rays, the intensity variations are due solely to the magnetic field fluctuations.

Consider the synchrotron intensities I coming from two points on the celestial sphere, separated by the angle θ. An angle between the large-scale field $\mathbf{B}$ and the vector connecting these two points is denoted by ϕ (Figure IV.5). The correlation

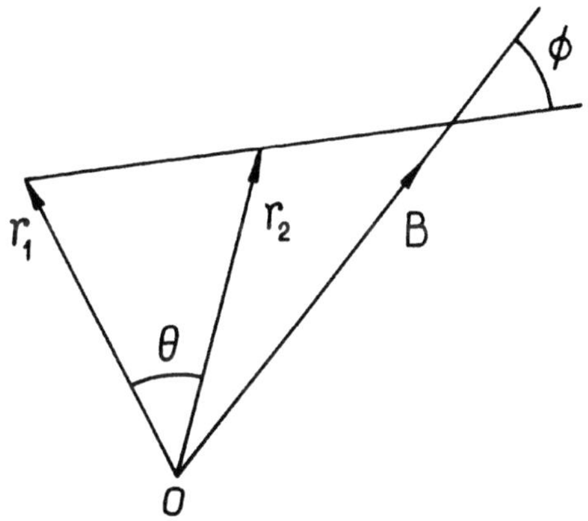

Fig. IV.5. The angles between chosen directions, that are determined by unit vectors $\mathbf{r}_1$ and $\mathbf{r}_2$, and the large-scale magnetic field $\mathbf{B}$.

function of intensity fluctuations on the sphere

$$\delta I^2 \equiv \langle (I(r_1) - \langle I \rangle)(I(r_2) - \langle I \rangle) \rangle$$

where $\mathbf{r}_1$ and $\mathbf{r}_2$ are unit vectors along chosen directions, can be represented as the sum of isotropic and anisotropic parts:

$$\delta I^2 = C_i(\theta) + C_a(\theta)\cos^2\phi. \tag{IV.36}$$

Chibisov and Ptuskin (1981) have shown (see Section III.2) that the anisotropic component C_a must have a maximum at $\theta_{\mathscr{L}} \approx \mathscr{L}/r$ where $\mathscr{L}$ is the characteristic scale of the field inhomogeneities, and r is the characteristic distance to the emitting region. The ratio of the isotropic part of the correlator to the minimal value of the anisotropic one gives the relative square of the fluctuation magnetic field:

$$\frac{(\delta B)^2}{B_\perp^2} \approx \frac{C_i(0)}{C_a(\theta_{\mathscr{L}})}. \tag{IV.37}$$

The correlation function reaches a maximum at $\phi = 0$, i.e. in the direction of the mean field.

The above relations are used by Dagkesamansky and Shutenkov (1987) to interpret their observations in the Galactic North Pole region (Figure IV.6). The observations were carried out at the frequency 102.5 MHz. The obtained correlation function shown in Figure IV.7 is in a good agreement with representation (IV.36). This confirms the existence of both the regular and the fluctuation fields. The correlation function shows a maximum which allows the large-scale field direction to be determined approximately as $l_0^{\mathrm{II}} = 90°$ (within 10° error), i.e. approximately along the spiral arm (see Figure IV.6 where the field direction is given by the arrow).

The characteristic size of inhomogeneities is $\theta_{\mathscr{L}} \simeq 8°$ which corresponds to a linear dimension of 140 pc at a distance of 1 kpc.

Extrapolation of the function $C_i(\theta)$ to $\theta = 0$ (Figure IV.8) yields $C_i(0) \simeq 1.7\ Jy^2$. The maximal value of $C_a(\theta)$ is 2.1 Jy^2 which gives, according to (IV.37) $\delta B/B \simeq 1$.

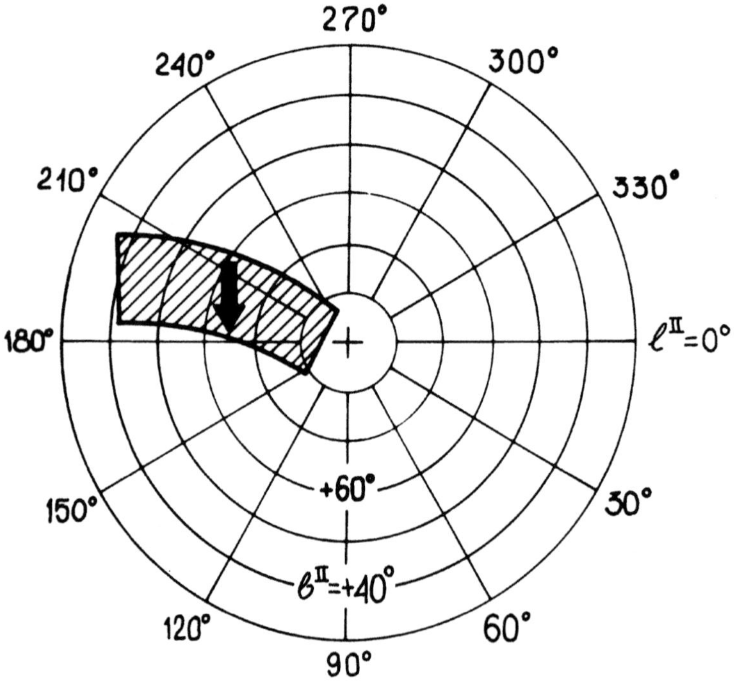

Fig. IV.6. The region (shaded), in galactic coordinates, in which Dagkesamansky and Shutenkov (1987) measured intensity variations of Galactic synchrotron emission. The arrow indicates the determined direction of the large-scale magnetic field.

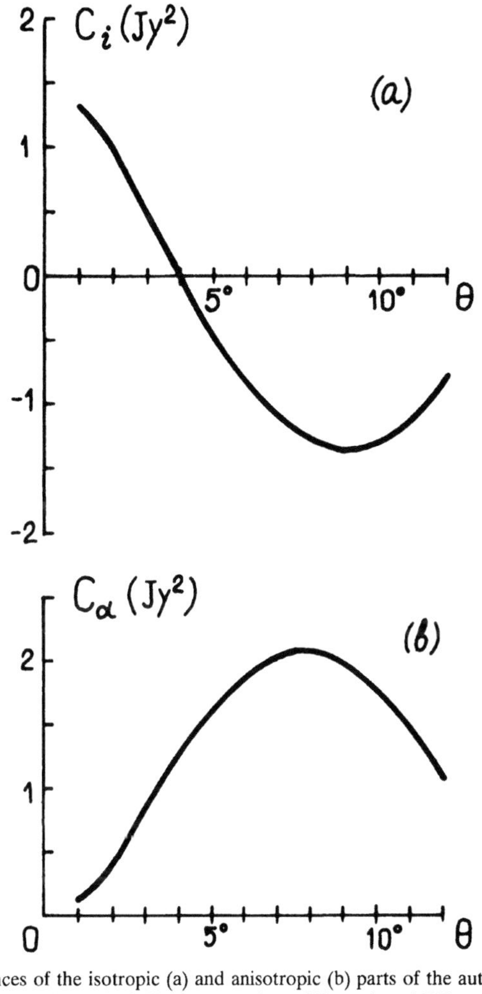

Fig. IV.7. Dependences of the isotropic (a) and anisotropic (b) parts of the autocorrelation function on the angle θ (Dagkesamansky and Shutenkov, 1987).

IV.7. Magnetic Fields in Nearby Spiral Galaxies

In this section we discuss possible global configurations proposed for the large-scale magnetic fields observed in external galaxies. Observations of sufficient quality to begin such an investigation have become possible only recently and only the first, pioneering results are available here.

The fields observed in nearby spirals can be crudely divided into two major classes: axisymmetric fields either concentrated within a ring at some distance from

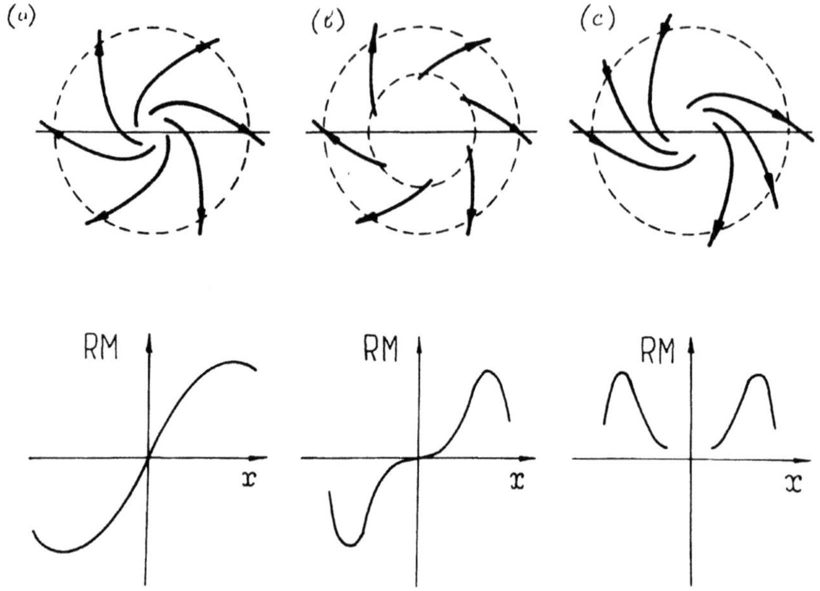

Fig. IV.8. Variations of the rotation measure and position angle along the major axis of a galaxy for three simplest structures of the large-scale magnetic field: (a) continuously distributed axisymmetric field; (b) axisymmetric ring; (c) bisymmetric structure.

the galactic center or smoothly distributed over the galactic disk; and non-axisymmetric (bisymmetric) structures (see Figure IV.8). This is, of course a very schematic division: both theoretical and observational evidences indicate the existence of more complicated configurations.

The most reliable method for determination of the magnetic field configuration in external galaxies is based on analysis of variations of the *intrinsic rotation measure* with the azimuthal angle in the galaxy plane (Tosa and Fujimoto, 1978; Sofue *et al.*, 1980; Sofue *et al.*, 1986). For the ring and continuous axisymmetric structures (b) and (a) in Figure IV.8, respectively, the intrinsic rotation measures RM_s in an observed galaxy vary according to

$$RM_s \propto \cos(\theta - i)$$

where θ is the azimuthal angle in the galactic plane measured from the major axis of the galactic image and i is the angle between magnetic lines and the azimuthal direction. For the bisymmetric structure, Figure IV.8(c), this law takes a markedly different form,

$$RM_s \propto \cos(2\theta - i - m) + \cos(i - m)$$

where m is the azimuthal angle at which the field is maximal.

The notable difference in the behaviour of RM_s with azimuthal variation periods differing by a factor of two enables to distinguish the axisymmetric structures from bisymmetric ones.

To determine RM_s over the whole galactic image, the observations with high sensitivity and resolution should be carried out at several wavelengths (cf. discussion above in this chapter). Such observational results are as yet rather rare, therefore simpler and thus less reliable methods are bravely applied. The major portion of the information on global configurations of magnetic fields in nearby spiral galaxies is obtained by investigation of *position angle* variations of the polarized non-thermal radiocontinuum with the azimuthal angle in a galaxy plane. It is presumed that these variations reflect the variations of RM_s with θ. The position angle variations with θ (averaged over 10° sectors) are shown in Figure IV.9 for the Andromeda Nebula for different distances to the galaxy center (Beck, 1982). The distinct cos θ behaviour can be seen in these plots, which indicates the axisymmetric structure of Figure IV.8(a) or (b) type. Analysis of the distribution of polarized intensity along the radius (see Figs. IV.10, VII.7) reveals a large-scale field concentration within the ring $7 \leqslant r \leqslant 14$ kpc. Note that the plots of Figure IV.9 for large galactocentric distances show the traces of bisymmetric structure (cos 2θ dependence) there as well.

Similar plots for the spiral galaxy NGC 6946 are shown in Figure IV.11 (Klein *et al.*, 1982). Here the θ-variation period for the position angle is about 180°, which is interpreted as evidence of the bisymmetric large-scale field in this galaxy.

Sofue *et al.* (1985) have proposed an elegant and still more simple method for determination of the global field configuration which requires measurements of the position angle variations along major and minor axes of a galaxy. The lower parts of Figure IV.8 show the expected variations of RM_s and the position angle along the major axis for three basic configurations. Of course, we cannot expect the observed dependences to be as neat as shown at these very schematic plots. However, the symmetry properties of these plots may be preserved in reality with the even symmetry corresponding to axisymmetric fields and the odd one, to bisymmetric structures. The required observations are comparatively simple and provide a way of rapid preliminary estimation of the global field configuration. However, this economical method sometimes gives unreliable results due to the ambiguity of an integer number of half-rotations in the definition of the position angle. As an example, we give two diametrically opposite interpretations of the position angle distributions along the major axis of NGC 2903 obtained by Sofue *et al.* (1985) and shown in Figure IV.12. The solid line (an even dependence) drawn through observed points corresponds to the two-armed spiral (bisymmetric) magnetic structure, (c) in Figure IV.8. However, addition of 180° to the angles on the right-hand side of the plot allows an equally successful interpretation of these data as showing an odd dependence (broken line), which corresponds to an axisymmetric field structure (Ruzmaikin *et al.*, 1986). We should note, however,

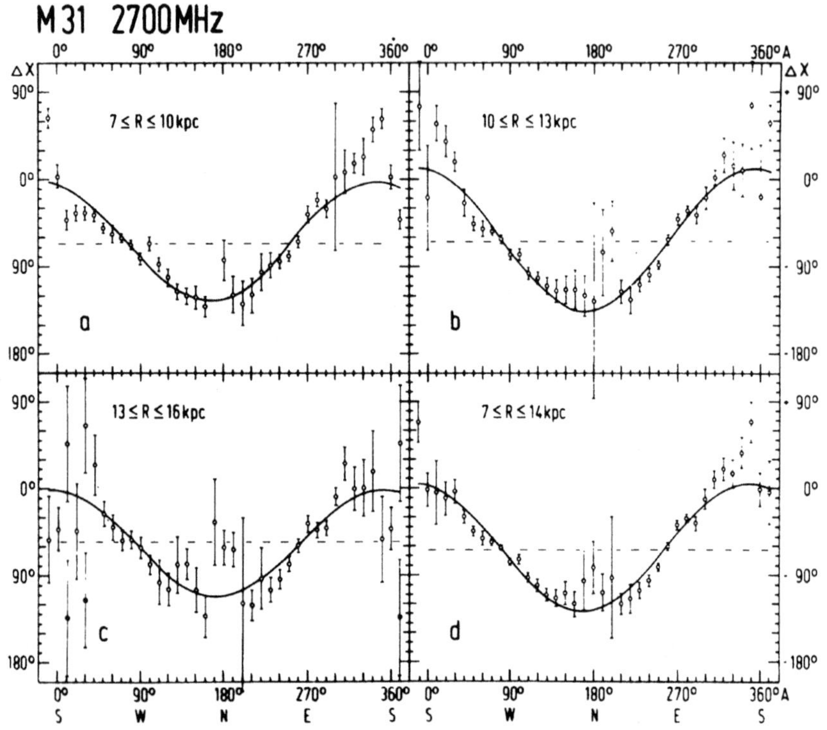

Fig. IV.9. Position angle of polarized radioemission of M31 versus the azimuthal angle in the galaxy plane for different galactocentric distance ranges. The observed dependence of cos θ type indicates the ring axisymmetric structure of the large-scale magnetic field (Beck, 1982).

that more detailed multifrequency observations confirm bisymmetric structure in NGC 2903 (Wieleblinski, 1986, private communication).

The two methods described above require measurements at only a single wavelength. However, the cost for economy of effort in observations is ambiguity of interpretation. For instance, it is difficult to establish the cause of azimuthal variations of the position angle: do they result from RM_s variations or are they due to variations in the intrinsic position angle?

It is no wonder therefore that observations at higher resolution and sensitivity give new information that corrects and deepens earlier results (cf. Beck *et al.*, 1985).

The very first pieces of information about the configuration of the large-scale magnetic fields have been obtained now for a dozen nearby galaxies. The majority of the results have been derived from single-frequency observations and they may be reconsidered in the future. The majority of galaxies have magnetic fields of

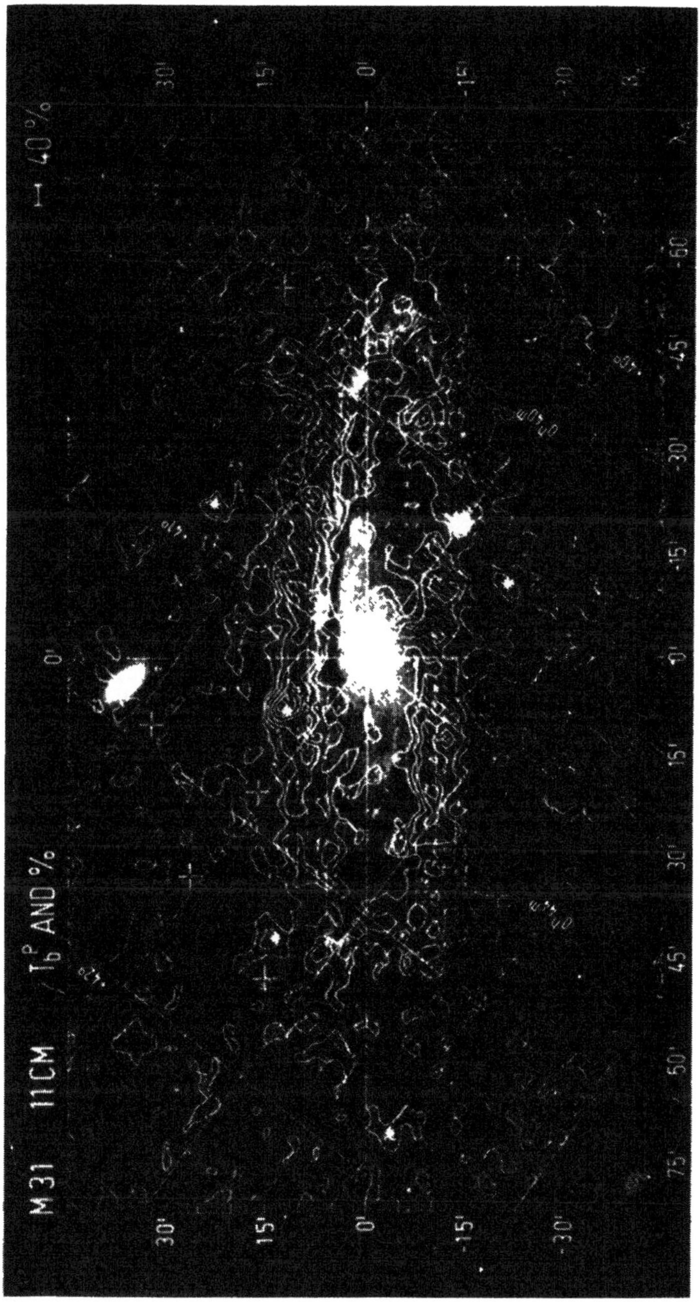

Fig. IV.10. Positions of synchrotron polarization planes and polarized intensity superposed on the optical image of M31 (Beck *et al.*, 1980). The large-scale magnetic field concentrates in a ring around $r = 10$ kpc.

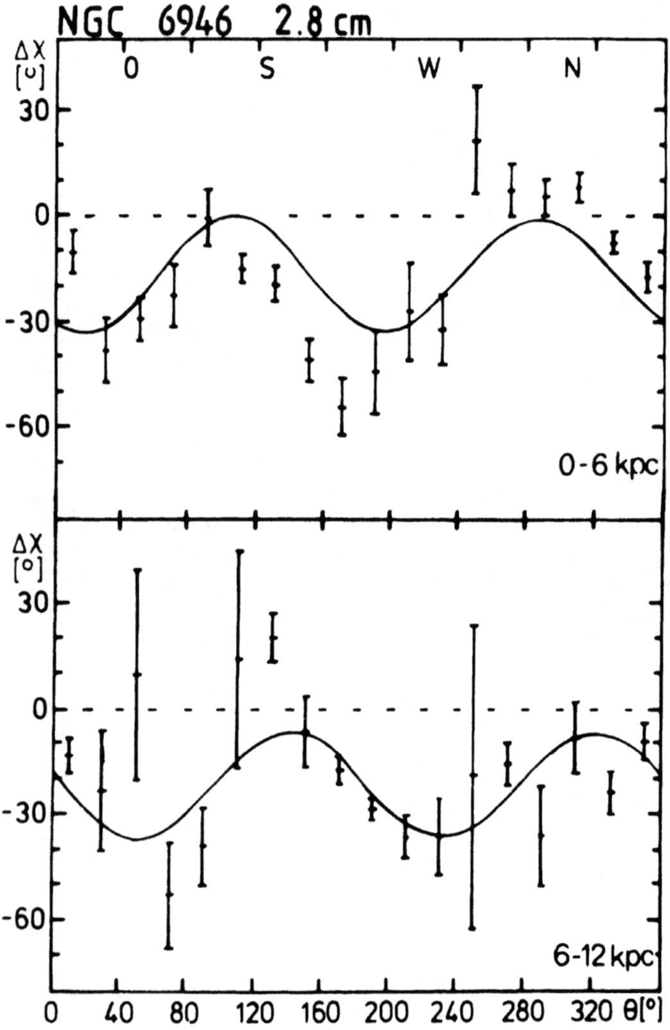

Fig. IV.11. The same as in Figure IV.9 but for the spiral galaxy NGC6946. The cos 2θ dependence indicates the bisymmetric large-scale magnetic structure (Klein *et al.*, 1982).

bisymmetric structure. The axisymmetric structures have been revealed in only two cases. The available observational information is compiled in Table IV.3 (see also Beck, 1983; Beck, 1986; Sofue *et al.*, 1986; Beck and Gräve, 1987). The corresponding theoretical models are developed by Ruzmaikin and Shukurov

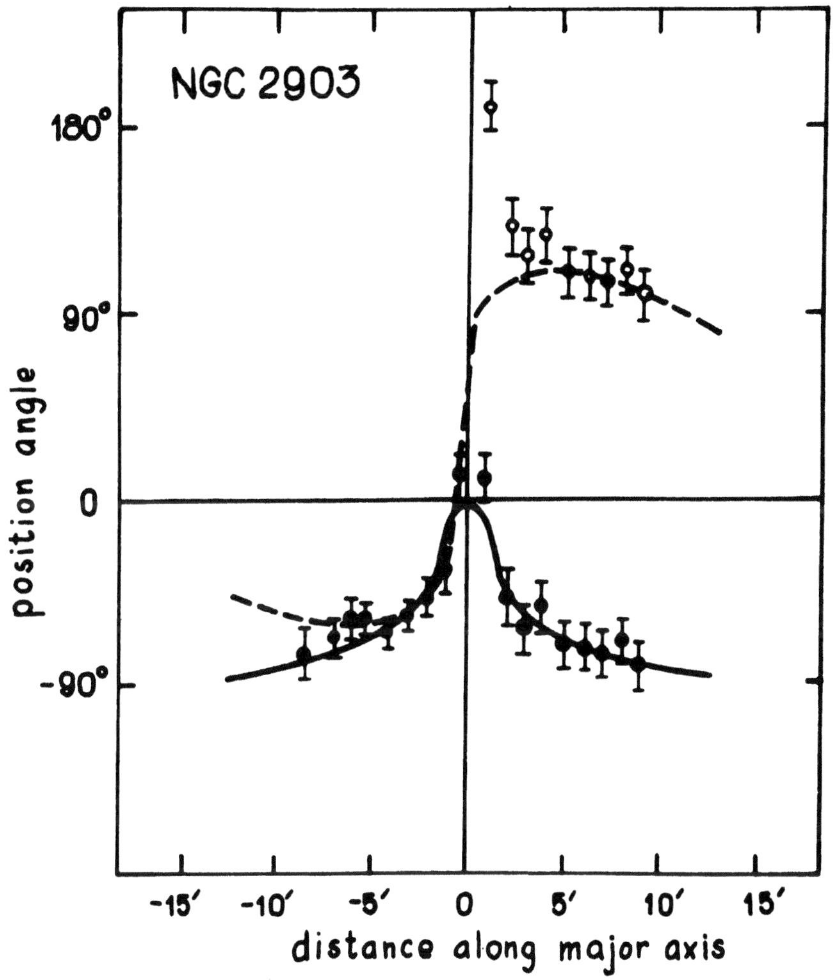

Fig. IV.12. Different interpretations of position angle variations along the major axis of a spiral galaxy NGC2903.

(1981), Ruzmaikin *et al.* (1985), Sawa and Fujimoto (1986), Baryshnikova *et al.* (1987b), Ruzmaikin (1987) and discussed in Chapter VII.

Observational investigation of magnetic fields in spiral galaxies is now in a state of rapid and fruitful development. The main results are due to observations in the radio range carried out at Effelsberg and the V. L. A. Quite recently optical polarimetric observations with CCD devices also began to give qualitatively new information. An example of such results is given in Figure III.3 which shows

TABLE IV.3
Magnetic fields in spiral galaxies according to radio observations

Galaxy	Morphological type	Large-scale magnetic field configuration	Large-scale magnetic field strength (μG)	Ratio of fluctuation to regular field	References
Milky Way	Sb	?	2—3	1.7	See the text
M31 (NGC 224)	Sb	Axisymmetric	2—4	0.4—0.5	Sofue and Takano, 1981; Beck, 1982
M33 (NGC 598)	Scd	Bisymmetric	1	2.5	Sofue and Takano, 1981; Buczilowski, 1985, 1987
M51 (NGC 5194)	Sc	Bisymmetric ?	5	1.7	Tosa and Fujimoto, 1978; Sofue *et al.*, 1980; Beck, 1983
M81 (NGC 3031)	Sb	Bisymmetric ? Bisymmetric	3	2.3	Sofue *et al.*, 1980 Beck *et al.*, 1985 Krause *et al.*, 1987
M83 (NGC 5236)	Sbc	Bisymmetric	?	?	Onderchen, 1985; Sofue *et al.*, 1986
IC 342	Scd	Axisymmetric	3	2	Sofue *et al.*, 1985; Gräve and Beck, 1987
NGC 253	Sc(p)	?	4	2.5	Sofue *et al.*, 1986
NGC 6946	Scd	Bisymmetric	5	3	Klein *et al.*, 1982
NGC 2903	Sbc	Bisymmetric	?	?	Sofue *et al.*, 1985
NGC 4258	Sbc	? Bisymmetric	5—20	? 2.5	Krause *et al.*, 1984 Hummel *et al.*, 1987

optical polarization vectors (presumably showing magnetic field vector orientations) over the galaxy M51. Unfortunately, optical data cannot give the direction of the magnetic field. Thus, they hardly can help to distinguish axisymmetric magnetic structure from bisymmetric. In this respect, observation between arms, where the field is not affected by spiral shock waves, may be helpful: in the bisymmetric structure there are sites (spiral lines) where the large-scale field strength is zero. Of course, these sites do not necessarily lie between the arms. At any rate, observations between arms seem to be most relevant for the purposes of comparison with theoretical predictions. For instance, the dynamo theory predicts that the pitch angle of magnetic lines (i.e. the angle between the magnetic lines and the circumferential direction) is 10°—20° between the spiral arms and thus is generally greater than the pitch angle of the spiral arms and the field within them (see Chapter VII). Observations in M31 (Loiseau *et al.*, 1987), IC 342 (Gräve and Beck, 1987) and M81 (Krause *et al.*, 1987) confirm the theory, giving consistent estimates of the magnetic pitch angles. However, it is important to detect observationally the difference of the magnetic pitch angles within the arms and between them.

Estimates of the large-scale magnetic field strength in galaxies can be obtained from intrinsic Faraday rotation measures derived from the observed amplitude of the polarization angle variations (Sofue and Takano, 1981). For example, the characteristic value of the intrinsic rotation measure in the Andromeda Nebula is 20 rad m^{-2} (Beck, 1982). For a disc thickness of 900 pc and thermal electron density 0.03 cm^{-3}, which are the values for our Galaxy, this gives an estimate of 2.4 μG for the large-scale magnetic field strength. Estimates of the field strength obtained for nearby galaxies in this way are given in Table IV.3.

The total strength of magnetic fields in spiral galaxies is estimated from the observed intensity of nonthermal radioemission under the assumption of equipartition between the cosmic ray and magnetic pressures. For M31, for example, this gives an estimate 4 ± 1 μG for the total field strength (Beck, 1982). Then the observed polarization of the non-thermal emission gives an estimate of the ratio of fluctuation and regular field components (see Section III.2). The model widely used for such estimates is that by Segalovitz *et al.* (1976), Equation (III.10) above. After extraction of the thermal component of radioemission, the polarization of non-thermal component is about 65% in M31 (Beck, 1982). According to the model of Segalovitz *et al.*, this corresponds to the ratio $\delta B/B \approx 0.5$ of the fluctuation to regular field. Shortcomings of the model of Segalovitz *et al.* are discussed in Section III.2. According to the model of Korchak and Syrovatskii (1961), the same polarization gives $\delta B/B \approx 0.4$ (see equation (III.7) above). Thus, the model of Segalovitz *et al.* somewhat overestimates the ratio $\delta B/B$. On the other hand, note that the estimate of the regular field in M31 which follows from the polarization degree and equipartition arguments, $B \simeq 0.93H \simeq 3.7$ μG, does not agree with the estimate $B = 2.4$ μG from the Faraday rotation observations.

Possibly, this indicates either a lower value of the thermal electron density (or other physical parameters) in M31 as compared with the Galaxy, or deviations from equipartition.

Estimates of the fluctuation fields given in Table IV.3 are obtained by the authors of the corresponding papers with use of the model of Segalovitz *et al.* Note also that linear resolution of observations used in compilation of the Table IV.3 generally deos not exceed 1—3 kpc (see Beck, 1983). Hence, the gradients of the regular field strength may contribute considerably to the estimated fluctuation field strength. New V. L. A. observations have reduced this figure to 200 pc for M31 (Loiseau *et al.*, 1987), which is close to the presumed value of the energy-range scale of interstellar turbulence, 100 pc.

CHAPTER V

ORIGIN OF MAGNETIC FIELDS

V.1. Introduction

In previous chapters we have shown how in many independent ways magnetic fields are revealed in the interstellar gas of galaxies, within the huge volumes of radiogalaxies, in quasars and in other objects. Details of magnetic field configurations and their strengths are not always reliable, but the general result is certainty of the presence in galaxies and galactic-type objects of magnetic fields that cover wide ranges of spatial scales and strengths. Let us now discuss the basic concepts of the origin of galactic magnetic fields.

Only 30 years ago the origin of magnetic fields at scales comparable to the size of the Galaxy seemed to be quite mysterious. Indeed, a tremendous potential difference of 3×10^{12} V is required in order to build up a field of 10^{-6} G strength over a 10 kpc scale over the period of 10^{10} years. The formidable enormity of this result provoked F. Hoyle (1958) to suggest that the Galactic magnetic field is an offspring of a field on a still larger, cosmological scale. Further development of this idea is due to Piddington (1964). We discuss this hypothesis in §2 below and the discussion of modern views on cosmological magnetic fields can be found in Section 3.

In fact, the hypothesis of the relic origin of galactic magnetic fields does not solve the problem but only shifts it to earlier stages of evolution of the Universe. The problem of the origin of galactic magnetic fields is simply replaced by the problem of the origin of cosmological magnetic field. Hoyle himself realized that this approach seems to be somewhat artificial. Therefore, attempts continued to find an alternative source of the magnetic field inside the Galaxy. In particular, the suggestion was considered that the global galactic field is produced from local magnetic fields ejected by stars (Michel and Yahil, 1973; see §4 below). Although this mechanism is incapable of explanation of the observed large-scale field, stellar ejections undoubtedly play an important role in the local field configuration.

A convincing explanation of the origin of magnetic fields in the Milky Way galaxy and other spiral galaxies is provided by the dynamo theory. This theory develops the suggestion of Sir Joseph Larmor (1919) who first conceived the idea that magnetic fields can be excited by motions of conductive fluid or gas (see Figure I.1 and the monographs of Moffatt, 1978; Parker, 1979; Krause and Rädler, 1980; and Zeldovich *et al.*, 1983).

The mechanism of the field maintenance is based on the electromagnetic induction law discovered by Michael Faraday. When a conductor moves in a magnetic field, an electromotive force is induced within it that drives an electric current, giving rise to a new magnetic field and thus modifying the initial one. But does this chain of events amplify the initial field or attenuate it? It would seem that very specific motions are required for amplification and maintenance of magnetic fields (cf. Cowling, 1934; Zeldovich, 1956; Zeldovich and Ruzmaikin, 1980). The beauty and fascination of the dynamo theory is due to the fact that it shows that in many classes of astronomical objects — from planets and stars to galaxies — the *natural* motions of conductive medium are present and are capable of exponentially fast amplification of some weak initial magnetic fields. In particular, turbulent motions lacking mirror symmetry are sufficient for the production and maintenance of a large-scale magnetic field by a hydromagnetic dynamo. The overall rotation and non-uniform gas distribution are in turn sufficient for violation of the mirror symmetry of small-scale motions — turbulence or convection. One more intriguing aspect is thus inherent in the generation of the large-scale fields: the magnetic fields of large spatial and temporal scales are maintained by random small-scale fluid motions. The order arises from the chaos.

The role of motions in magnetic field evolution is determined, in comparison with dissipative effects, by the magnetic Reynolds number

$$R_m = VL/\nu_m.$$

Here V and L are the characteristic velocity and scale of the motions, and the magnetic diffusivity ν_m is determined by the electric conductivity σ through

$$\nu_m = \frac{c^2}{4\pi\sigma}.$$

The main specific feature of interstellar magnetic fields is their extremal spatial extent. Typical scales range from 100 pc for chaotic fields to several kiloparsecs for the mean field. Turbulent motions in the interstellar medium have typical velocities $v \approx 10$ km s^{-1}. The regular motion (differential rotation) within galaxies is characterised by velocities of the order of 200 km s^{-1}. Dissipation of electric currents in partially ionised interstellar gas provides magnetic diffusivity of order $\nu_m \approx 10^{21}$ cm^2 s^{-1}. Hence, the magnetic Reynolds number is 3×10^5 for chaotic motions, while it is still larger, $R_m \simeq 5 \times 10^7$, for the regular motion.

Extremely high values of the magnetic Reynolds number indicate a prevalent role of motions in the evolution of interstellar magnetic fields. Dissipation is essential only for the fields of very small scales of the order of tenth and hundredth of parsec (McIvor, 1977; Ruzmaikin and Shukurov, 1982).

In other words, the concept of magnetic fields frozen into the medium can be widely used in qualitative description of the influence of the ambient medium on interstellar magnetic fields. This provides a very convenient way of describing the magnetic lines that move, stretch and entangle together with streamlines. Magnetic lines can be usefully considered as elastic ropes that move and extend together with the ambient medium (cf. Alfvén, 1950, §4.4).

The frozen-in field is completely determined by the velocity field: $\nu_m = 0$. The field in a non-moving medium $\mathbf{V} = 0$ is subject to magnetic diffusion alone: an initial distribution $\mathbf{H}_0(\mathbf{r})$ spreads out with a characteristic time L_H^2/ν_m where L_H is the scale of the spatial distribution of the field. When the ambient conductive medium is made turbulent, the large-scale magnetic field is destroyed by turbulent diffusion; for its description the diffusivity ν_m can be replaced by the turbulent magnetic diffusivity

$$\beta \simeq \tfrac{1}{3} lv,$$

where l and v are the characteristic scale and velocity of turbulent motions, respectively.

V.2. The Relic Field Hypothesis

The frozenness of galactic magnetic fields suggests a pleasingly simple solution for the origin of the large-scale galactic magnetic field. If the protogalactic gas cloud had been permeated by an external magnetic field (perhaps of cosmological origin), the cloud contraction in the course of the formation of the galaxy would amplify the field by compression, while the galactic differential rotation would finish the process of formation of the present large-scale magnetic field (Hoyle, 1958; Piddington, 1964). Now this idea is eagerly advocated by Piddington (1970, 1974, 1975a, 1981a) who, moreover, ascribes to magnetic fields a decisive role in establishing the morphological diversity of galaxies (see also Wasserman, 1978). The relic field hypothesis, in the form developed by Piddington, has provoked some discussion. Parker (1973a, b) has proposed forcible arguments against the hypothesis; contrary ideas have been given in a series of papers by Piddington (1970, 1972, 1973, 1975a, b, 1981b). The following result is important: magnetic fields currently observed in galaxies cannot be of pre-galactic origin because the relic fields are rapidly (in a period of the order of 10^8 years) destroyed by the turbulent magnetic diffusion. In order to provide a prolonged, over ten billion years, existence of galactic magnetic fields, a continuously operative regeneration mechanism is required. The arguments in favour of this view are not very simple because the relic field is not only destroyed by the turbulent diffusion, but also simultaneously amplified by differential rotation. Therefore, the hypothesis of the relic field needed more careful attention. Note, however, that differential rotation can only provide a linear temporal growth of magnetic field while diffusion causes its exponential decrease in time. Hence, over long periods the destructive influence of diffusion would always dominate. The problem is to decide whether galaxies are sufficiently old from this point of view.

Consider the evolution of an external, initially uniform magnetic field $\mathbf{H}_0$ which permeates the differentially rotating conductive disc of a spiral galaxy. The field behaviour is determined by the competition of differential rotation, which amplifies the field by stretching magnetic lines, and magnetic diffusion which entangles the field, reducing its scale, and thereby promoting Ohmic dissipation and magnetic line reconnections.

Details of the field evolution depend on its orientation with respect to the rotation axis. Evolution of magnetic field $\mathbf{H}$ in a continuous conducting medium moving at velocity $\mathbf{v}(\mathbf{r})$ is described by the induction equation

$$\frac{\partial \mathbf{H}}{\partial t} = \mathrm{rot}(\mathbf{V} \times \mathbf{H}) - \mathrm{rot}(\nu_m \,\mathrm{rot}\, \mathbf{H}), \tag{V.1}$$

which is a consequence of Maxwell's equations and Ohm's law. Considering the magnetic field in a differentially rotating galaxy we adopt

$$\mathbf{V} = \mathbf{\Omega} \times \mathbf{r}, \qquad \mathbf{\Omega} = \mathbf{\Omega}(\mathbf{r}).$$

As long as the induction equation (V.1) is linear in magnetic field, it is sufficient to consider two cases separately, firstly when the external field $\mathbf{H}_0$ is orthogonal to the rotation axis and the secondly for the parallel field. A general case reduces to a superposition of these two.

We begin with a uniform magnetic field and a differentially rotating conductive disc embedded in the field with the rotation axis orthogonal to the field. When the inner regions rotate faster than the outer ($\partial\Omega/\partial r < 0$, as is the case in spiral galaxies), the frozen-in field takes, after a few rotations, the form of a trailing two-armed spiral (Figure V.1), akin to the magnetic configuration deduced for some spiral galaxies (see Section IV.7). However, the galaxies undergo a few tens of rotations during their lifetimes and such similarity is very transient. It is well known that the magnetic field is effectively excluded from regions occupied by flows with closed streamlines, surviving only in thin skin layers (R. Parker, 1966; Weiss, 1966; see also Moffatt, 1978, §§3.8 and 3.9; E. Parker, 1979, Chapter 16).

At the initial stage the field strength grows linearly with time, but the distance between neighbouring spiral turns of oppositely directed magnetic fields (Figure V.1) decreases at the same rate:

$$\Delta r \simeq r_0/\Omega_0 t, \tag{V.2}$$

where Ω_0 and r_0 are the characteristic angular velocity of rotation and its scale, respectively: $|\partial\Omega/\partial r| \simeq \Omega_0/r_0$. Approach of the regions with oppositely directed fields enhances magnetic dissipation and the linear growth of the field comes to the end when the diffusion effects become essential.

With the scale of the field given by (V.2), the diffusion time is

$$\tau_d \simeq \frac{(\Delta r)^2}{\beta} = \frac{r_0^2}{\beta}\,\frac{1}{(\Omega_0 t)^2}\,.$$

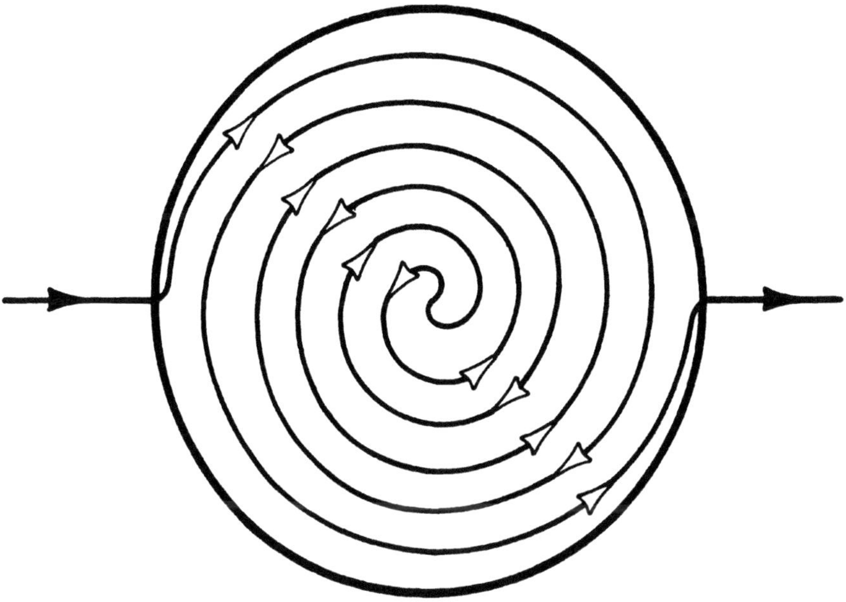

Fig. V.1. Differential rotation distorts the magnetic field lying orthogonally to the rotation axis (Parker, 1973b).

The linear growth lasts until $t \lesssim \tau_d$, i.e.

$$\Omega_0 t < \Omega_0 t_0 \simeq \left(\frac{\Omega_0 r_0^2}{\beta} \right)^{1/3}$$

(Parker, 1963, 1973b). Note that the dimensionless combination $\Omega_0 r_0^2/\beta$ is just the turbulent magnetic Reynolds number based on rotational shear. During this period the field has time to grow up to the strength

$$H_{\max} \simeq \left(\frac{\Omega_0 r_0^2}{\beta} \right)^{1/3} H_0 .$$

What follows is a rapid decay of the field in a major part of the disc volume. In galactic discs $r_0 \simeq 3$ kpc, $\Omega_0 \simeq 10^{-15}\ \mathrm{s}^{-1}$ and $\beta \simeq 10^{26}\ \mathrm{cm}^2\ \mathrm{s}^{-1}$, which gives

$$t_0 \simeq 3 \times 10^8 \text{ years}, \qquad H_{\max} \simeq 10\, H_0 .$$

For $t \gg t_0$ the disc is free of magnetic fields, except skin layers whose thickness is of the order

$$\delta r \simeq \left(\frac{\Omega_0}{\beta} \right)^{-1/2} \simeq 100 \text{ pc}.$$

Hence, the component of intergalactic field parallel to the disc is excluded from the disc rapidly indeed.

Consider now the external magnetic field parallel to the rotation axis (see Figure V.2 where the disc thickness is exaggerated to make details visible). In this case the exclusion effect typical of the orthogonal configuration is absent: the wound-up field is directed uniformly in neighbouring turns (except the region near the equator) and the fatal enhancement of dissipation does not occur. Let us write out the azimuthal component of induction equation (V.1) with turbulent diffusivity in polar cylindrical coordinates (r, φ, z) with the z-axis coincident with $\mathbf{\Omega}$, i.e. $\mathbf{\Omega} = (0, 0, \Omega)$ and $\mathbf{H}_0 = (0, 0, H_0)$:

$$\frac{\partial H_\varphi}{\partial t} = r\,\frac{\partial \Omega}{\partial z}\,H_0 + \beta(\Delta\mathbf{H})_\varphi, \tag{V.3}$$

where the last term on the right-hand side is the azimuthal component of the Laplacian. Differential rotation affects the external field only when angular velocity varies along the field. Note that in observed galactic discs the angular velocity

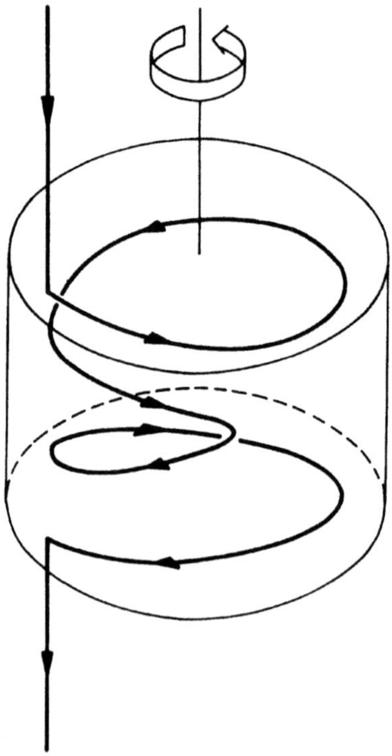

Fig. V.2. Distortion of the magnetic field parallel to the rotation axis with angular velocity varying along the axis.

varies predominantly along the radius, $|\partial\Omega/\partial z| \ll |\partial\Omega/\partial r|$. However, vertical variations of Ω could be considerable at the stage of contraction of a protogalactic gas cloud. In this case Figure V.2 can be viewed more literally. Consider for simplicity the field evolution in stationary conditions, although the time of the protogalactic cloud evolution (free fall time $(G\rho)^{-1/2}$) may be comparable with the field evolution time. Taking account of the cloud contraction does not change the final result.

At the initial stage the diffusion term can be neglected in equation (V.3) which gives

$$H_\varphi = \frac{r\,\mathrm{d}\Omega}{\mathrm{d}z} H_0 t.$$

Again, the field initially grows linearly.

When Ω is symmetric with respect to the equatorial plane, H_φ is an anti-symmetric function, $H_\varphi(z) = -H_\varphi(-z)$.

The linear growth of the azimuthal field lasts until diffusion becomes essential, i.e. up to the moment, say z_0^2/β. During this time the azimuthal field becomes

$$H_\varphi \simeq H_0\Omega_0 \frac{z_0^2}{\beta} \qquad \text{(V.4)}$$

where we have assumed that the vertical z_0 and radial r_0 scales are comparable. For the periods much exceeding the diffusion time the stationary state is reached which is determined by a balance of field amplification by differential rotation (the first term on the right-hand side of (V.3)) and attenuation by turbulent diffusion (the second term). The stationary field strength is given by the order-of-magnitude estimate (V.4).

Thus, in the case when the protogalaxy rotation axis is parallel to the magnetic field differential rotation provides the azimuthal field whose stationary strength exceeds the initial poloidal field by a factor of the order of a number of protogalaxy revolutions in diffusion time.

Let us now turn to the possible strength of the initial field H_0. Observations give a stringent upper limit on the possible intergalactic field

$$H_{\mathrm{IG}} < 10^{-9}\ \mathrm{G}$$

(see the next section). Isotropic contraction of the gas from intergalactic density $\rho_{\mathrm{IG}} \simeq 10^{-29}$ g cm^{-3} up to interstellar density $\rho \simeq 10^{-24}$ g cm^{-3} with frozen-in intergalactic field gives

$$H_0 < H_{\mathrm{IG}}(\rho/\rho_{\mathrm{IG}})^{2/3} \simeq 2 \times 10^{-6}\ \mathrm{G}. \qquad \text{(V.5)}$$

Since $\Omega_0 z_0^2/\beta > 1$ in (V.4), it seems that the resulting field could explain the present galactic magnetic field. Let us show why this is not true.

First, Parker (1973b) notes that under such strong compression, magnetic lines would be very strongly bent (see Figure V.3 where $\phi \simeq \arctan(\rho_{IG}/\rho)^{1/3} \simeq 1°$). Near the neutral surface marked by the broken line, rapid reconnection occurs (when the gas pressure is negligibly small) which diminishes the total magnetic flux through the galacatic disc. In other words, due to the specific topology of the magnetic lines, field dissipation is enhanced during formation of a galaxy which invalidates estimate (V.5). The reconnections proceed at the rate εV_A where $V_A = H(4\pi\rho)^{-1/2}$ is the Alfvén velocity and $\varepsilon \simeq 0.1$—1.0. For $H_{IG} = 10^{-9}$ G and $\rho_{IG} = 10^{-29}$ g cm^{-3} we obtain $V_A \simeq 1$ km s^{-1} which means that in 10^{10} years reconnections change the magnetic field within $2V_A t \simeq \varepsilon \times 20$ kpc. Of course, the gas pressure makes the rate of reconnections smaller but the derived estimate very much exceeds the thickness of the present galactic disc.

Second, the resulting azimuthal component of galactic field is antisymmetric with respect to the galactic equator, $H_\varphi(z) = -H_\varphi(-z)$. Meanwhile, the azimuthal field observed in the Galaxy is probably symmetric over the disc thickness (see Chapter IV). Hence, observations in the Milky Way galaxy exclude the possibility that the present large-scale magnetic field is a result of compression of the relic

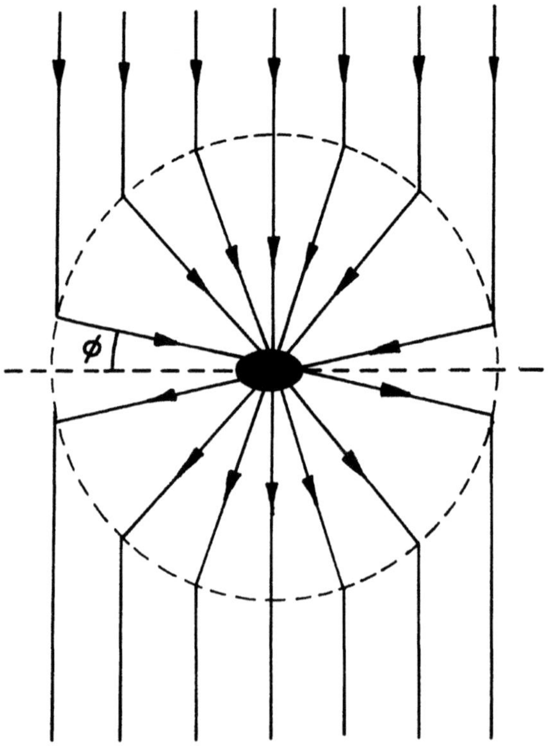

Fig. V.3. A highly distorted configuration of magnetic lines arising when protogalaxy contracts in an external magnetic field.

field parallel to the rotation axis. As we have seen above, the field orthogonal to the rotation axis even cannot survive against differential rotation and dissipation.

There is one more worrying problem associated with the relic field hypothesis. According to this hypothesis, the field within galaxies exists due to some external source. If the external large-scale field decays, e.g. owing to turbulent diffusion, the field inside a galaxy would disappear in approximately 10^8 years. Discussing the fate of the external field within the disc, we have considered only the azimuthal component (V.3) of the induction equation that includes the z-component of the external field. The z-component of the induction equation reduces to the diffusion equation for H_0. As long as turbulent motions are present in intergalactic gas and within galactic gaseous coronae, the large-scale intergalactic field becomes tangled and the large-scale component aligned with the rotation axis decays exponentially. Accordingly the azimuthal field decays within the galaxy.

We have devoted so much space to the discussion of the hypothesis of the primordial origin of galactic magnetic fields because recent observations of bisymmetric magnetic structures have provoked, in the absence of an alternative theory, a suggestion of the relic origin of these galactic magnetic fields. The basis for such an interpretation is an outward resemblance of bisymmetric structures to the magnetic lines of a uniform external field stretched by differential rotation, only, however, after a few revolutions. The spiral galaxies have made several tens of revolutions during their lifetimes. The bisymmetric magnetic structures are only mildly twisted and thus they could not exist in differentially rotating galactic discs if they are not presently maintained by some mechanism. The situation here is quite similar to that encountered in the theory of the spiral structure: a relatively large inclination of spiral arms tells us that they cannot be a permanent star condensations but are rather waves travelling along the disc. Bisymmetric magnetic structures also can be explained as dynamo-waves propagating along the azimuth (see Section VII.9). Bisymmetric structures cannot serve as an indication of the relic origin of the galactic magnetic field.

V.3. Cosmological Magnetic Fields

As we have described above, the hypothesis of a magnetic field existing at the largest conceivable scale in the Universe has been suggested in connection with the problem of the Galactic magnetic field origin. This suggestion by Hoyle (1958) was a reaction to the absence, at that time, of the theory of the large-scale magnetic field of the Galaxy. The presence of such a field has been anticipated from views on cosmic ray confinement and its direct manifestation was the polarization of starlight. In the years since that time a dynamo theory has been developed that explains the existence of the large-scale Galactic field by internal causes, i.e. by plasma motions inside the Galaxy. However, the problem of the cosmological magnetic field, once suggested, lives its own life. Apart from purely applicational importance, it is of general scientific interest. Magnetic fields have been revealed in all classes of astronomical objects, from planets and stars to galaxies and their clusters. Can that unique and universal object, the Universe, be magnetic?

A uniform magnetic field existing in the Universe *ab initio* does not violate its homogeneity but makes anisotropic the cosmological expansion (Zeldovich, 1965; Doroshkevich, 1965; Thorne, 1967; see also Zeldovich and Novikov, 1975, §19.3). As long as the field is uniform, its curl is zero, rot $\mathbf{H} = 0$, and no electric currents are required to maintain it. A uniform magnetic field within a finite body is maintained by currents flowing over its surface. In an infinite Universe such currents are absent and therefore the magnetic field dissipation is absent regardless of the conductivity of the medium. The uniform magnetic field is exactly frozen into the expanding cosmological plasma and changes only due to transverse Hubble expansion,

$$\frac{\mathrm{d}H}{\mathrm{d}t} = -2\mathscr{H}_\perp H,$$

where $\mathscr{H}_\perp$ is the Hubble constant in the direction orthogonal to the field.

The magnetic field changing in time induces the vortex electric field $\mathbf{E} = (\mathscr{H}_\perp / c)\mathbf{r} \times \mathbf{H}$. Due to the frozenness of the magnetic field the electric field vanishes in a co-moving reference frame.

In principle, the magnetic field could be given as an initial condition and formed together with the Universe. This approach seems, however, to be artificial, as Hoyle, the creator of this hypothesis, has realized. This calls for a mechanism capable of magnetic field creation at early stages of cosmological evolution. A large-scale magnetic field could be created at the early stages, when $z > 10^{14}$ (z is the redshift), due to the alignment of magnetic moments of baryons and leptons (Brecher and Blumenthal, 1970). This would give the present intergalactic field $H_0 \lesssim 10^{-14}$ G. Vilenkin and Leahy (1982) propose a mechanism of magnetic field generation in the early Universe based on parity non-conservation in weak interactions (see also Leahy and Vilenkin, 1981). Magnetic fields could be generated by fluxes of charged particles that arise due to asymmetric neutrino emission by primordial black holes or asymmetric decay of unstable particles whose spins are oriented when they are emitted by rotating primordial black holes. When rotation axes of the black holes are randomly oriented and they make a minor contribution to the total density, Vilenkin and Leahy predict

$$H_0 < 2 \times 10^{-9}\ \mathrm{G}$$

presently at galactic scale. Note that the fields thus produced are inhomogeneous and their strength decreases toward larger scales. A possibility of generating inhomogeneous magnetic fields in the early Universe during the first-order phase transition is considered by Hogan (1983) who predicts a possible present intergalactic field $H_0 < 10^{-13}$ G.

The cosmological magnetic field breaks the isotropy of expansion of the Universe making anisotropic the energy—momentum tensor: expansion along the field is decelerated and that transversely to the field it accelerates. Cosmological astronomical observations indicate a very isotropic character of the Universe expansion at the present time (see Zeldovich and Novikov, 1975; Weinberg, 1972). This allows us to place limits on the possible cosmological magnetic field.

Firstly, anisotropy in the expansion rate results in an anisotropic temperature of relic microwave background radiation. Frequent Thomson scatterings of relic photons by free electrons keep the radiation isotropic as long as the Universe is sufficiently hot and dense so as to provide a sufficient density of free electrons. However, after the Universe becomes transparent, anisotropy in the redshift causes anisotropy in the temperature of the microwave background while the radiation spectrum remains the black-body one. If $A(t)$ and $W(t)$ denote the transverse and longitudinal expansion scale factors, and $\mathscr{H}_\perp = \dot{A}/A$ and $\mathscr{H}_\parallel = \dot{W}/W$ are the respective Hubble constants, the temperature anisotropy is given by

$$\frac{\Delta T}{T} \simeq \left.\frac{A - W}{W}\right|_{z = z_L},$$

where z_L is the redshift corresponding to the moment when the intergalactic gas became transparent (see Zeldovich and Novikov, 1975, Section 19.3). Expansion becomes more isotropic with time, therefore the anisotropy of relic radiation is preserved from the moment of the last scattering. More exactly, the temperature distribution over the sky is as follows:

$$T(\theta) = T_0 \left[\frac{1}{A^2} \sin^2 \theta + \frac{1}{W^2} \cos^2 \theta \right]^{-1/2},$$

(see e.g. Thorne, 1967), where T_0 is the mean temperature, θ is the angle measured from the longitudinal direction W, and the normalization adopted is $A = W = 1$ presently. In the framework of the simplest quasi-Euclidean, weakly anisotropic, pressure-free cosmological model with uniform magnetic field (Doroshkevich, 1965; Thorne, 1967) the last expression reduces to

$$T(\theta) \approx T_0 \left[1 + \frac{3}{2\pi} (1 + z_L) \frac{H_0^2}{\rho_0 c^2} \cos^2 \theta \right],$$

where H_0 is the present strength of the magnetic field and ρ_0 is the present matter density.

Anisotropic expansion caused by a magnetic field results in the quadrupole pattern of the relic radiation temperature. Observations give the following upper limit of the quadrupole component of the angular temperature variations:

$$a_2 < 2 \times 10^{-5}$$

(Strukov and Skulachev, 1986). The corresponding limit on the cosmological magnetic field strength is given by

$$H_0 < 2 \times 10^{-8}\, h \left(\frac{1 + z_L}{1300}\right)^{-1/2} \left(\frac{a_2}{10^{-5}}\right)^{1/2} \mathrm{G}$$

(Juszkiewicz and Shukurov, 1986), where $h = \mathscr{H}/100$ km s^{-1} Mpc^{-1} is the present value of the Hubble constant. The value of the redshift z_L of the last Thomson scattering depends on the thermal history of the Universe after recombination (see Zeldovich and Novikov, 1975, Chapter 8; Weinberg, 1972, Section 15.5). In any case, $10 \lesssim z_L \lesssim 1300$ which gives the following upper limit:

$$H_0 < (5\text{–}50) \times 10^{-8}\, \mathrm{G}.$$

Other effects can also contribute to the quadrupole anisotropy of the relic radiation, density fluctuations among them. Hence, the sign 'much less than' should be used in the last inequality.

Cosmological magnetic fields can be also estimated from observations of the polarization of the relic radiation (Milaneschi and Fabbri, 1985). However, in contrast to the anisotropy degree, the polarization degree is very sensitive to the ionization history of the Universe (see, e.g., Basko and Polnarev, 1980; Negroponte and Silk, 1980), which factor is rather uncertain.

The other way to restrict the cosmological magnetic field follows from nucleosynthesis of light elements (H, ^{4}He, etc.) that had occurred in the period $1 \lesssim t \lesssim 100$ s after the Big Bang (Zeldovich and Novikov, 1975, Chapter 6). Seventy per cent of the mass of matter, unprocessed in stars, comprises hydrogen, while thirty per cent is helium. Such abundances require a highly isotropic expansion during nucleosynthesis (Thorne, 1967; Zeldovich, 1969). The condition of small anisotropy for $t > 1$ s gives the following limit on the magnetic field:

$$H_0 < 3 \times 10^{-7}\, \mathrm{G}$$

(Zeldovich, 1969), which is equivalent to the requirement for the magnetic energy density not to exceed one per cent of the relic radiation energy density.

A more stringent limit of cosmological magnetic field strength follows from the observed abundance of ^{4}He if the neutrino has non-zero rest mass. Neutrino spin precession in a magnetic field would not affect the abundance of ^{4}He provided

$$H_0 < \frac{2 \times 10^{-10}\, \mathrm{G}}{(m_\nu/20\ \mathrm{eV})}$$

at the present time (Shapiro and Wasserman, 1981; Beech, 1985b). According to available estimates, the neutrino rest mass m_ν lies in the range $14 \leqslant m_\nu \leqslant 46$ eV (Lubimov *et al.*, 1980). Detection of neutrino pulse from the supernova SN 1987a in the Large Magellanic Cloud has allowed Hillebrandt *et al.* (1987a, b) to obtain the estimate of the neutrino rest mass close to the given lower limit.

The high degree of isotropy of the Universe places a severe limit also on the universal rotation characterized by the pseudovector of angular velocity. The observed isotropy of the relic microwave radiation (see above) implies that the Universe has undergone no more than 10^{-6} of full rotation during its lifetime (Barrow *et al.*, 1985). The presence of a global distinguished direction could manifest itself in the anisotropy of properties of distant radiogalaxies. The claimed observations of such anisotropy (Birch, 1982, Sanders, 1984b) are not confirmed by a more careful analysis (see Bietenholz and Kronberg, 1984; Kapahi *et al.*, 1985 and references therein).

At later stages of the evolution of the Universe a weak cosmological field could reveal itself through correlation between orientations of rotation axes of galaxies in clusters (Reinhardt, 1971; Reinhardt and Roberts, 1972). However, observations do not reveal the required correlation (see, e.g., Helou and Salpeter, 1982).

Intergalactic magnetic fields could be discovered through the corresponding Faraday rotation of radioemission from extragalactic radiosources. Numerous attempts to detect such intergalactic rotation (Sofue *et al.*, 1968; Kawabata *et al.*, 1969; Reinhardt and Thiel, 1970; Reinhardt, 1972; Kolobov *et al.*, 1976; Andreassian, 1982) have failed to provide a reliable detection of a global intergalactic magnetic field. The observations only give more and more stringent upper limits. The arguments in favour of cosmological magnetic field seem to be unsatisfactory (Brecher and Blumenthal, 1970; Kuznetsova, 1976, Vallée, 1983c).

Since analysis of Faraday rotation measures of extragalactic radiosources (see Chapter IV) provides the most valuable limit on the strength of hypothetical intergalactic magnetic field, we discuss these results in more detail. The observed rotation measures can be divided into additive contributions from the magnetic field of the Milky Way, from the plasma within the source, and from the intergalactic medium along the propagation pathlength. Each component can be distinguished in statistical data processing because they depend differently and specifically on such parameters as galactic coordinates, the redshift of (the distance to) the source, etc. For instance, the Galactic rotation measure varies with the angle between the line of sight and the local magnetic field but is independent of the source redshift. The intrinsic rotation measure accumulated within the source depends on the redshift but not on galactic coordinates. The contribution from a hypothetical intergalactic magnetic field must increase with the distance to the source sample and must be anisotropic, being zero for the sources seen transversely to the field. Statistical analysis of rotation measures for 85 sources with the redshifts $z \lesssim 2.2$ yields the estimate

$$H_0 \lesssim 5 \times 10^{-10}\, \rho_{cr}/\rho_0 \text{ Gauss}$$

(Ruzmaikin and Sokoloff, 1977a), where ρ_0 is the present density of the universe and $\rho_{cr} = 3\mathscr{H}^2/8\pi G = 2 \times 10^{-29}$ g cm^{-3}. Analysis of data for a larger sample (142 sources) in the same range of redshifts $z \lesssim 2.2$ but situated in the southern Galactic hemisphere where the Galactic field is more regular than in the northern one, has allowed Vallée (1983c) to obtain an even stronger restriction,

$$H_0 \lesssim 3 \times 10^{-11}\ \text{G}.$$

This result is the most stringent from all available. It places possible intergalactic fields far below the level of 10^{-8}—10^{-9} G required by the hypothesis of the primordial origin of galactic magnetic fields (Piddington, 1981a). A weakness of this method is due to our uncertain knowledge of the free electron number density between galaxies.

Virtually all conceivable ideas were tested in the search for cosmological magnetic fields. Kronberg *et al.* (1977) have studied the possibility of a purely chaotic field at cosmological scales. Such a field cannot produce a non-zero rotation measure averaged over many sources. It shows itself in a dependence of the variance of RM on the redshift. Observations of quasars at high galactic latitudes, where the Galactic contribution is small, have shown the absence of any difference in RM variances for 31 quasars with $1.0 < z \leqslant 2.5$ and 34 quasars with $z < 1.0$. This gives the limit

$$H_0 < 5 \times 10^{-9} (\rho_{cr}/\rho_0)\ \text{G}$$

for the field at the scale 1 Mpc which corresponds to protogalaxies (Kronberg *et al.*, 1977). Earlier observational restrictions on primordial magnetic fields at various scales are discussed by Rees and Reinhardt (1972). With a decrease in scale, the restrictions, of course, become weaker.

Altogether, observations indicate that there are no magnetic fields in the Universe that are required by the hypothesis of the relic origin of magnetic fields of the Milky Way and other galaxies. Decisive restrictions follow from the analysis of Faraday rotation measures which 'seem to have nailed the coffin shut', as Verschuur (1979) puts it. Table V.1 compiles all available observational limits on the strength of the cosmological field given by rotation measures (see Beech, 1985b). It can be clearly seen how the resulting estimate becomes more and more stringent with growth in the number of investigated sources, i.e. with improvement in statistics. Very weak fields still allowed by observations cannot affect any processes in the early Universe (Beech, 1985a).

The available observational data readily allow for magnetic fields at the scales exceeding galactic dimensions — large-scale fields of the strength of few micro-gauss in galactic gaseous coronae (Kronberg and Perry, 1982; Welter *et al.*, 1984) and turbulent fields of the strength of order 0.1 μG and scale $\gtrsim 20$ kpc in intergalactic gas of galaxy clusters (Dennison, 1979; Lawler and Dennison, 1982). However, these can hardly be of cosmological origin (see Chapter VIII). In

TABLE V.1
Restrictions on cosmological magnetic field strength from analysis of Faraday rotation measures

Year	Authors	Magnetic field strength (G)	Number of investigated sources
1968	Sofue *et al.*	10^{-9}	40
1969	Kawabata *et al.*	10^{-8}–2×10^{-9}	60
1970	Reinhardt and Thiel	10^{-8}–2×10^{-9}	88
1972	Reinhardt	$< 10^{-8}$–10^{-9}	98
1975	Vallée	10^{-6}–10^{-9}	44
1976	Kolobov *et al.*	10^{-9}	139
1976	Kronberg and Simard-Normandin	10^{-8}	108
1977a	Ruzmaikin and Sokoloff	$< 5 \times 10^{-10}$	94
1977	Kronberg *et al.*	$< 5 \times 10^{-9}$	65
1983	Vallée	$< 3 \times 10^{-11}$	142

principle, intergalactic fields can be generated *in situ*, without any relation to the primordial field. From the observational point of view, such a possibility was noted by Kolobov *et al.* (1976) (see also Sokoloff, 1975; Ruzmaikin and Sokoloff, 1977a).

Available observational results show that even if cosmological magnetic fields exist they are too weak to provide a basis for the hypothesis of the relic origin of galactic magnetic fields. Nevertheless, a possibility of weak ($\lesssim 10^{-11}$ G) intergalactic magnetic fields still remains open. Such fields can serve as seed fields for the hydromagnetic dynamo in galaxies.

V.4. Stellar Ejections

Let us discuss the hypothesis which is opposite, in a certain sense, to that considered in the previous two sections: whether the Galactic magnetic field may be the product of ejections from more compact objects, say stars? We begin the discussion with the well understood fields of the Sun and in interplanetary space. Following is the role of stars with much more matter outflow e.g. supernova explosions. The role of the Galactic nucleus concludes the discussion.

It is well known that, like the Sun, the majority of stars have internal magnetic fields that can be expelled into interstellar space with the stellar wind. Direct measurements reveal fields of 50 μG strength in the solar wind at the Earth orbit. It is not clear, however, how and to what extent interplanetary magnetic fields penetrate the shock wave which isolates the heliosphere from interstellar space at the distance 200 AU from the Sun. Michel and Yahil (1973) consider a number of plausible scenarios. We discuss one of them.

Owing to the rotation of the Sun, magnetic lines in the solar wind acquire a spiral form; the radial field component $B_r \propto 1/r^2$ drops more rapidly than the azimuthal one, $B_\varphi \propto 1/r$, which is continuously supplied by differential rotation. Near the Earth's orbit $B_r \simeq B_\varphi \simeq 35\ \mu$G, i.e. magnetic lines are inclined at approximately 45° to the radial direction and $B = (B_r^2 + B_\varphi^2)^{1/2} \simeq 50\ \mu$G. Thus, the interplanetary field is predominantly azimuthal in the outer regions of the solar system. The solar wind velocity near the Earth is $u = 500$ km s^{-1}. The wind brings magnetic energy to the outskirts of the solar system. When dissipation in the shock wave that isolates the solar wind region from interstellar space is neglected, the rate of magnetic energy supply is estimated as

$$\frac{\mathrm{d}E_m}{\mathrm{d}t} = \int \frac{B_\varphi^2}{8\pi}\, u\, \mathrm{d}s \simeq \tfrac{1}{2} B_\varphi^2 u r^2,$$

where integration is carried out over a sphere with the radius, for instance, equal to the Earth orbital radius $r = 1$ AU. Suppose that this energy subsequently fills, more or less uniformly, the volume $V = 10$ pc^3, the average vollume per star in the galactic neighbourhood of the Sun. The field strength within this volume is estimated from

$$\frac{B^2}{8\pi} V \simeq T \frac{\mathrm{d}E_m}{\mathrm{d}t},$$

where $T \simeq 5 \times 10^8$ years is the Sun's lifetime. This brings us to the estimate

$$B \simeq 0.3\ \mu\mathrm{G}$$

of the field averaged over the Galactic volume which is ejected by individual stars similar to the Sun. Analysis of other possible processes of magnetic field penetration from the interplanetary space into interstellar medium (based on presumptions of adiabatic isotropic expansion with shock-wave compression neglected, of conservation of the total wind energy in the shock wave with subsequent equipartition, and others) shows that the resulting average field lies presumably in the range

$$3 \gtrsim B \gtrsim 0.3\ \mu\mathrm{G}$$

(Michel and Yahil, 1973). Under the approximation of adiabatic isotropic expansion the value 0.08 μG is derived, which can be considered as the absolute lower limit since it does not take into account field intensification in the shock wave and anisotropy of expansion.

Thus, the *strength* of the resulting field may be close to that observed in the Galaxy. This picture of the origin of the Galactic magnetic field would seem to be quite self-consistent if stars of solar type could eject into interstellar space, together with the magnetic field, gas in sufficient quantity to fill the whole Galaxy at the observed density. However, the solar rate of mass loss, $\simeq 10^{-14}\ M_\odot$ year^{-1}, is so small that even 10^{11} such stars cannot contribute considerably into the interstellar gas density. The radius of a sphere filled by an 'old' solar wind is as small as 10^3 AU $\simeq 5 \times 10^{-3}$ pc (Lüst, 1970). The corresponding volume is far below the average volume per star, 10 pc^3.

On the other hand, it is known that early-type, young stars lose matter at a much higher rate. Mass loss from O and B stars is of order $dM/dt \simeq 10^{-16}\ M_\odot$ $year^{-1}$, and the stellar wind velocity reaches $u \simeq 1500$ km s^{-1} (see Pottash, 1970). In principle, such stars can fill the whole Galaxy by ejected gas and magnetic field. Considerable contribution comes from M-type supergiants and stars of other types (Pottash, 1970).

The field estimate for this case can be deduced in the following manner (Bisnovatyi-Kogan *et al.*, 1973). First of all, let us estimate the magnetic field B_s near the star's surface. The equipartition estimate may be adopted with the stellar wind density

$$\rho_s \simeq \frac{dM/dt}{4\pi R^2 u},$$

where $R \simeq 5 \times 10^{11}$ cm is the typical radius of the O type star, and the thermal velocity

$$v_T = (kT_{\text{eff}}/m)^{1/2},$$

where $T_{\text{eff}} \simeq 4 \times 10^4$ K is the effective stellar radiation temperature. With these estimates, the equipartition magnetic field is estimated as

$$B_s \simeq 4\pi\rho_s v_T^2 \simeq 10\ \text{G}.$$

In fact, the outflow velocity grows outside both in the solar and stellar winds, and ρ_s is underestimated above. This justifies the presumption of equipartition which usually overestimates the field. Now we use the isotropic adiabatic expansion law to obtain the lower estimate of the average field in interstellar space:

$$B \simeq B_s(\rho/\rho_s)^{2/3} \simeq 0.3\ \mu\text{G},$$

where $\rho \simeq 10^{-24}$ g cm^{-3} is the interstellar gas density.

Supernovae are no less a powerful source of interstellar gas enrichment by magnetic fields (Syrovatskii, 1970). Suppose that supernova explosions occur once every 30 years in the Galaxy (Shklovsky, 1976; Lozinskaya, 1980). This gives a total number of 3×10^8 supernovae which have exploded in the Galaxy. Since the volume of the magnetized Galactic disc is

$$V = 2\pi h R^2 \simeq 5.7 \times 10^{11}\ \text{pc}^3$$

($h \simeq 400$ pc and $R \simeq 15$ kpc), the volume per supernova is 2×10^3 pc^3. The direct estimate of the magnetic field 3×10^{-4} G is available e.g. for the Crab nebula (Woltjer, 1972; Shklovsky, 1976, Section 13), a young supernova remnant of radius 0.8 pc. For a simple estimate suppose that magnetic field expansion to the volume 2×10^3 pc^3 occurs adiabatically as $(\text{volume})^{-2/3}$; this gives the following estimate of magnetic field produced by supernovae in the interstellar medium:

$$B \simeq 3 \times 10^{-4}\,\mathrm{G}\,(2 \times 10^3\,\mathrm{pc}^3/2\,\mathrm{pc}^3)^{-2/3} \simeq 3\,\mu\mathrm{G}.$$

Magnetic field estimates for other supernova remnants give (0.1—3) $\times$ 10^{-4} G (Woltjer, 1972). The characteristic scale of the resulting field is close to the maximal size of supernova remnants. The remnants a few million years old have radii of 50 pc (see Woltjer, 1972). This gives the scale $l \simeq 100$ pc.

Enrichment of interstellar gas by magnetic fields from supernova remnants is especially interesting because the remnants merge to give rise to the hot rarefied coronal phase of the interstellar medium which may occupy a large fraction of the Galactic disc (see Section II.3).

Outflows from young hot stars and supernova explosions are capable of producing in the Galaxy the magnetic field of average strength comparable to the observed value of few microgauss. But can the regular, large-scale magnetic field be produced in this way? The fields ejected by stars form a chaotic cellular structure and the above estimates give only the root mean square value of the field. The azimuthal dimension of a cell of course becomes larger due to differential rotation. The fields within individual cells are orientated randomly and the statistically possible large-scale component has a much smaller strength than local fields. In Section VII.14 we estimate the ratio of the small-scale to the large-scale field ejected from stars to be

$$\delta B/B \simeq 10^3.$$

Meanwhile, observations give $\delta B/B \simeq 1$—2 (see Chapter IV). The resulting large-scale field $B \sim 10^{-3}$ G can only serve as a seed field for the dynamo. Note that this source of seed fields seems to be the principal one in spiral galaxies.

Stellar ejections cannot directly explain the existence of the large-scale magnetic field of the Galaxy. This factor makes a noticeable contribution into the local structure of interstellar magnetic fields. Magnetic field distortions by the shells possibly formed by supernova explosions or by a strong mass outflow from young stars are directly seen in the distribution of Faraday rotation measures over the sky (Vallée and Bignell, 1983; Vallée *et al.*, 1984; Vallée, 1984; Broten *et al.*, 1985; see Figure IV.4). Supernova remnant shells are notable for their enhanced synchrotron emission which indicates magnetic fields of 10^{-5}—10^{-4} G within them (Woltjer, 1972; Spoelstra, 1972). This is probably the interstellar magnetic field compressed in an expanding shell. The field components normal and tangential to the shell are often comparable in strength (Woltjer, 1972). Although such fields cannot be considered as being ejected from stars, such prominent radiosky objects as the Crab Nebula, North Polar Spur, loops in Cygnus, Monoceros and others, are all intrinsically connected with supernova explosions or stellar mass outflows. As a matter of fact, chaotic fields of magnetic bubbles constitute an independent, separate component of the Galactic magnetic field. Apart from supernova shells, the shells formed by the stellar winds from massive young stars can also have considerable mass ($10^4\ M_\odot$) and radius (50 pc) (McKee *et al.*, 1984). Properties of magnetic bubbles around massive ($> 10\ M_\odot$) protostars in molecular clouds are discussed by Draine (1983).

Disturbances produced by stars undoubtedly play an important role in the local structure of the interstellar magnetic field. Supernova explosions and stellar outflows, which provide the source of interstellar turbulence (Section VI.3), enrich the interstellar gas by small-scale magnetic fields as well. The scale $l \simeq 100$ pc determined by diameters of supernova remnants agrees well with both the basic scale of interstellar turbulence and the correlation scale of fluctuation magnetic fields, as indicated by many independent methods (see Chapter IV and Section VII.13).

To conclude this section, we note another hypothetical possibility of formation of the Galactic magnetic field from the field of smaller scale. This is the field ejection from the Galactic center (Hoyle, 1969). The presumption that a strong field of the nucleus, whose mass is thought to be $10^8\ M_\odot$, expands adiabatically over the Galactic disc volume $V \simeq 10^{67}$ cm^3 requires the presence in the nucleus of the enormous field 10^9 G. In this case the magnetic energy of the nucleus would be more than ten per cent of the total rest energy! The presence of such a strong field in the Galactic nucleus contradicts observations. Moreover, due to the high electric conductivity of interstellar gas the field should be ejected together with the gas: even with turbulent diffusion, the field diffuses only 1 kpc in 10^{10} years. The field strength in the central 1 kpc of the Galaxy does not exceed 2×10^{-5} G (Sanders and Wrixon, 1973; Stephens, 1981; see Section IX.1). It is clear that this possibility seems to be very unrealistic, at least for our Galaxy and other normal galaxies.

V.5. The Dynamo

Large scales which are encountered in galaxies, make predominant the role of motions in the evolution of magnetic fields, as compared with the Ohmic dissipation. Indeed, for the scales $l \simeq 100$ pc and velocities $v \simeq 10$ km s^{-1} typical of the interstellar medium, the magnetic Reynolds number

$$R_m = lv/\nu_m$$

is of order of 10^5 for ambipolar diffusion with $\nu_m \simeq 3 \times 10^{21}$ cm^2 s^{-1} (see Section VII.13).

We have seen in Section 2 above how fatal are motions for relic fields. Turbulent, chaotic motions of interstellar gas trap magnetic lines and entangle them. This is the turbulent diffusion. The action of turbulence on the large-scale magnetic field enhances electric resistivity pumping magnetic energy toward smaller scales where it rapidly transfers into heat. Let on its own, the large-scale magnetic field of the Galaxy would become $e = 2.7\ldots$ times weaker only in 10^8 years. But the large-scale fields are actually observed in galaxies! This means that some processes are continuously acting in galaxies that replenish the large-scale field and counterbalance turbulent diffusion.

Curiously enough, the creative role is played by the very same motions. The generation mechanism of magnetic field is based on Faraday's electromagnetic induction law. When a conductor, e.g. a wire loop, moves in a magnetic field, an electromotive force arises. The resultant electric current flowing along the loop produces its own magnetic field. This idea works in a dynamo, now more frequently called a generator. In the cosmic medium, there are no loops and electromagnets prepared by man. Nature itself organizes motions of conductive media in such a way that electric currents, arising due to some weak initial magnetic field, amplify and maintain this field.

The galactic dynamo must be very powerful. As we have mentioned in the introduction to this chapter, the electromotive force of 3×10^{12} V must permanently act during the galactic lifetime.

Now we intend to show how this electromotive force is being produced by motions of interstellar gas. The large-scale magnetic fields in spiral galaxies exist due to differential rotation of galaxies and helicity of interstellar turbulence. The effects of differential rotation have already been considered in discussion of the relic field hypothesis (§2). Let us now make clear the origin of another ordering factor, the helicity.

The density of interstellar gas decreases with distance above the galactic central plane. In combination with galactic rotation this brings about a very specific and delicate violation of symmetry of turbulent motions. In a common laboratory turbulence, turbulent vortices rotate in completely random directions and the number of clockwise rotating vortices is on average the same as the number of counter-clockwise rotating vortices. But let us follow an individual vortex cell (cloud) of interstellar gas that ascends, due to random forces, above the galactic plane (see Figure V.4). As the cell goes up, the ambient gas density decreases and the cell expands. Due to conservation of angular momentum, this results in an additional clockwise rotation of the cell (viewed from above, i.e. from the direction of the galactic north pole). Due to such rotations added to the random rotations of all turbulent cells, the number of clockwise rotating eddies no longer coincides with the oppositely rotating vortices. This means violation of the *mirror symmetry* of the turbulence. In other words, the turbulence becomes *helical.* Turbulent eddies acquire an additional helical, or screw-like component of motion. It is important that this additional helical motion has the same sense of screwedness for all eddies ascending in a given galactic hemisphere. It can be easily checked that for descending vortices both the velocity $\mathbf{v}$ and its curl rot $\mathbf{v}$ simultaneously change signs, so that the *helicity* $\mathbf{v} \cdot \text{rot}\ \mathbf{v}$ has identical signs both for ascending and descending motions.

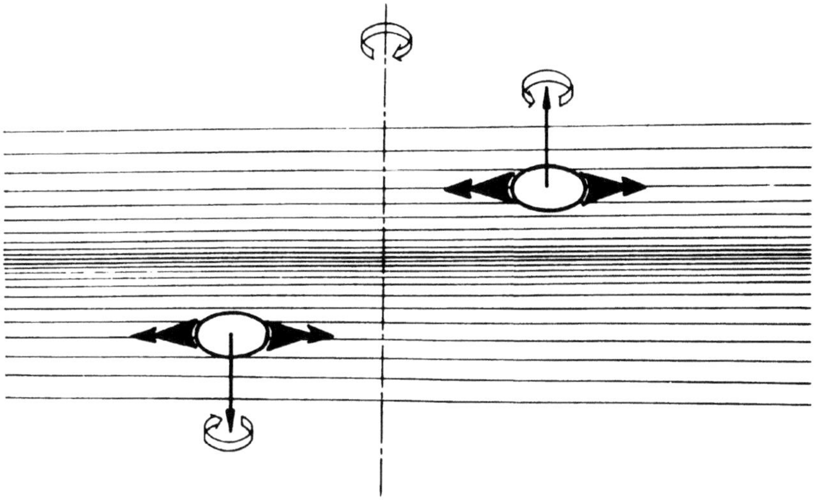

Fig. V.4. A cloud of interstellar gas moving away from the galactic plane expands due to ambient density gradient. Owing to conservation of the angular momentum, expanding clouds rotate oppositely to the galactic rotation in the upper half-space and in the direction of galactic rotation in the lower half-space. This is how the mirror symmetry of interstellar turbulence breaks down and the turbulence becomes helical.

Violation of the mirror symmetry of the turbulence is typical not only of galactic discs but also of all rotating bodies whose density varies along the angular velocity direction – convective zones of the Sun and stars, molten cores of planets and gaseous discs arising due to accretion in double stellar systems.

Discussion of the mirror symmetry properties reminds us of an accident met by a careless photographer, a friend of one of the authors. While printing some photographs of a group of war veterans and being in a hurry, he arranged the negative incorrectly and then obtained mirror-image prints. Were his clients ordinary men, they could have overlooked the mistake. But richly decorated by medals, which strongly violated the mirror symmetry, the veterans immediately noticed the mistake and a profitable job turned into a great problem.

Fortunately, deviations from mirror symmetry do not always cause losses. In a turbulent conductive medium, they ensure amplification of magnetic fields. Note a specific aspect of the large-scale, ordered magnetic field generation as an example (one of the first to be recognized) of emergence of the order from chaos in physical systems. We call attention to a curious coincidence: spiral patterns play an important role in mythological concepts of the origin of the Cosmos, in ordering of the primeaval Chaos. From ancient times, the spiral has been one of the most important spiritual symbols. A very well known example of the spiral in cosmic mythological symbolism is provided by Yin Yang 'whose interlocking spirals are a symbolic cross section through the spherical vortex' (Purce, 1974). A more recent example is provided by a drawing from the notebooks of Paul Klee (Figure V.5). The spiral pattern is also not accidental in the well known project of the III International monument by Vladimir Tatlin (Figure V.6). The ascending spiral

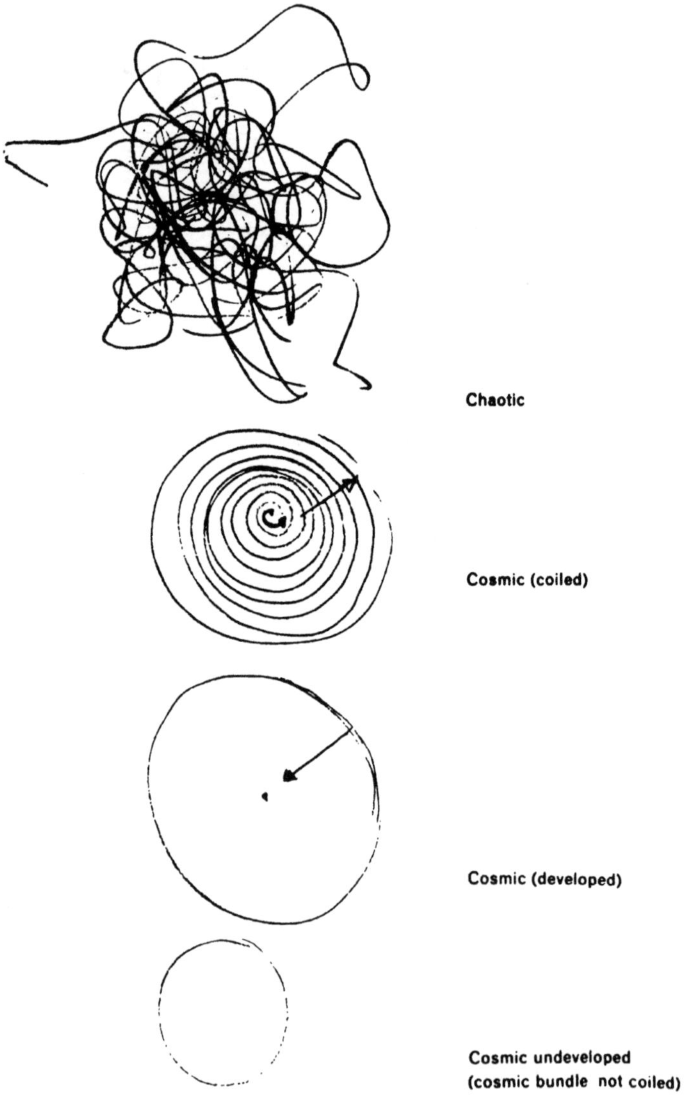

Fig. V.5. A drawing from notebooks of Paul Klee presents an example of an artistic view of the role of helical motions in the Universe, in the emergence of the ordered Cosmos from uninspired Chaos (Klee, 1961).

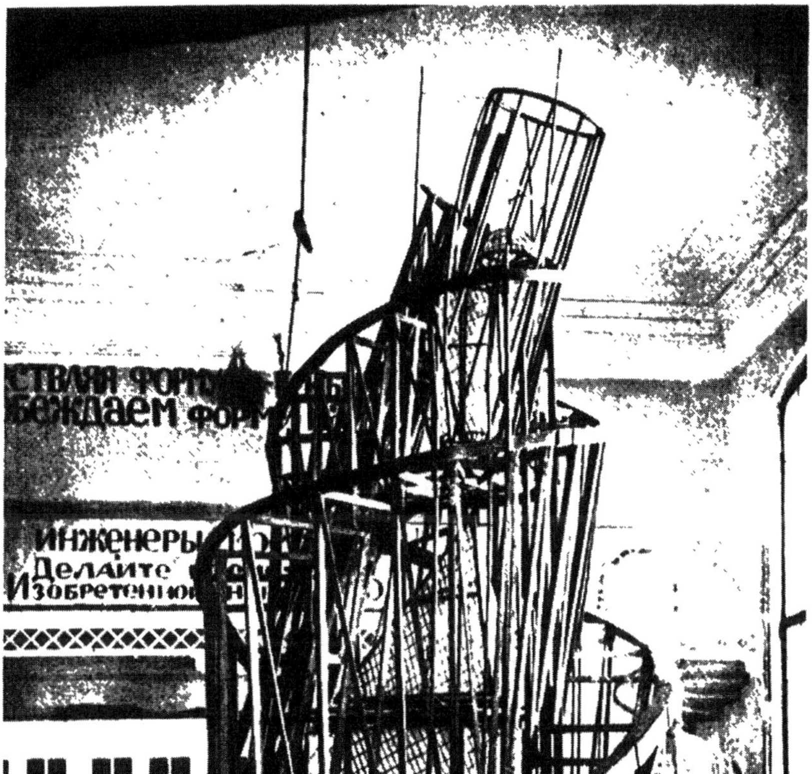

Fig. V.6. A two-armed helical design renders dynamism to the project of the Third International monument by Vladimir Tatlin symbolizing the birth of the new world and new order (Punin, 1922).

symbolizes here the emergence of a new order, creation of the new world. Being guided by a 'common sense', we usually connect the order and harmony with the presence of a certain symmetry. Excitation of the ordered magnetic field turns out to be associated with *violation* of the mirror symmetry of turbulence. This peculiarity is in accord with the ideas of Louis Pasteur who considered symmetry violation a distinctive sign of a living thing.

To illustrate how the galactic dynamo works, consider a quadrupole magnetic line which lies initially in a meridional plane of a highly conductive disc in differential rotation (Figure V.7). In this disc the points lying at different distances from the center rotate over different angles during the same period. The inner parts rotate faster than the outer and the points residing initially at the same azimuthal direction are connected, after a time, by a deformed and stretched

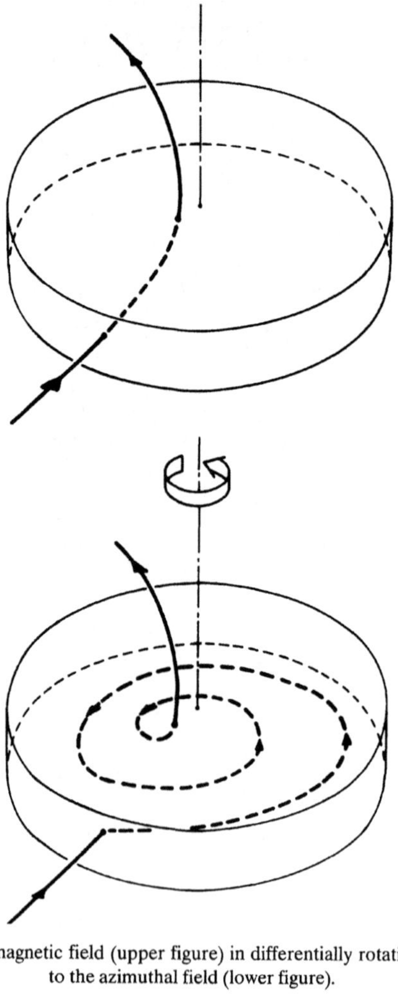

Fig. V.7. The meridional magnetic field (upper figure) in differentially rotating galactic disc gives rise to the azimuthal field (lower figure).

magnetic line, because the magnetic field is frozen into the disc and is trapped by matter (the interstellar gas in galaxies). Thus, due to differential rotation and the frozenness of the magnetic field the initially purely meridional field is stretched into a combination of meridional and azimuthal fields. In other words, the *azimuthal* field is generated by *differential rotation.* It is clear that with every rotation the azimuthal field strength is doubled at a given point (provided the meridional field does not change with time) and the differential rotation alone could ensure only linear growth in time of the azimuthal field; if there were not turbulent diffusion which destroys the total large-scale field together with its meridional component. It is necessary to maintain also the meridional field. When the meridional component is maintained at the expense of the azimuthal one, the total field can exist without any external sources. At this point a weak violation of the *mirror symmetry* of interstellar turbulence, also associated with the disc rotation, plays a decisive role.

Consider now the azimuthal field and how helical turbulent motions modify it (see Figure V.8). Due to the field frozenness, the turbulent vortices (gas clouds) ascending from the equatorial plane distort locally the azimuthal magnetic line into the shape of the capital Greek ‘omega’. Conservation of the angular momentum of expanding cells makes them rotate about a vertical axis, local magnetic loops being rotated as well. Oppositely directed fields at the base of a loop are rapidly removed by electric resistivity, small loops becoming independent of the parent magnetic line and coalescing with each other, again due to the electric gas resistivity essential at small scales. As long as the turbulence is mirror-asymmetric, the numbers of oppositely screwed loops are not identical to one another. Hence, after merging of the loops, small-scale fields are not cancelled completely, and a magnetic line of a new meridional field is the result of the coalescence. It can be easily seen, even from Figure V.8, that the meridional field thus growing within the disc has the same sense as required for further amplification of the azimuthal field through differential rotation.

The *radial* component of the large-scale magnetic field is directed identically above and below the galactic equator, while the *vertical* field points in the opposite directions. In combination, these components produce a meridional field with four poles, rather than two poles of the dipole field of the Earth. Such field is called *quadrupolar.*

Let us estimate the electromotive force E produced in spiral galaxies by helical turbulent motions. Let $\alpha/r = \langle \mathbf{v} \cdot \operatorname{rot} \mathbf{v} \rangle / v^2/l$ be the fraction of velocity associated with the organized property of the turbulence, dominance of the left-handed screwed cells over right-handed ones. Then

$$\mathscr{E} = |\oint \mathbf{E}\, d\mathbf{l}| \simeq \frac{\alpha}{c} B_\varphi \cdot 2\pi R.$$

In spiral galaxies $\alpha \simeq 1$ km s^{-1} (see Section VI.4), $B_\varphi \simeq 2\ \mu$G and $R \simeq 10$ kpc,

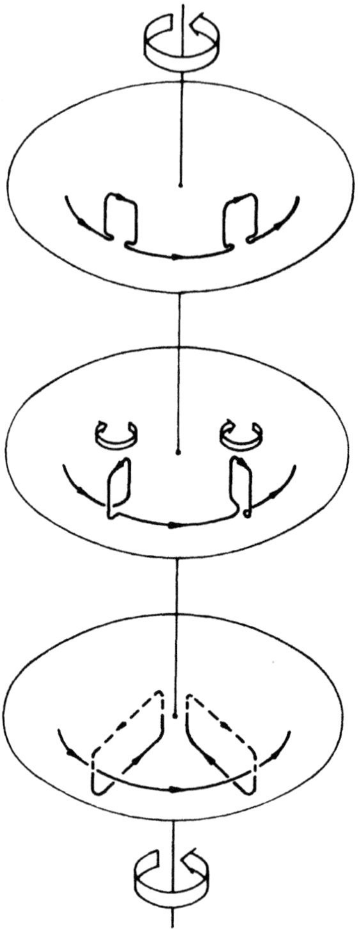

Fig. V.8. Influence of the helical turbulent motions on the azimuthal magnetic field results in the formation of the meridional field inside the disc. The upper part of the meridional magnetic line (shown dashed in the lowest figure) leaves the disc. Only the upper half-space is shown in all figures. In the lower half-space of the middle panel, mirror-antisymmetric events occur.

which gives

$$\mathscr{E} \simeq 3 \times 10^{14}\,\mathrm{V}.$$

This crude estimate even exceeds the required electromotive force. This means that the turbulent dynamo regenerates the galactic magnetic field in a time two orders of magnitude shorter than the galactic lifetime. We should emphasize that this enormous electromotive force is created not by a single source but by thousands of small 'batteries' the size of a turbulent cell (100 pc), connected in series.

The mechanism of formation of magnetic field by differential rotation was proposed by Elsasser (1946) in application to the Earth's magnetic field. The possibility of generation of a large-scale meridional field by helicity of small-scale motions was envisaged by Parker (1955). Helical small-scale motions also appear as an element of the theory of geomagnetic field generation developed by Braginsky (1964a, b). The concepts of generation action of the helical turbulence were developed and justified by Steenbeck *et al.* (1966) (see the English translation of their series of papers by Roberts and Stix, 1971).

CHAPTER VI

GALACTIC HYDRODYNAMICS

VI.1. Rotation

Rotation is the essential property of spiral galaxies which determines their appearance (spiral structure) and shape (oblateness). Curiously enough, rotation of the spiral galaxies M81 and NGC 4594 was discovered by M. Wolf and V. Slipher as early as 1914, three years before the extragalactic origin of spiral nebulae was recognized.

Differential rotation of interstellar gas is an important element of the galactic hydromagnetic dynamo. In the majority of galaxies and in a major portion of the volume of every galaxy, the helical turbulence alone cannot maintain the observed magnetic fields. The role of differential rotation is therefore of paramount importance. The radial distribution of rotational shear, $r\,\mathrm{d}\Omega/\mathrm{d}r$, largely determines the radial distribution of magnetic field in a galaxy: in those regions where the shear is small (see Chapter VII for a quantitative estimate), large-scale magnetic fields cannot be generated and maintained.

The rotation law of a galaxy is closely connected with the mass distribution: the gravitational force within the galactic disc is balanced primarily by the centrifugal force, which gives

$$V^2(r)/r = -\partial\Psi/\partial r,$$

where $V(r)$ is the rotational velocity at distance r from the axis, and $\Psi(r)$ is the gravitational potential which is obtained in terms of the distribution of mass, comprised mainly of stars, through the Poisson equation

$$\Delta\Psi = 4\pi G\rho_{\mathrm{tot}}(\mathbf{r})$$

The dependence $V(r)$ is called the *rotation curve.*

The observational study of galactic rotation uses Doppler shifts of spectral lines at different distances from the galactic center.

The rotation curve of the Milky Way galaxy is shown in Figure VI.1. The full curve for the inner Galaxy, $r < 10$ kpc, shows smoothed data from the 2.6 mm emission line of the ^{12}CO molecule and the 21 cm HI line (Burton and Gordon, 1978). Coefficients of the corresponding polynomial approximation are given in

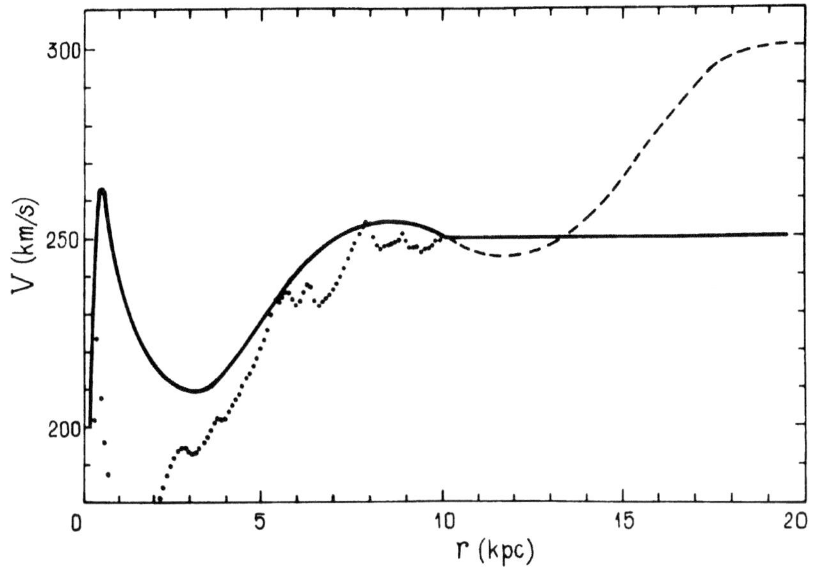

Fig. VI.1. Rotation curve of the Milky Way galaxy. The polynomial approximation of Burton and Gordon inside the solar orbit and the flat rotation curve in the outer Galaxy are shown by the solid line. The broken line is based on data of Blitz *et al.* (1980) and the dots mark the rotation curve of Haud (1979; 1984) which takes into account the radial gas velocities.

Table VI.1. The broken line for the outer Galaxy is based on observations of the optical HII regions by Blitz *et al.* (1980).

Dots show the rotation curve of Haud (1979) who has taken into account the radial motions of neutral hydrogen. The dispersion of observational points is due to chaotic gas motions and perturbations from spiral arms rather than observational scatter. The spiral pattern produces wavy perturbations of the rotation curve at the scale 1—2 kpc with amplitudes up to 15 km s^{-1} in outer parts.

TABLE VI.1

Polynomial approximation for the Galactic rotation curve at $r < 10$ kpc: $V(\mathrm{km\ s^{-1}}) = \sum_{n=0}^{5} C_n r^n$ for $b^{II} = 0°$ (Burton and Gordon, 1978)

Range	Coefficient					
	C_0	C_1	C_2	C_3	C_4	C_5
$r < 0.8$ kpc	0.0	1573.4	−3543.6	3893.5	−2283.9	595.4
$0.8 < r < 4.0$ kpc	284.6	−70.11	31.887	−10.2669	1.96392	−0.142689
$4 < r < 10$ kpc	266.3	−63.31	20.574	−2.4789	0.12820	−0.002437

Observations in the outer Galaxy seem to show that the rotation curve does not fall down at great distances, but even rises up to 300 km s^{-1} at $r \gtrsim 16$ kpc (Jackson *et al.*, 1979; Blitz, 1979; Blitz *et al.*, 1980; Schneider and Terzian, 1983; Chini and Wink, 1984; Clemens, 1985; Kolesnik and Yurevich, 1985; Petrovskaya and Teerikorpi, 1986; Rohlfs *et al.*, 1986; Hron, 1987). However, this rise can be attributed to a systematic error in data processing due to unaccounted radial gas motion (Haud, 1984). The line-of-sight component of rotational velocity is small near the anti-center direction $l^{\rm II} = 180°$ and even a slow radial flow with velocity 5 km s^{-1} may strongly affect the result producing a seemingly rising rotation curve from an actually flat one that is shown by the full curve at $r > 10$ kpc in Figure VI.1. A better determination of distances to HII regions used in the derivation of the rotation law can also make flatter the measured rotation curve (Avedisova, 1985). In our models of the dynamo in the Milky Way we use the flat rotation curve outside the solar circle.

The widely accepted value of rotational velocity near the Sun is $V_\odot = 250$ km s^{-1}, the galactocentric distance of the Sun is $r_\odot = 10$ kpc. Recently indications have appeared that $V_\odot = 220$ km s^{-1} and $r_\odot = 8.5$ kpc (see Graham, 1979; Knapp, 1979; Einasto *et al.*, 1979 and references therein).

A complicated form of the Galactic rotation curve is explained by the multi-component mass distribution in the Galaxy. The inner maximum of $V(r)$ at $r \simeq 0.5$ kpc is due to the bulge whose radius is about 0.2 kpc. The minimum at $r \simeq 2$—3 kpc is attributed to the 'hole', or density deficiency in the stellar disc at $r \lesssim 4$ kpc (Kormendy, 1977; Barabanov and Zasov, 1979; Einasto *et al.*, 1980; Caldwell and Ostriker, 1981; Rohlfs and Kreitschmann, 1981). Absence of a decrease in rotational velocity at large distances from the center of the Galaxy (as well as in the majority of other spiral galaxies, see Bosma, 1981; Rubin *et al.*, 1978) is usually interpreted as an indication of unseen massive spheroidal halos with density distribution $1/r^2$ that surround galaxies (Einasto *et al.*, 1974; Ostriker *et al.*, 1974). Some authors have attempted to explain the flatness of rotation curves without invoking the hypothesis of the massive halo (Pryor and Lecar, 1983; Kalnajs, 1983), sometimes replacing it by very radical assumptions (R. H. Sanders, 1984a). The nature of the dark matter has been discussed, e.g., by Blumenthal *et al.* (1984).

The quantity essential for the large-scale magnetic field generation is the rotational shear,

$$r\frac{\partial\Omega}{\partial r} = \frac{\partial V}{\partial r} - \frac{V}{r}.$$

This is twice the opposite Oort constant A. Figure VI.2 shows the radial distribution of rotational shear in the Galaxy. A weak maximum near $r = 10$ kpc, which plays an important role in the galactic dynamo, is more pronounced in the radial

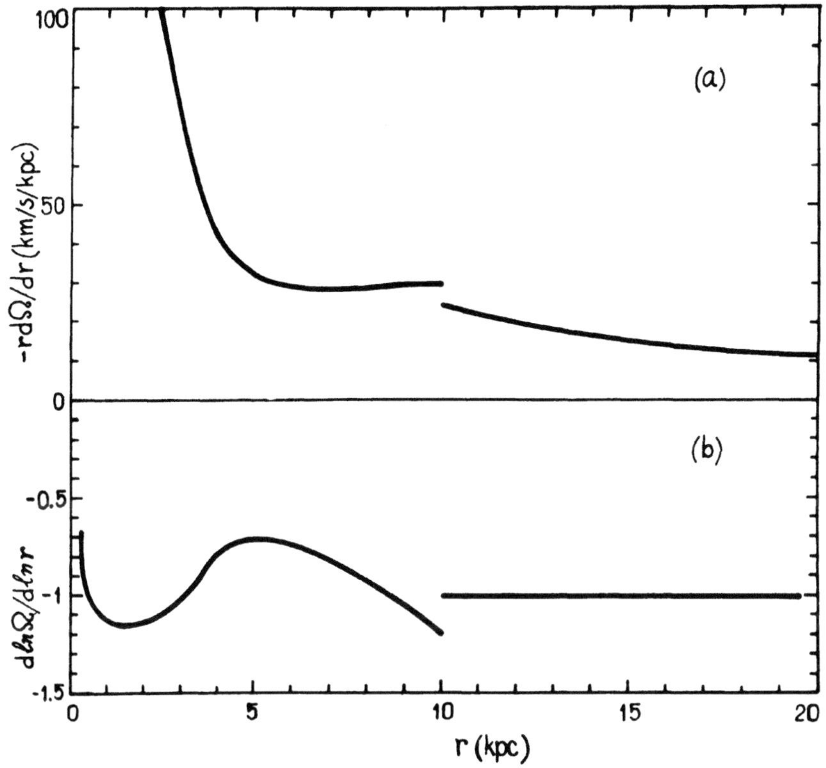

Fig. VI.2. Radial variations of the rotational shear, $r\,d\Omega/dr$ and $d \ln \Omega/d \ln r$ in the Galaxy from the rotation curve shown in Figure VI.1 by the solid line. Discontinuities at $r = 10$ kpc are artefacts of poor analytical approximation of observational data.

distributed of the dimensionless quantity $d \ln \Omega/d \ln r$, shown in Figure VI.2b. Looking at these figures, two regions can be distinguished. The shear is relatively weak in the regions $4 \lesssim r \lesssim 6$ kpc and in the outer parts. As we shall see in Chapter VII, this can weaken the large-scale magnetic field generation in these regions to such an extent that the ring-like magnetic field distribution may arise in the Galaxy.

Rotation curves have been measured for a few hundreds of spiral galaxies. The catalogue of Baiesi-Pillastrini *et al.* (1983) gives references to data on 271 galaxies. Kyazumov (1984) cites 116 galaxies, some of them not mentioned by Baiesi-Pillastrini *et al.* A few principal types of rotation curves of normal galaxies can be distinguished that are given in Figure VI.3 (see also Zasov and Kyazumov, 1983). Variations of rotation curves with morphological type, luminosity and size are

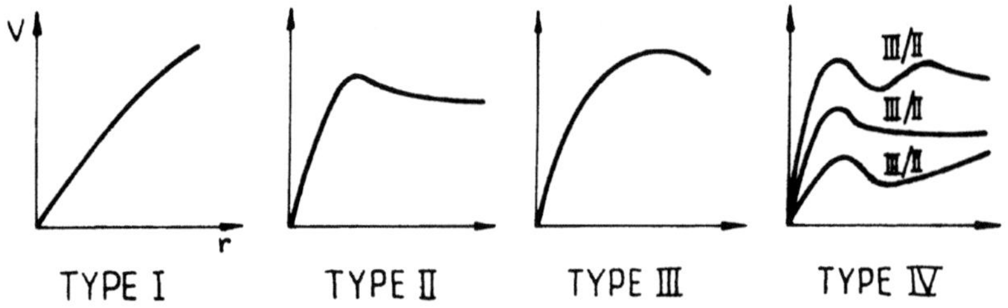

Fig. VI.3. Typical rotation curves of spiral and irregular galaxies according to Kyazumov (1984). See also Table VI.2.

discussed for relatively large samples of spiral galaxies by Rubin *et al.* (1978, 1980, 1985 and references therein) and by Zasov and Kyazumov (1983).

We note a statistical correlation between the radii of the region of solid-body rotation, r_t, and of the region occupied by spiral arms, r_s. In the region of solid-body rotation, the generation of the large-scale magnetic field is weakened (see the next chapter), while spiral arms trace the region of intense star formation, where interstellar gas and cosmic rays are concentrated. Therefore, one can expect that those galaxies where $r_t \gtrsim r_s$ would show relatively weak magnetic fields and polarization of non-thermal emission. Distributions of the r_t/r_s ratio for 100 spiral galaxies of various morphological types are shown in Figure VI.4. For later-type galaxies, $T = 6 \ldots 9$, one can expect a statistical decrease in non-thermal emission, especially its polarized component which is an indicator of the large-scale magnetic field.

The distribution of spiral galaxies over types of rotation curves is given in Table VI.2 for various morphological types. Any type of rotation curve can be met among galaxies of both earlier and later types. However, later-type galaxies $T > 5$ from the investigated sample seldom have multi-peaked rotation curves. These galaxies have a relatively simple mass distribution: the spheroidal component is weak and the disc density monotonously decreases with radius. The galaxies whose rotation curves fall down at great distances from the centers have, as a rule, a close companion (Zasov and Kyazumov, 1983).

We give here rotation curves of a few spiral galaxies for which we shall discuss, in the next chapter, the generation of magnetic fields. Figure VI.5 illustrates the rotation of the Andromeda Nebula (M31), the nearest spiral galaxy, whose magnetic field has been studied theoretically and observationally in considerable detail. The rotation curve of M31 resembles that of the Milky Way galaxy. There also exists a region with weakened non-uniformity of rotation in the main part of the disc of M31. (In addition, between $r = 2$ kpc and $r = 3$ kpc the angular velocity of the Andromeda Nebula even grows with radius.) As we shall see in the next chapter, this makes the magnetic field concentrated in a ring around $r \approx 10$ kpc in this galaxy.

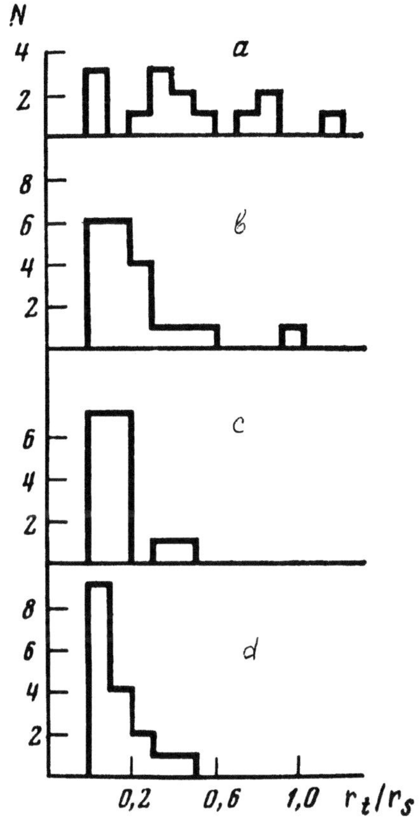

Fig. VI.4. Distributions of the ratio r_t/r_s for the galaxies of different morphological types T in numerical notations of de Vaucouleurs *et al.* (1976): $T = -3$ corresponds to SO, $T = 9$ — Irr. (a): $T = -3 \ldots 3$; (b): $T = 4$; (c) $T = 5$; (d): $T = 6 \ldots 9$.

TABLE VI.2
The number of galaxies with the indicated type of rotation curve (see Figure VI.3) among galaxies of various morphological types (de Vaucouleurs *et al.*, 1976): $T = -3$ corresponds to SO, $T = 9$ to Irr (after Zasov and Kyazumov, 1983)

Type of rotation curve	Number of galaxies of indicated type			
	$T = -3 \ldots 3$	$T = 4$	$T = 5$	$T = 6 \ldots 9$
I	3	5	13	11
II	9	9	10	6
III	4	3	4	8
IV	5	5	6	none
Total	21	22	23	25

A useful analytical approximation for the rotation curve of M31 is given by Deharveng and Pellet (1975) and Haud (1981) as

$$V(\mathrm{km\ s^{-1}}) = \begin{cases} [2808625\,r^2 - 6036750\,r^3 - 16861250\,r^4 + 43293750\,r^5]^{1/2} \\ \qquad\qquad \text{for} \quad r < 0.30\ \mathrm{kpc}, \\ (15472/r)^{1/2} \qquad \text{for} \quad 0.30 < r < 0.85\ \mathrm{kpc}, \\ 407.16 - 508.5\,r + 253.125\,r^2 - 37.0875 \\ \qquad\qquad \text{for} \quad 0.85 < r < 2.75\ \mathrm{kpc}, \\ [4624\,r^2 - 672.5\,r^3 + 35\,r^4 - 0.625\,r^5]^{1/2} \\ \qquad\qquad \text{for} \quad 2.75 < r < 14\ \mathrm{kpc}, \\ 220\exp(-4.9475 + 1.3128\,r - 0.04925\,r^2) \\ \qquad\qquad \text{for} \quad r > 14\ \mathrm{kpc}, \end{cases}$$

where the radius is measured in kiloparsecs.

A similar double-peaked form has the rotation curve of M81 (Figure VI.6). However, the minimum in the differential rotation is not sufficiently pronounced in this case so as to weaken the generation of magnetic field.

The majority of galaxies have more simple rotation curves, similar to those of M33 (Figure VI.7) or M51 (Figure VI.8). In this case nearly solid-body rotation is observed only near the center.

Rotational properties of elliptical galaxies differ very much from those of spirals. Although images of ellipticals are flattened, this is not due to rotation. Ellipticals rotate much more slowly than do spirals. Rotational velocities in the majority of elliptical galaxies do not exceed 20—30 per cent of the values required to produce the observed flattening (Illingworth, 1977; Binney, 1976, 1980). The shape of ellipticals is explained by anisotropy of stellar velocities dispersion (Binney, 1976, 1982 and references therein). Stars in ellipticals rotate at velocities not exceeding 50—150 km $\mathrm{s^{-1}}$ which is below the stellar velocity dispersion in the central parts; the stellar rotation curve rapidly grows within 2—10 core radii and slowly decreases or remains constant at greater distances from the centers (see Illingworth, 1983).

Data on gas rotation in elliptical galaxies are rather inconclusive. The interstellar gas is very rarefied and observations are difficult. Observations indicate some most unexpected situations, e.g. when gaseous disc rotates oppositely to the stars; in ellipticals gaseous discs are often inclined to the major axis of star distribution (Knapp, 1983).

To conclude, a few words should be said about variations of rotational velocity with height above the Galactic disc. Only very general considerations are now available in this field, which seem to be applicable to all spiral galaxies.

A simplified consideration of hydrostatic gaseous coronae based on assumption of barotropic equation of state $P = f(\rho)$, where P is the pressure and ρ is the gas density, shows that the gas must have a rotation curve which is independent of z (York *et al.*, 1982). Indeed the radial balance equation with gas pressure included

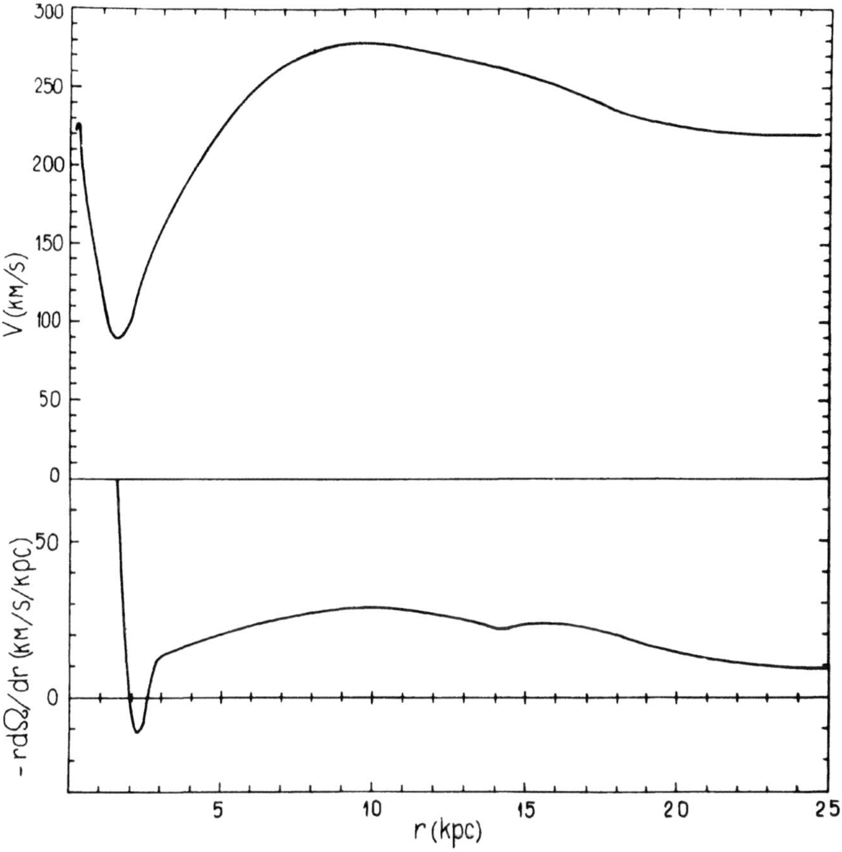

Fig. VI.5. The rotation curve of the Andromeda nebula (M31) (upper panel) and rotational shear (lower panel) according to Deharveng and Pellet (1975) and Haud (1984). The local minimum near $r = 14$ kpc is an artefact of poor analytical approximation.

is given by

$$-\frac{1}{\rho}\nabla P + \frac{V^2}{r}\mathbf{e}_r = -\nabla\Psi,$$

where Ψ is the gravitational potential, V is the rotational velocity, and $\mathbf{e}_r$ is the cylindrical radius unit vector. Notice that in the disc, where the thermal speed ($\simeq 10$ km s^{-1}) is much smaller than the rotation velocity ($\simeq 200$ km s^{-1}), the contribution of the gas pressure to the radial balance equation can be neglected in comparison with centrifugal and gravitational forces. In contrast, gas pressure cannot be neglected in the coronae where the gas temperature is of the order of 5×10^5 K and thermal and rotation velocities are therefore comparable.

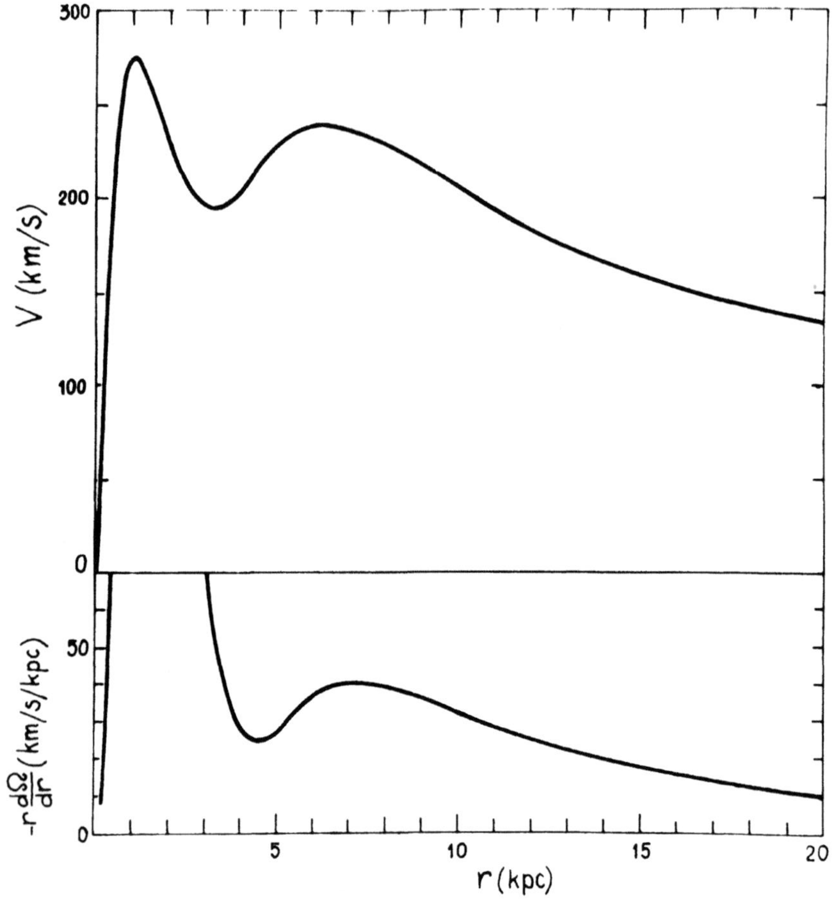

Fig. VI.6. The rotation curve (Goad *et al.*, 1979) and differential rotation of the spiral galaxy M81.

For barotropic gas, the vertical and the radial components of the balance equations are

$$\frac{\mathrm{d}}{\mathrm{d}z}(G(\rho) - \Psi) = 0; \qquad \frac{\mathrm{d}}{\mathrm{d}r}(G(\rho) - \Psi) = \frac{V^2}{r}$$

with

$$G(\rho) = \int \frac{1}{\rho}\frac{\mathrm{d}P}{\mathrm{d}\rho}\,\mathrm{d}\rho$$

in cylindrical polar coordinates (r, φ, z) with the origin at the Galactic center. The

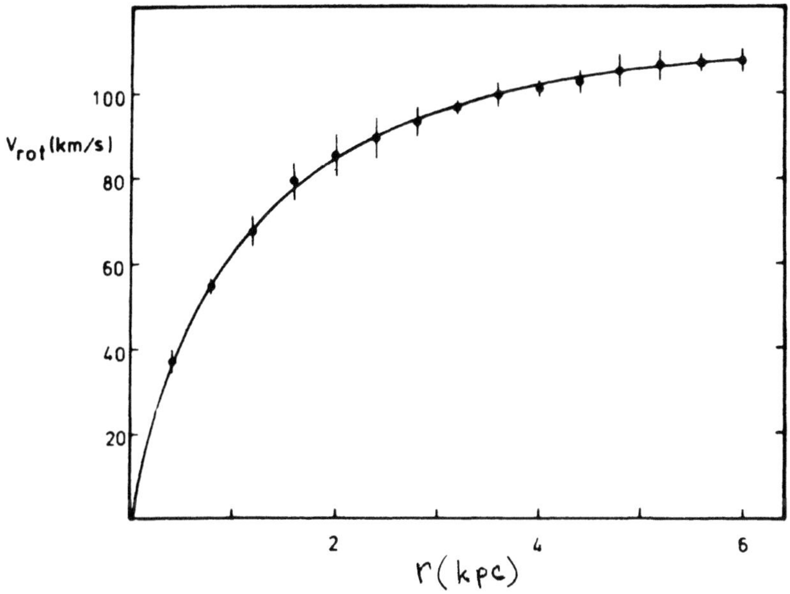

Fig. VI.7. The rotation curve of the spiral galaxy M33 (Unwin, 1980a).

first equation implies that $G(\rho) - \Psi$ does not depend on z, which means, in turn, by the second equation, that the rotation velocity is a function of r only. Arguments in favour of the co-rotating corona (z-independent rotation) also gives Weisheit (1978) basing on the presumption of the large-scale magnetic field in a corona that is an extension of the disc magnetic field. Magnetic coupling presumed by Weisheit would effectively equalize rotational velocities of the disc and corona. Considering this possibility, we can say that the large-scale magnetic field which may be present in the corona can hardly be closely connected with the disc one: buoyant fields rising from the disc have a relatively small scale of the order of 1 kpc (see Parker, 1979) and a possible large-scale field in corona must be generated *in situ* (see Section VIII.1). Therefore, magnetic coupling of Weisheit can hardly be sufficiently effective.

As for the argument of York *et al.* (1982), one should bear in mind that the rotating gas cannot be barotropic: under the influence of the Coriolis force (with Ω dependent on z), level surfaces of pressure and density inevitably become inclined, $\nabla P \times \nabla \rho \neq 0$. Deviations from co-rotation arise consistently with deviations from the barotropic equation of state.

Gas rotation above the gravitating disc has been considered by Gvaramadze and Lominadze (1986) in the opposite limiting case of the dust (pressure-free) medium $P = 0$. When the gravitating mass distribution is chosen so as to

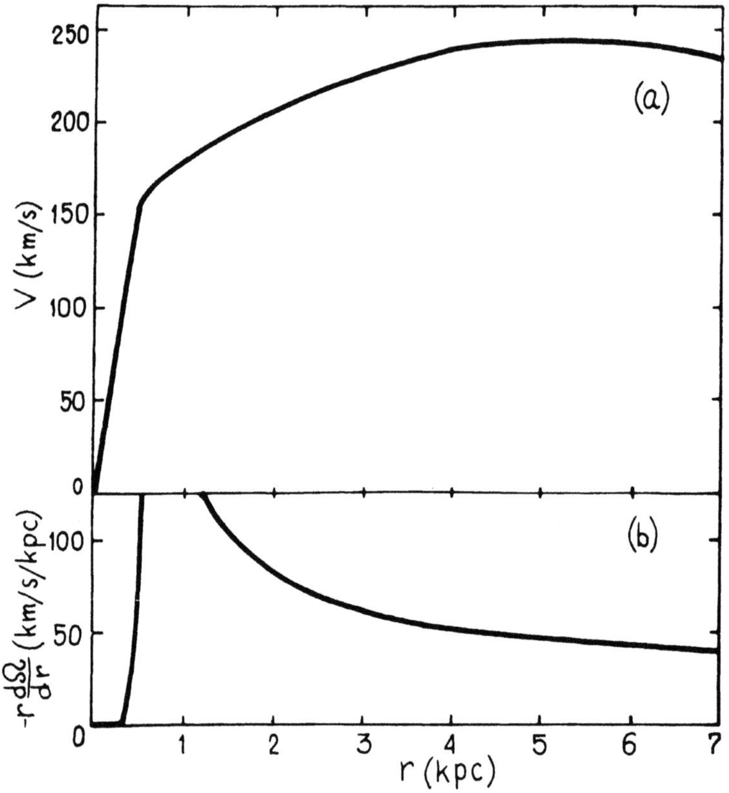

Fig. VI.8. The rotation curve (a) (Tully, 1974) and rotational shear (b) in the spiral galaxy M51.

reproduce the Galactic rotation curve in the disc, the angular velocity very weakly depends on z for $z \lesssim 2$ kpc and decreases at greater heights down to $\frac{1}{3}$ of its disc value at approximately $z \simeq 15$ kpc. It is interesting that the calculated rotation curve is single-peaked at $z \gtrsim 2$ kpc, while at smaller z, the double-peaked form is preserved. An attempt to take the gas pressure into account shows that it only slightly affects rotation immediately above the disc but can even make the rotation curve decreasing beyond the disc edge, at $r \gtrsim 10$ kpc.

Observations with the IUE (International Ultraviolet Explorer) satellite have made possible an investigation of the gas in the Galactic corona which produces ultraviolet absorption lines in spectra of extragalactic sources and stars at great heights above the disc. Difficulties in interpretation of these observations are due to uncertainties in the distances to the absorbing gas. Earlier investigations which were based on data along few lines of sight have been compatible with the co-rotating corona (Savage and de Boer, 1981; York *et al.*, 1982). However, a broader observational basis has shown that rotation velocity actually decreases with the distance from the disc and peculiar motions are considerable (de Boer and Savage, 1983; Albert, 1983; Morton and Blades, 1986; Savage and Massa, 1987). HI observations indicate a possible asymmetry of rotation with respect to the Galactic plane (Bania and Lockman, 1985).

VI.2. Shape of the Gaseous Disc

Properties of magnetic fields maintained by the dynamo strongly depend on the geometrical shape of a region where the generation proceeds. The generated large-scale magnetic field is dipolar in spherical regions, as in the convective zone of the Sun and in the outer core of the Earth or planetary cores. A thin spherical shell — the convective zone of the Sun — generates the oscillatory field (the 22-year activity cycle), while the Earth's core produces a quasi-stationary magnetic field over time intervals of 10^5 years. Meanwhile, the very same dynamo mechanism prefers the fields with quadrupole symmetry in thin discs — gaseous discs of spiral galaxies and accretion discs near black holes. The specific shape of the disc influences considerably the radial distribution of large-scale magnetic fields in spiral galaxies. We postpone discussion of the nature of this influence to Chapter VII, turning now to observational and theoretical data on the shapes of galactic discs.

Magnetic fields are generated in regions occupied by conductive gas. The major proportion of interstellar gas in the Galaxy is concentrated within the disc near the equatorial plane, the layer of ionized gas being thicker than the denser neutral gas layer. Discs of spiral galaxies become thicker with galactocentric distance.

The regions where dense neutral gas concentrates reveal themselves directly by powerful emission and absorption in radiolines of neutral hydrogen and interstellar molecules. In contrast, the electron gas can be detected only indirectly, through its effect on passing radiowaves. Study of dispersion measures and Faraday rotation of pulsar emission shows that half-thickness of the ionized gas layer is

$$h_0 \approx 400 \text{ pc}$$

in the solar vicinity (Falgarone and Lequeux, 1973; Ruzmaikin and Sokoloff, 1977b; see also Guélin, 1974). This is the so-called equivalent half-thickness, i.e. the half-thickness of a uniform disc which contains the same mass of gas as the actual disc and has the same gas density at $z = 0$. Equivalent thickness is simply the ratio of the surface density to the central density. When the density (and magnetic field strength) actually decreases with distance from the equatorial plane, the actual disc thickness exceeds the equivalent one.

Many different characteristics of vertical gas distribution are used. Of course, when observing galactic discs it is much more convenient to use, e.g. the height at which the density or some other quantity decreases to the half-maximum value. Below we list some ways to characterize the thickness of a gaseous disc based on different models of gas distribution. For the Gaussian distribution

$$\rho(z) = \rho_0 \exp(-z^2/2h_g^2)$$

one obtains

$$h_g = (2/\pi)^{1/2} h_0 \approx 0.8\, h_0 \approx 320 \text{ pc}.$$

Meanwhile, the height at which the density is half the central value is

$$h_{1/2} = (2 \ln 2)^{1/2} h_g \approx 1.2\, h_g \approx 380 \text{ pc}.$$

The first moment of the density distribution is also often used:

$$\bar{h} = \int_0^\infty z\, \rho(z)\, \mathrm{d}z \Big/ \int_0^\infty \rho(z)\, \mathrm{d}z = (2/\pi)^{1/2} h_g \approx 255 \text{ pc}.$$

For the exponential distribution,

$$\rho = \rho_0 \exp(-z/h_0),$$

the scale height coincides with the equivalent half-thickness and

$$h_{1/2} = h_0 \ln 2 \approx 0.7\, h_0,$$
$$\bar{h} = h_0.$$

Parameters of the disc of thermal electrons can also be derived from an analysis of scattering of radioemission of pulsars and extragalactic sources by electron density inhomogeneities. The width of the pulsars distribution is smaller than the thickness of the electron layer near the Sun. Thereby properties of electron distribution are extracted from observations by matching the parameters of electron distribution models (cf. Harding and Harding, 1982).

The estimates are strongly affected by the choice of a pulsar sample: removing pulsars which are seen through HII regions leads to greater estimates of electron layer thickness. Estimation of distances to pulsars is also far from being definite. For example, Taylor and Manchester (1977) have probably used overestimated distances to pulsars which has resulted in an overestimate of the scale height (see Hall, 1980 and Table VI.3). Note also that widening of pulsar impulses (measured by $\int n_e \, \mathrm{d}l$) and scattering of radiowaves and scintillations may be products of different phases in the interstellar gas: delay in the arrival of impulses may occur in the warm phase while scattering occurs on fluctuations of electron density in the hot tunnels (Shapirovskaya and Bocharov, 1986a, b).

Various estimates of the half-thickness of the disc of thermal electrons in the solar vicinity are compiled in Table VI.3. Individual estimates differ considerably, but they seem to converge to

$$h_0 = 400\text{—}600 \text{ pc}.$$

TABLE VI.3
Observational estimates of equivalent half-thickness of the discs of free electrons and intercloud hydrogen in the solar vicinity

Method of determination	Equivalent half-thickness, pc	Reference
	I. *Free thermal electrons*	
Faraday rotation and pulsar dispersion measures	405 ± 115	Falgarone and Lequeux, 1973
	400	Ruzmaikin and Sokoloff, 1977b
Distribution of pulsar dispersion measures	152	Prentice and ter Haar, 1969
	$>400 \ldots 500$	Falgarone and Lequeux, 1973
	>200	Terzian and Davidson, 1976
	1800 ± 1500	Taylor and Manchester, 1977
	1000	Harding and Harding, 1982
Free-free absorption of 10 MHz radiowaves	300 ± 100	Bridle, 1969
Comparison of distributions of pulsar dispersion measures and of neutral hydrogen	>800	Bridle and Venugopal, 1969
Scattering of radiowaves	$430 \ldots 670$	Readhead and Duffet-Smith, 1975
	>500	Cordes *et al.*, 1985
	II. *Intercloud neutral hydrogen*	
λ 21 cm observations	292 ± 49	Falgarone and Lequeux, 1973
	300	Jackson and Kellman, 1974
		Baker and Burton, 1975
		Lockman, 1984
Absorption by OVI ions	300^{+200}_{-150}	Jenkins, 1978
L_α absorption	$440 \ldots 500$	Bohlin *et al.*, 1978

The number density of thermal electrons at $z = 0$ near the Sun lies between 0.025 and 0.048 cm^{-3} and a widely adopted estimate is

$$n_e = 0.03 \text{ cm}^{-3}.$$

The distribution of interstellar gas may be crudely described by a sequence of enclosed layers. The thinnest is a layer of molecular clouds. The intensity of CO molecule emission, a tracer of molecular gas, falls to half the central value at the height of 75 pc (D. B. Sanders *et al.*, 1984) in the solar vicinity. The disc occupied by the young stellar population, supernovae and HII regions, has nearly the same thickness. The layer of neutral hydrogen clouds is somewhat thicker, $h_{1/2}$ ranges from 95 pc to 155 pc according to different estimates (Baker and Burton, 1975;

Falgarone and Lequeux, 1973). The layer occupied by the intercloud warm neutral hydrogen is still thicker, $h_{1/2}$ ranging from 140 pc according to Baker and Burton (1975) to 274 pc as estimated by Falgarone and Lequeux (1973). The thickest is the disc of thermal electrons and ions. Its half-thickness, as we have discussed above, is at least 400 pc.

Stratification of various phases of the interstellar gas is revealed not only by observations of neutral hydrogen (Falgarone and Lequeux, 1973; Baker and Burton, 1975; Quiroga, 1980), but also by analysis of the scattering of emission of pulsars and extragalactic sources in the interstellar medium. According to this analysis, free electrons are concentrated in two distinct layers: extended, with the scale height exceeding 500 pc and relatively low level of fluctuations, and the thinner one, with scale height 50—100 pc, which is probably associated with HII regions (Harding and Harding, 1982; Rao and Ananthakrishnan, 1984; Cordes *et al.*, 1985).

What determines the thickness of the electron disc? It is not clear to which phase of the interstellar medium can be attributed the gas of thermal electrons and ions high above the galactic plane. It may be a separate completely ionized warm component of the interstellar gas (Terzian and Davidson, 1976; Quiroga, 1980); or an ionized part of a warm intercloud component (Falgarone and Lequeux, 1973). In the latter case the greater effective thickness of an ionized layer, as compared with the layer of neutral intercloud gas, is explained by the growth of the degree of ionization with decreasing density, i.e. with height. The recombination rate is approximately proportional to the squared density, while the ionization rate varies as the density itself. Therefore, the equilibrium ionization grows with distance from the equatorial plane (Falgarone and Lequeux, 1973). According to calculations of thermal and ionization equilibrium of interstellar gas, in the intercloud medium with $n_{\rm HI} \lesssim 0.2\ {\rm cm}^{-3}$, dependence of the degree of ionization on the density gives the following approximate relation:

$$\log n_e \simeq 0.4 \log n_{\rm HI} + {\rm const}$$

which holds both for heating by hard ultraviolet radiation (Bergeron and Souffrin, 1971) and soft x-rays (Grewing and Walmsley, 1974). This implies that for a Gaussian distribution of neutral intercloud gas the distribution of electrons is also Gaussian but with the width $0.4^{-1/2} \approx 1.6$ times larger. For the equivalent half-thickness of 290 pc of neutral intercloud gas (Falgarone and Lequeux, 1973) one thus obtains

$$h_0 \approx 460\ {\rm pc}$$

for the free electron layer, which is close to the estimate of 400—600 pc following from observations (see Table VI.3).

Closeness of the vertical distributions of free electrons and the warm intercloud medium can be considered as an indication of their intrinsic connection. Distribution of the 2×10^5 K gas, which is revealed through absorption by OVI ions, is approximated by exponential law with $h_0 = 300\ (+200, -150)$ pc (Jenkins, 1978).

Analysis of interstellar absorption in the L_α line shows that intercloud neutral hydrogen has the density 0.16 cm^{-3} at $z = 0$ and half-thickness $h_g \simeq 350$ pc ($h_0 \approx 500$ pc for exponential approximation) (Bohlin *et al.*, 1978). A weak wing in the gas distribution extending farther than 1 kpc from the disc can be traced in λ 21 cm observations. The density of this component is extrapolated as 0.05 cm^{-3} at $z = 0$, its scale height is $h_0 \simeq 480$ pc (Lockman, 1984; Lockman *et al.*, 1986); the layer of thermal electrons may be directly associated with this component of neutral gas.

Consider now the dependence of the thickness of a multi-layer gaseous disc of the Galaxy on the galactocentric distance. Observations in λ 21 cm line show that the neutral hydrogen disc becomes thicker — in contrast to the stellar disc — with distance from the galactic center (Westerhout, 1957; Lozinskaya and Kardashev, 1963; Jackson and Kellman, 1974). The dependence of the half-thickness of the diffuse neutral hydrogen layer on galactocentric distance is shown in Figure VI.9

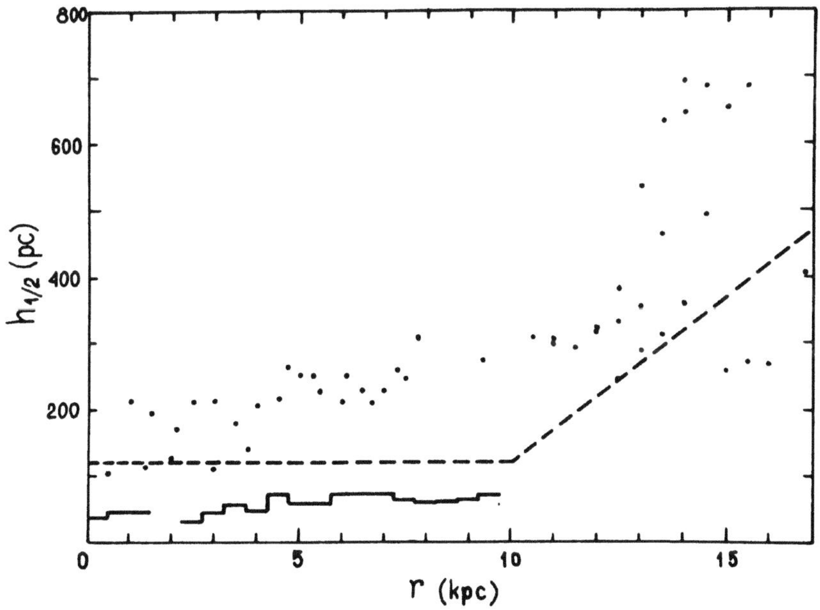

Fig. VI.9. The half-maximum half-thickness for the distributions of various components of interstellar gas as a function of the galactocentric distance: diffuse neutral hydrogen (dots, Jackson and Kellman, 1974), dense HI clouds (dashed line, Knapp *et al.*, 1978), and molecular clouds (the histogram is from Sanders *et al.*, 1984). Large scatter of the points for the diffuse HI is probably due to local variations of the thickness. For instance, anomalously small thickness is observed at $r > 10$ kpc, $160° \gtrsim l^{II} \gtrsim 110°$.

(Jackson and Kellman, 1974; van Woerden, 1979). The half-thickness of the layer of dense HI clouds is somewhat smaller; its radial dependence can be approximated as

$$\frac{h}{r_\odot} = \begin{cases} 0.012 & \text{for} \quad r < r_\odot, \\ 0.012 + 0.05(r/r_\odot - 1) & \text{for} \quad r > r_\odot \end{cases} \tag{VI.1}$$

with $r_\odot = 10$ kpc the galactocentric distance of the Sun (Knapp *et al.*, 1978). This dependence is given by the broken line in Figure VI.9. It seems, nevertheless, that the neutral hydrogen disc becomes somewhat thinner in the inner Galaxy, down to 50—100 pc half-thickness at $r < 4$ kpc (see van Woerden, 1979).

The distribution of neutral hydrogen across the disc is intermediate between Gaussian and exponential and is the same at all radii (Kulkarni *et al.*, 1982).

Also expanding, as $r^{1/2}$, toward the edge is the layer of molecular clouds whose shape is given by the histogram in Figure VI.9 (D. B. Sanders *et al.*, 1984).

Of similar shape is the disc of free electrons. Statistical analysis of pulsar dispersion measures indicates that its thickness decreases toward the center. According to Harding and Harding (1982), the scale height drops down by a factor of ten between $r = 10$ kpc and $r = 4$ kpc. Accordingly the density of electron gas grows: from 0.01—0.02 cm^{-3} at $r > 10$ kpc to 0.03 cm^{-3} near the Sun and 0.06—0.1 cm^{-3} at $r = 7$—9 kpc (Shapirovskaya and Bocharov, 1986a, b). Some intensification of radiowave scattering is registered in the ring between $r = 4$ and 6 kpc (Rao and Ananthakrishnan, 1984).

Observations of the Galactic synchrotron background radioemission also show that the thickness of the magnetic field and cosmic rays disc grows towards the edge approximately as exp[(r − 10 kpc)/8 kpc], while the scale height of the synchrotron disc is $h_{1/2} \simeq 850$ pc near the Sun (Phillipps *et al.*, 1981). The level lines of synchrotron emmissivity are given in Figure VI.10 according to a model of Phillipps *et al.*

Thickening of the disc of interstellar gas with galactocentric distance is naturally explained by a decrease in the vertical component of the gravitational force which is produced by stars whose number density in the disc decreases outwards. Dependence of the HI disc half-thickness on the galactocentric distance under hydrostatic equilibrium conditions is given by

$$h_{1/2} \simeq \left(\frac{1.386}{\pi G} \right)^{1/2} \frac{Q}{\rho_{\text{tot}}^{1/2}} \tag{VI.2}$$

(Kellman, 1972; see also Parker, 1979, Chapter 22) if the quantity

$$Q^2 = v_z^2 + H^2/8\pi\rho + P_{\text{c.r.}}/\rho$$

and all the summands separately do not depend on z. Here v_z^2 is the root mean

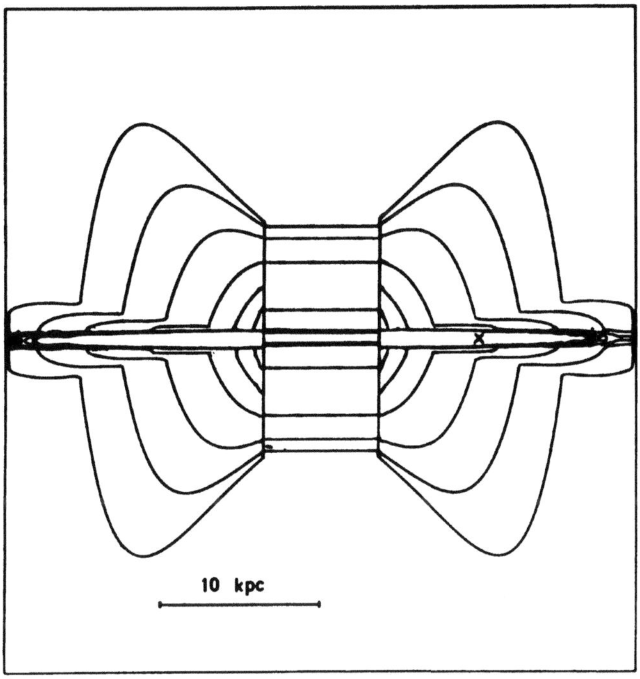

Fig. VI.10. Meridional cross section of level lines of the synchrotron emissivity in the Galaxy. The contours correspond to emissivities 0.1, 0.2, 0.4, 0.8, 1.2, 1.6, and 2.0 K kpc^{-1}. Increase in the radiodisc thickness toward the edges is clearly seen. The Sun position is marked by a cross. From Phillipps *et al.* (1981).

square vertical component of the turbulent gas velocity, $H^2/8\pi$ and $P_{c.r.}$ are magnetic and cosmic rays pressures respectively, ρ is the gas density, ρ_{tot} is the total mass density (all values are taken at $z = 0$), and G is the gravitational constant. The difference in vertical distributions of various components of the interstellar medium are explained by the difference in supporting pressure, i.e. the value of Q.

Let us estimate $h_{1/2}$ for the warm component of the interstellar gas whose density at $z = 0$ is $n_{HI} \simeq 0.2$ cm^{-3}, i.e. $\rho \simeq 3.4 \times 10^{-25}$ g cm^{-3} near the Sun's orbit. As far as magnetic pressure is produced by the field with approximately equal regular and chaotic components $B \simeq b \simeq 2$ μG, we have $H^2/8\pi \simeq 2B^2/8\pi \simeq 0.3 \times 10^{-12}$ dyn cm^{-2}. Cosmic ray pressure can be estimated as 0.5×10^{-12} dyn cm^{-2}. Adopting $v_z^2 = \frac{1}{3}(10 \text{ km s}^{-1})^2$ and the stellar mass density $\rho_{tot} = 0.15\ M_{\odot}$ pc^{-3} we obtain $Q \simeq 16.4$ km s^{-1} and

$$h_{1/2} \simeq 445 \text{ pc}.$$

The vertical gas distribution can actually be described neither by an exponential nor by a Gaussian law. In the gravitational field of an isothermal star disc of finite thickness and infinite radius the gas density distribution is given by

$$\rho(z) = \rho_0 \, ch^{-q}(z/z_0),$$

where

$$q = 2\pi G \rho_{\mathrm{tot}} z_0^2 / Q^2,$$

and z_0 is the equivalent half-thickness of the star disc (Celnick *et al.*, 1979). For $z \ll z_0$ the gas distribution is close to the Gaussian with $h_g = z_0 q^{-1/2}$, while for $z \gg z_0$ it approaches the exponential law with $h_0 = z_0/q$. According to Celnick *et al.* (1979), the Galactic HI disc can be described, on average, by

$$z_0 \,(\mathrm{pc}) = 11.8\, r - 0.6,$$

$$q = 0.14\, r - 0.03$$

for $r > 3$ kpc, where r is measured in kiloparsecs.

The expressions for the vertical distribution of the interstellar gas given above are obtained under assumption that the pressures of magnetic field, cosmic rays and the turbulent pressure are all identically distributed along the z-coordinate. It is, however, clear that the magnetic field is generated in an ionized gas that is distributed wider than the neutral hydrogen; cosmic rays are confined in a much more extended region than the Galactic disc. Hence, the magnetic pressure and the cosmic ray pressure must decline with z slower than the neutral gas density. A detailed analysis of the hydrostatic gas equilibrium in the Galaxy and comparison of inferred distributions of magnetic fields and cosmic rays with observed distributions of the synchrotron background radioemission confirms this prediction (Badhwar and Stephens, 1977; Fuchs and Thielheim, 1979). With account for this effect, the hydrostatic equilibrium models satisfactorily describe the vertical distribution of the interstellar gas.

A good agreement of the hydrostatic equilibrium models and observations of the interstellar gas is actually fraught with important consequences for the interstellar gas dynamics. The point is that the hydrostatic equilibrium of the system of the gas and quasi-uniform magnetic field may be strongly unstable with respect to the Parker instability (a magnetic analogue of the Rayleigh—Taylor instability): in those places where magnetic field is enhanced, the pressure and density of the gas decrease below the average value since the total (thermal + magnetic) pressure is uniform; as a result, gas parcels with stronger magnetic fields rise buoyantly above the disc (evacuating, by the way, the magnetic fields from the disc), while the gas slides down the magnetic lines, further diminishing the density at the crest (see Parker, 1979, Ch. 13; Elmegreen, 1982). A typical wavelength of the growing disturbances is of the order of 1 kpc, the growth time is comparable with the free-fall time for the scale height of the gas distribution ($\simeq 10^7$ years for the Galactic disc). The Parker instability seems to develop very rapidly in spiral galaxies and it is known that neither chaotic magnetic fields nor the disc rotation are able to suppress it. Only recently it has become clear that in galactic discs the Parker instability can be suppressed by cosmic rays due to diffusive character of their propagation (Kuznetsov and Ptuskin, 1983; Kuznetsov, 1986; Ptuskin, 1986 and references therein) and owing to anisotropy of their pressure (Nelson, 1985).

The surface density of stellar discs in spiral galaxies decreases with galactocentric distance, to the first approximation, as $\exp(-kr)$ with $k \simeq \frac{1}{4}$ kpc $-$ 1 for the Galaxy (Freeman, 1970). Since observations do not reveal a corresponding variation in thickness of inner ($r < 10$ kpc) parts of the gaseous disc, this means that the supporting pressure (i.e., Q) grows inward (cf. Jackson and Kellman, 1974). Meanwhile, observations in external galaxies show that the velocity dispersion v_z is fairly constant at all radii (see, e.g., van der Kruit, 1981). Therefore, this very weak variation of the gaseous disc thickness in the inner Galaxy may be interpreted as an indication of inward growth of magnetic and/or cosmic ray pressure.

The dominant contribution to the mass of the Galaxy at $r > 10$ kpc comes from the unseen massive halo with $\rho_{\mathrm{tot}} \approx r^{-2}$. The linear growth of the gaseous disc thickness beyond the solar orbit is thus compatible with existence of a massive halo with the r^{-2} density distribution. Since flat rotation curves are also attributed to the influence of the halo, the linear growth of the disc thickness turns out to be intrinsically connected with flatness of the rotation curve. In addition, the linear growth of h indicates that Q is approximately constant in the outer Galaxy up to $r \approx 16$ kpc (see also Jackson and Kellman, 1974; Paul *et al.*, 1975; Kulkarni *et al.*, 1982).

Equilibrium configurations of interstellar gas in the gravitational field of uniform spheroids, which model the distribution of stars, were studied by Abrahamian and Kaplan (1975), Abrahamian (1975), Abrahamian and Oganessian (1977), and Kaplan *et al.* (1977). According to these studies, the shape of the gaseous disc can be approximated as the surface of a parted hyperboloid of rotation,

$$h \propto [1 + (r/r_0)^2]^{1/2},$$

at small distances from the galactic center and by the surface of a spheroid at large distances. For $r \ll r_0$ the disc thickness little varies with radius and for $r \gg r_0$ the thickness grows nearly linearly.

Growth of the disc thickness towards the edge seems to be typical for spiral galaxies. Behaviour of the disc thickness is usually inferred from indirect evidence. Velocity dispersion of interstellar gas observed in the λ 21 cm HI line is, as a rule, remarkably constant along the radius in spiral galaxies. From the rotation curve the mass distribution can be deduced, which allows estimation of the gaseous disc thickness through formulae analogous to (VI.2) (e.g., Emerson, 1976). As long as rotation curves are flat at large distances from galactic centers, the thickness thus obtained grows with radius. It was obtained in this way that the half-thickness of the HI disc in the Andromeda Nebula grows linearly from approximately 100 pc at $r = 6$ kpc to 750 pc at $r = 25$ kpc (Emerson, 1976; Brinks and Burton, 1984). Similarly, in M33, $h_{1/2}$ (0.5 kpc) $\simeq$ 150 pc and $h_{1/2}$ (5 kpc) $\simeq$ 4000 pc (Warner *et al.*, 1973). Note that the value of velocity dispersion in the inner parts of M33 is overestimated by Warner *et al.*; the thickness is overestimated correspondingly (see next section). In M51 the half-thickness grows from 60 pc in the central part to $\simeq$270 pc at $r = 3$ kpc (Tully, 1974). In the galaxy IC 342 the HI disc thickens linearly from 200 pc at $r = 5$ kpc to 900 pc at 40 kpc distance (Newton, 1980b).

These estimates of disc thickness do not take into account a considerable contribution to the total pressure from magnetic fields and cosmic rays. Although velocity dispersion is probably nearly the same in all galaxies, magnetic fields and cosmic rays pressure may differ considerably from one galaxy to another (see Beck, 1983).

Direct observation of the shape of gaseous discs in external galaxies is a formidable task. Such observations were performed by Sancisi and Allen (1979) in the spiral galaxy NGC 891 seen edge-on. The disc of neutral hydrogen becomes markedly thicker toward the edge. At $r < 8$ kpc its half-thickness is less than 500 pc, and at 20—24 kpc radius it is as large as 500—1000 pc. We also note that in NGC 891 the synchrotron radioemission is distributed wider than neutral gas, just as in the Galaxy. Further analysis of the distribution of neutral hydrogen in NGC 891 has shown that the observed shape of the disc is compatible with radially invariant dynamic pressure (Q from Equation (VI.2)) and with the observed distribution of stars (van der Kruit, 1981). The half-thickness of the HI disc is approximated as

$$h_{1/2}\,(\mathrm{pc}) = 220\,\exp(r/2r_0),$$

while the emission from old population stars in the disc is given by

$$L \propto \exp(-r/r_0)$$

with $r_0 = 4.9$ kpc.

In other galaxies, as in our Galaxy and NGC 891, the synchrotron disc has a greater thickness than the neutral gaseous disc. For example, the half-intensity half-thicknes of the synchrotron disc in NGC 253 is 700 ± 100 pc (Hummel *et al.*, 1984b).

VI.3. Interstellar Turbulence

Interstellar gas is being kept in a state of intense chaotic motions. Interstellar turbulence markedly differs from that observed in a laboratory. Effects of magnetic fields, compressibility, and self-gravitation are pronounced in interstellar gas; observed velocities often exceed sound velocities. The source of turbulent energy is also unusual: it is a random force whose correlation time is possibly even shorter than the revolution time of the turbulent cell. In this sense interstellar turbulence is much closer to δ-correlated random flow, so beloved by theoreticians, than are laboratory flows that are turbulized by stationary grids of quasi-stationary shear stresses.

Hydrodynamic and magnetohydrodynamic turbulence is a phenomenon associated with high Reynolds numbers. The great value of Reynolds number is a necessary condition for the onset of turbulent motion. For instance, the turbulence arises in a flow between concentric cylinders at $Re \gtrsim 1500$ (see Di Prima and Swinney, 1981). Free shear layers become turbulent when $Re \gtrsim (3–5) \times 10^3$ (see Ho and Huerre, 1984).

It is easy to estimate the Reynolds number for galactic gaseous discs. The regular global rotation is characterized by the velocity $V \simeq 200$ km s^{-1} and the scale $L \simeq 1$ kpc. A kinematic viscosity estimated for fully ionized gas, $\nu = 1.2 \times 10^{-16} T^{5/2} \rho^{-1}$ cm^2 s^{-1}, gives for $T = 10^4$ K and $\rho = 2 \times 10^{-24}$ g cm^{-3}

$$Re \simeq 10^{11}.$$

This value is so enormous that some arbitrariness in choice of characteristic parameters cannot keep one from the final decision about the turbulent character of interstellar gas motions.

The criteria for turbulent transition between cylinders and in shear layers mentioned above refer to turbulization of a laminar flow by internal instabilities. However, the origin of interstellar turbulence is associated with a quite different source: the driving force is provided by explosions of supernovae, which occur randomly in time and at random places within the disc. Thus, even if the turbulence in the usual sense would not be present in the galactic disc, random flow driven by supernovae is surely present. Note that the flow property required for the turbulent dynamo is actually simply its randomness (Molchanov *et al.*, 1985).

An important property of the interstellar medium are magnetic fields whose energy density is estimated as approximately equal to the energy of chaotic motions,

$$\frac{H^2}{8\pi} \simeq \tfrac{1}{2}\rho v^2 \simeq 10^{-12}\,\text{erg cm}^{-3}.$$

The interstellar medium is characterized not only by a great hydrodynamic Reynolds number, the magnetic Reynolds number is also very high. Thus, interstellar turbulence is the *magnetohydrodynamic turbulence.*

High values of Reynolds numbers are understood to be an indication that viscous and Ohmic dissipation of kinematic and magnetic energies plays a generally negligible role in the formation and evolution of turbulent flow over a wide range of scales. The energy is conserved in interactions of motions of different scales and cascades from the largest scale towards the smallest one where it is transferred into heat.

Under the random phase approximation, the isotropic and homogeneous turbulent flow is conveniently described by the spectral densities of kinetic, $E(k)$, and magnetic, $M(k)$, energies, which characterize the distribution of energy over scales, or wave numbers $k = (\text{scale})^{-1}$:

$$v_0^2/2 = \int_0^\infty E(k)\,\mathrm{d}k; \qquad v^2(k) \sim 2kE(k);$$

$$a_0^2/2 = \int_0^\infty M(k)\,\mathrm{d}k; \qquad a^2(k) \sim 2kM(k).$$

Here $a(k) = b(k)/(4\pi\rho)^{1/2}$ is the Alfvén velocity in magnetic field $b(k)$ and v_0 and a_0 are velocities at the greatest, basic scale. Spectra do not continue to zero and infinite wave numbers, but are rather restricted by certain wave numbers $k_0 = 2\pi/l_0$ and $k_d = 2\pi/l_d$ corresponding to the basic, l_0, and dissipation, l_d, scales. As long as $Re \gg 1$ and $R_m \gg 1$, these scales differ considerably, $l_0 \gg l_d$. Spectral densities are usually decreasing functions of wave number. Therefore, the dominant contribution to the energy comes from the neighbourhood of the basic scale, which is therefore called the energy-range scale. The specific value of the energy-range scale is determined by properties of the source of turbulent motions, while the dissipation scale depends on the rate of energy dissipation and on the energy flux towards smaller scales.

Note that these spectral energy distributions provide an adequate description of isotropic and homogeneous random fields only if the phases of modes of the random field are random themselves.

A widely known and relatively well studied type of turbulent flow is the hydrodynamic vortex turbulence where the cascade of kinetic energy toward smaller scales is due to the interaction described by the non-linear term $(\mathbf{v}\nabla)\,\mathbf{v}$, inertial force, in the Navier—Stokes equation. To see how this term can produce the energy cascade, imagine the velocity field $\mathbf{v}$ made up initially by a single mode $\sin(kx)$. Then $(\mathbf{v}\nabla)\,\mathbf{v} \sim \sin 2kx$, that is non-linearity produces the driving force for the second harmonic and what follows is the cascade of motions with wave numbers k, $2k$, $4k$, $8k$, etc. The kinetic energy is conserved in the course of such scale halvings and the spectrum can be obtained from conservation of energy flux along wave number axis:

$$\frac{kE(k)}{\tau(k)} = \varepsilon,$$

where ε is a constant which does not depend on k and is determined by the power of the source of turbulent energy. It is clear that high Reynold numbers are required for the existence of a developed range of wave numbers $k_0 \gg k \gg k_d$ where the energy cascade occurs.

In magnetohydrodynamic turbulence a similar cascade of magnetic energy occurs duc to thc curl of electric field, $\mathrm{rot}(\mathbf{v} \times \mathbf{H}) = -(\mathbf{v}\nabla)\mathbf{H} + (\mathbf{H}\nabla)\mathbf{v}$.

In hydrodynamic vortex turbulence the characteristic time of spectral energy transfer, $\tau(k)$, is essentially the lifetime of a vortex of a scale $2\pi/k$ and is estimated as a revolution time of the vortex:

$$\tau(k) \sim l/v(l) = 1/kv(k).$$

With this relation we can easily obtain, from the energy flux conservation, the *Kolmogorov spectrum* of incompressible, isotropic, homogeneous hydrodynamic vortex turbulence

$$\begin{aligned} E(k) &= \varepsilon^{2/3} k^{-5/3} \\ v(k) &= (\varepsilon k^{-1})^{1/3} = v_0(k/k_0)^{-1/3} \end{aligned} \tag{VI.3}$$

(Kolmogorov, 1941; Obukhov, 1941), where constants of order unity are dropped. Since the energy transfer rate ε does not depend on k, it is conveniently determined at the energy-range scale as

$$\varepsilon = k_0 v_0^3.$$

The spectral interval $k_0 > k > k_d$, where both energy sources and energy sinks can be neglected, is called the *inertial range*. Termination of the inertial range is found at the wave number, k_d, where the spectral transfer time, $\tau(k)$, becomes as long as the dissipation time $(\nu k^2)^{-1}$:

$$k_d = k_0 Re^{3/4},$$

where $Re = v_0/\nu k_0$ is the kinematic Reynolds number.

Magnetic field strongly affects turbulence. When densities of magnetic and kinetic energies are close to each other turbulence takes the form of an ensemble of hydromagnetic waves of various lengths moving at nearly the Alfvén velocity for which the equipartition

$$E(k) = M(k)$$

approximately holds. In the inertial range the rate of the energy transfer toward greater wave numbers is conserved again:

$$\frac{kM(k)}{\tau_m(k)} = \frac{kE(k)}{\tau_m(k)} = \varepsilon.$$

However, the time of spectral energy transfer, $\tau_m(k)$ is no longer equal to the eddy lifetime $1/kv(k)$ but is essentially the timescale for the non-linear interaction of random waves of the scale k^{-1}. For the Kolmogorov turbulence $(kv)^{-1}$ is the only relevant combination with dimension of time. However, when magnetic field appears, sole dimension arguments become inadequate and more complicated physical considerations are to be invoked. Conservation of energy and momentum dictate that the three-wave interaction prevails in an ensemble of random Alfvén waves when non-linearity is weak. For this interaction

$$\tau_m^{-1}(k) \simeq \omega_k \frac{kM(k)}{U}$$

(see Sagdeev and Galeev, 1969; Kaburaki and Uchida, 1971) where $\omega_k = ka_0$ is the Alfvén frequency and U is a homogeneous function of dimension $(\text{cm/s})^2$. The simplest such function which reflects properly the role of magnetic fields is $U = a_0^2$. This gives

$$\tau_m^{-1}(k) = \frac{k^2 M(k)}{a_0} . \tag{VI.4}$$

Energy flux conservation now yields the spectrum of an ensemble of Alfvén waves in the inertial range

$$\begin{aligned} M(k) &= F(k) = (a_0 \varepsilon)^{1/2} k^{-3/2}, \\ a(k) &\simeq v(k) \simeq (a_0 \varepsilon k^{-1})^{1/4} = v_0 (k/k_0)^{-1/4} \end{aligned} \tag{VI.5}$$

which was determined by Iroshnikov (1963) and Kraichnan (1965).

The equipartition $M(k) \approx E(k)$ is maintained as long as the Alfvén frequency $\omega_k = ka_0$ exceeds the inverse dissipation time (Kraichnan and Nagarajan, 1967) that is for $k < k_A = a_0/\nu_m$. In the interstellar medium $k_d < k_A$ and thus in the dissipation range the equipartition is still maintained.

In the interstellar medium the inequality $R_m < Re$ holds and, therefore, the turbulence dissipation scale is determined by magnetic viscosity from the equation $\tau_m(k) = \nu_m k^2$ which gives

$$k_d = k_0 R_m^{2/3} (v_0/a_0)^{1/3},$$

where possible departures from equipartition $v_0 \neq a_0$ are allowed at the energy-range scale.

Theoretical views on interstellar fluctuation magnetic fields are further discussed in Section VII.13.

The spectrum (VI.5) is derived for isotropic, homogeneous, and incompressible turbulence. Properties of interstellar gas are far from being so simple. Large-scale magnetic fields violate isotropy. Co-existence of different phases of the interstellar medium, spiral arms, and other factors infringe homogeneity. Closeness of observed turbulent velocities to the sound velocity may be an indication of considerable effects of compressibility. Interstellar shock waves lead to energy dissipation in all scales. Large masses of gas involved in turbulent motions suggest the possible importance of self-gravitation. However, the observed spectrum of interstellar turbulent motions is surprisingly close to the inertial one (VI.5) (Armstrong *et al.*, 1981) and the mentioned deviations from symmetry probably can be accounted for as being small corrections.

We illustrate the role of compressibility in turbulence by the example of acoustic turbulence which arises when compressible gas or plasma is driven by random potential forces (Zakharov and Sagdeev, 1970; Kadomtsev and Petviashvili, 1973). In the absence of dispersion, finite-amplitude acoustic waves are breaking, shock waves are forming and kinetic energy dissipates at all scales. An energy cascade cannot be established. The spectrum of random shock waves is given by

$$E(k) \propto k^{-2}.$$

In a weak acoustic turbulence, where shock waves do not appear and the spectral energy cascade is established, the spectrum is

$$E(k) \propto k^{-3/2}.$$

A principal difference between these two spectra lies in the fact that in the latter case all modes have random, uncorrelated phases while in an ensemble of shock waves phase correlation is prominent between different harmonics, which decay simultaneously due to energy dissipation at shock fronts. A useful phenomenological modification of the turbulent cascade concept can be applied: the spectral energy transfer rate $\varepsilon(k)$ is thought to be not a conserved quantity, but is described by the equation

$$\frac{\partial}{\partial t} E(k) + \frac{\partial}{\partial k} \varepsilon(k) = -\mu\varepsilon(k)/k$$

(Kadomtsev and Petviashvili, 1973), where, quite similarly to the Alfvén-wave turbulence

$$\varepsilon(k) = \frac{kE(k)}{\tau(k)} = \frac{k^3E^2(k)}{c_0}$$

(cf. (VI.4) with Alfvén velocity a_0 replaced by the sound velocity c_0) and μ is a constant. In stationary state, $(\partial/\partial t)E(k) = 0$, this gives $\varepsilon(k) \propto k^{-\mu}$. For an ensemble of shock waves $E(k) \propto k^{-2}$ and, therefore, $\mu = 1$ and

$$\varepsilon(k) \propto k^{-1}.$$

Thus, the spectral energy transfer rate decreases with the scale in the presence of shock waves, which is manifested as steepening of the spectrum.

Turbulence of random shock waves is considered by Kaplan (1966), Bykov and Toptygin (1981, 1985), and Bykov (1982). At scales below the energy-range one, the spectrum of shock waves is steeper than those associated with energy cascades.

Therefore, energy cascade probably would prevail over a wide range of scales and only at small scales below e.g. the Coulomb free path of thermal particles (3×10^{17} cm in the hot coronal phase of interstellar medium) random shock waves can be dominant. Bykov and Ushakov (1986) consider evolution of an ensemble of random shock waves in a cloudy interstellar medium. Their important result is that acoustic and enthropy disturbances are damped by the shock waves while the vortex mode can be amplified. Inclusion of magnetic fields into their theory remains a demanding problem.

Note that the shock-wave turbulence does not require an anomalously powerful energy source: the energy transfer rate does not grow catastrophically, $\varepsilon(k) \leqslant \varepsilon(0) \simeq v_0^2 k_0$ and the time of energy transfer along the spectrum remains $\tau_0 \simeq (k_0 v_0)^{-1}$ for $v_0 = c_0$. Enhanced dissipation in shock waves leads to a steepening of spectrum but not to disastrously high required energy supply.

Now we turn to a discussion of the sources of interstellar turbulence and its observed properties.

We have already mentioned that the energy-range scale and the energy transfer rate are determined by properties of the energy source which drives the turbulence. Observationally, the correlation scale of interstellar turbulence can be estimated by a number of independent methods. Among them are statistical analysis of the fluctuations of starlight polarization (Section III.4) and non-thermal radiobackground intensity (Section IV.6) and direct observations of the velocity field of the interstellar gas (see below). All these methods give

$$l_0 \simeq 50\text{—}150 \text{ pc}.$$

The corresponding velocity is

$$v_0 \simeq 10 \text{ km s}^{-1},$$

and the fluctuation magnetic field

$$b_0 \simeq (1\text{—}1.7)B \simeq 2\text{—}5\ \mu\text{G}$$

while the mean, large-scale magnetic field has the strength $B \simeq 2\text{—}3\ \mu$G. The observed correlation scale of the fluctuation magnetic fields is the same as the velocity correlation scale in accordance with predictions of the dynamo theory (Artamonova and Sokoloff, 1986; Kleeorin *et al.*, 1986).

Thus, sources of interstellar turbulence maintain the energy transfer rate per unit mass

$$\varepsilon = 2v_0^3/l_0 \simeq 6 \times 10^{-3} \text{ erg g}^{-1}\text{ s}^{-1},$$

or

$$\varepsilon_V = 2\rho v_0^3/l_0 \simeq 10^{-26} \text{ erg cm}^{-3}\text{ s}^{-1}$$

per unit volume (factor 2 reflects a contribution of magnetic energy).

The random force that drives interstellar turbulence is produced by supernova explosions and energetic outflows from young hot stars, in which phenomena great amounts of kinetic energy are released. This source is very different from common sources of laboratory turbulence. It produces a non-stationary velocity field whose renovation time is of the order or less than the lifetime of a dominant eddy (see below).

Let us begin with the estimation of a contribution of supernova explosions to turbulence energetics. A shell formed due to explosion expands until the inner pressure drops down to the ambient value. After this the shell disintegrates dissolving in interstellar gas (see, e.g., Woltjer, 1972; Shklovsky, 1976; Chevalier, 1977). A major portion of the explosion energy, which is estimated as 3×10^{50} erg (Lozinskaya, 1980), is lost by radiation in the early stages of the supernova remnant's evolution. When the shock wave propagates at a speed exceeding the ambient sound velocity, the ambient gas remains practically free of density and velocity fluctuations from the wave; the kinetic energy transfer to the interstellar medium is basically null. The energy transfer becomes effective when the shock front propagation velocity approaches the velocity of sound: 10 km s^{-1}. According to calculations of the evolution of supernova remnants, no more than 5% of the total energy is transferred to the kinetic energy of interstellar gas (see Chevalier, 1977, §4.3), which amounts to 1.5×10^{49} erg. Assuming that supernovae explode once in 30 years in the Galaxy, we obtain an estimate of the energy supply rate as 2×10^{40} erg s^{-1}. Divided by the total mass of interstellar gas in the Galaxy, $4 \times 10^9 \, M_\odot = 8 \times 10^{42}$ g, this gives an estimate

$$\varepsilon = 2.5 \times 10^{-3} \, \mathrm{erg \, g^{-1} \, s^{-1}},$$

which is nearly half the energy flux required for maintenance of turbulence.

The pressure inside a supernova remnant, whose explosion energy is 7.5×10^{50} erg, equalizes with ambient pressure 10^{-12} dyn cm^{-2} when the outer diameter of the shell reaches 120 pc and the expansion velocity falls down to approximately 15 km s^{-1} (Smith, 1977). These values are essentially the dominant energy-range scale and velocity of interstellar turbulence.

An additional source of the turbulence is provided by expanding stellar winds from young O stars (Kaplan and Pikelner, 1970, §23) and stellar radiation pressure (Abbott, 1982). The average distance between expanding HII regions is a few hundreds parsecs, close to the dominant turbulent scale. The number of O stars in the Galaxy is 10^4, their evolution time and correspondingly the lifetime of the HII regions is 3×10^6 years. Therefore, HII regions appear every 300 years. Every star ionizes $10^3 \, M_\odot$ of the gas and the expansion pushes $2–5 \times 10^4 \, M_\odot$ of the gas up to the velocity ~ 10 km s^{-1}, which is equivalent to the energy 3×10^{49} erg (see Kaplan and Pikelner, 1970, Section 23). This gives the energy flux

$$\varepsilon = \frac{3 \times 10^{49}\ \text{erg}}{300\ \text{years} \times 4 \times 10^{9}\ M_{\odot}} \simeq 0.4 \times 10^{-3}\ \text{erg g}^{-1}\ \text{s}^{-1}.$$

Outflows from young stars provide approximately ten per cent of the total interstellar turbulent energy. In Table VI.4 we give kinetic energy supply rates by radiation pressure and winds from Wolf—Rayet stars and massive O stars and from supernovae. The table is based on the results of Abbott (1982) who has analyzed the situation in the region of a 3 kpc radius around the Sun (see also van Buren, 1985). As Abbott notes, these energy sources show spatial fluctuations at 100 pc scale. Analogous estimates have been made by Tarrab (1983) for some external galaxies.

TABLE VI.4
Kinetic energy supply to interstellar medium by Wolf-Rayet stars, O stars and supernovae

	Stellar radiation	Stellar wind	Supernova explosions	Total
ε erg g^{-1} s^{-1}	$\lesssim 2 \times 10^{-3}$	2×10^{-3}	5×10^{-3}	$\simeq 8 \times 10^{-3}$

One more possible energy source of interstellar turbulence is the differential rotation of the Galaxy. This mechanism of turbulence generation is more habitual for hydrodynamics. The energy transfer from regular flow to turbulent motions occurs through shear stresses. Instability of shear flow at large Reynolds number may arise, bringing about the turbulence despite the fact that the specific angular momentum in the Galaxy grows outward, $\Omega r^2 \propto r$ ($\Omega \propto 1/r$ in outer parts $r > 10$ kpc). In addition to theoretical considerations (Zeldovich, 1981) there are experimental results (Kruglyak *et al.*, 1980) that indicate the possibility of transition to a turbulent state in a flow between concentric cylinders when the outer cylinder rotates while the inner is in rest, i.e. when the specific angular momentum grows with distance from the rotation axis.

The energy transfer rate in this case can be estimated as

$$\varepsilon \simeq \beta \left(r \frac{\mathrm{d}\Omega}{\mathrm{d}r} \right)^2 \simeq \tfrac{1}{3} v_0 l_0 \Omega^2$$

(Shakura *et al.*, 1978), where $\beta \simeq \frac{1}{3} l_0 v_0$ is the turbulent viscosity and $\mathrm{d}\Omega/\mathrm{d}r = -\Omega/r$ for $\Omega \propto 1/r$. The quantity $r\,\mathrm{d}\Omega/\mathrm{d}r$ is the velocity shear $\mathrm{d}V/\mathrm{d}r$ at the scale l_0 in the reference frame rotating with the local angular velocity $\Omega = V/r$. For the Kolmogorov spectrum $v_0 \simeq (\varepsilon l_0)^{1/3}$ and with $l_0 \simeq 100$ pc we obtain

$$v_0 \simeq 0.7\,\Omega l_0 \simeq 2\ \text{km s}^{-1}$$

and

$$\varepsilon \simeq 2.7 \times 10^{-5}\ \text{cm}^{-3}\ \text{s}^{-1}.$$

For an ensemble of random Alfvén waves $v_0 \simeq (a_0 \varepsilon l_0)^{1/4}$ and in equipartition $a_0 \simeq v_0$ the result is the same.

Thus, it appears that differential rotation of the Galaxy can provide only one per cent of the required turbulent energy.

Figure VI.11 shows the radial distribution of turbulence sources in the Galaxy. Supernova remnant distribution is given in Figure VI.11a. Giant HII regions (see Figure VI.11b) are associated with star formation and are concentrated in the same areas as the interstellar gas, that is in the ring extending from 4 to 10 kpc radius (see Figure II.6). One would expect the corresponding enhancement of turbulence in gas-rich regions. Distribution of rotational velocity shear (Figure VI.2) is dictated by the distribution of stars in the Galaxy (see §1). Supernovae and young stars are highly concentrated around the galactic equator and the half-width of their vertical distribution does not exceed 70—100 pc (Ilovaisky and Lequeux, 1972).

Distribution of type-II supernovae in other spiral galaxies also has a ring-like form with a maximum at $r/R = 0.3$—0.4, where R is the radius of a galaxy (Guseinov *et al.*, 1980). Rings in distribution of HII regions are observed in early-type spirals, while in spiral galaxies of later types their surface density falls monotonously toward the edge (Hodge and Kennicutt, 1983; see also Hodge, 1974).

Overall, the source of interstellar turbulent motions is primarily a random force associated with kinetic energy release by supernovae and young stars. The spatial correlation scale is 50—150 pc and the corresponding revolution time of a turbulent cell is

$$\tau = \frac{\pi l_0}{v_0} \simeq 3 \times 10^7 \text{ years}.$$

Now we can estimate the renovation time of the Galactic turbulent flow. This can be determined as the time after which the sum of volumes of supernova remnants reaches the total volume of the galactic disc,

$$2\pi h R^2 \simeq 1.5 \times 10^{11} \text{ pc}^3,$$

where $h \simeq 100$ pc is the characteristic half-thickness of the (neutral) gaseous disc where supernovae concentrate and $R \simeq 16$ kpc is the outer radius of the supernovae distribution (see Burton, 1976). Since the volume of an average remnant is

$$\tfrac{4}{3}\pi\ 50 \text{ pc})^3 \simeq 5 \times 10^5 \text{ pc}^3,$$

the required number of supernovae is

$$\frac{1.5 \times 10^{11}}{5 \times 10^5} \simeq 3 \times 10^5.$$

Provided one supernova explodes every 30 years, their remnants fill the whole

(a)

(b)

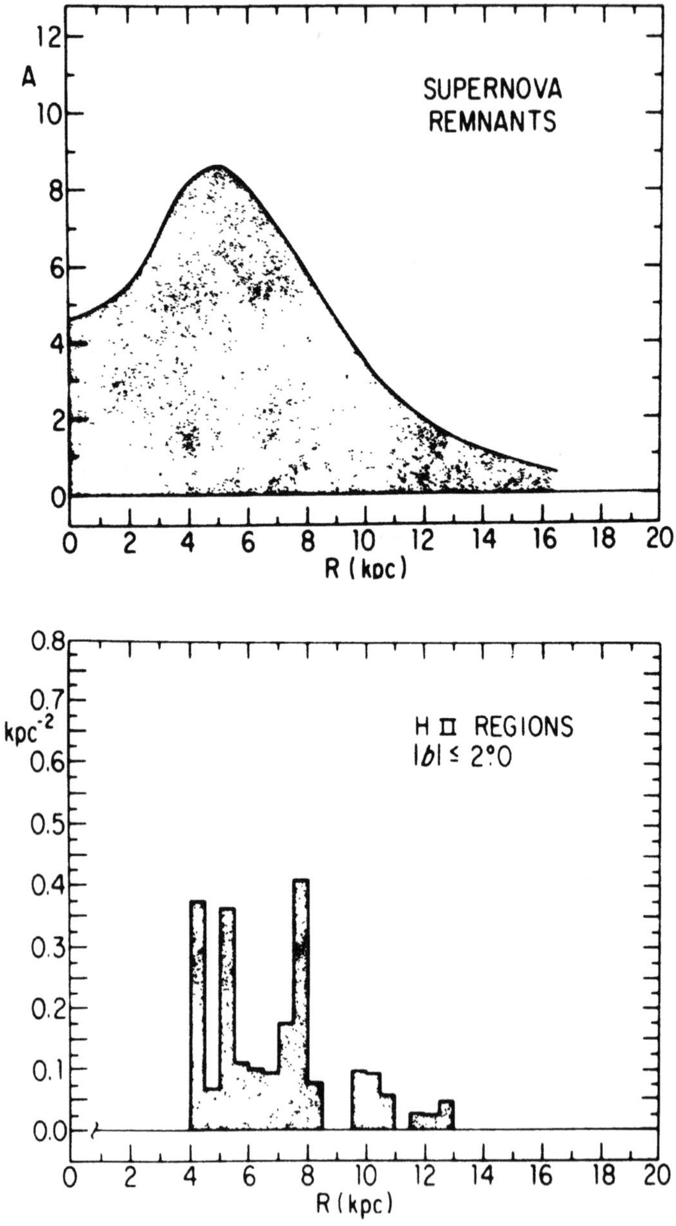

Fig. VI.11. Radial distributions in the Galaxy of (a) supernova remnants (Kodaira, 1974) and (b) surface density of giant HII regions (Burton *et al.*, 1975). From Burton (1976).

galactic disc in

$$\tau_0 \simeq 10^7 \text{ years.}$$

This is the time in which the turbulent velocity field changes completely in the Galaxy. The renovation time τ_0 is, thus, of the order or even less than the lifetime of the turbulent cell! It is this property which makes the interstellar turbulence very different from laboratory flows where turbulence is produced, e.g. by a grid. However, despite non-stationarity, interstellar turbulence has the usual correlation function which depends only on the time difference since the moments of flow renovation are distributed randomly (Molchanov *et al.*, 1985). Note also that supernovae are strongly concentrated around the galactic equator. Therefore the renovation time is even smaller near the central plane and approximation of short time-correlations can be adequately applied (see Chapter VII).

Let us turn now to the interpretation of observational data on interstellar turbulence. Description of interstellar gas motions in terms of Kolmogorov's theory of turbulence was first suggested by von Weizsäcker (1951). Observational study had begun with the determination of autocorrelation and structure functions by von Hoerner (1951) and Pikelner (1954). Autocorrelation function is given by

$$C(r) = \langle \mathbf{v}(\mathbf{r}')\mathbf{v}(\mathbf{r}' + \mathbf{r})\rangle \tag{VI.6}$$

where angular brackets $\langle \ldots \rangle$ denote the ensemble average. This function depends only on the separation $r = |\mathbf{r}|$ in homogeneous isotropic turbulence. Another characteristic, the structure function

$$D(r) = \langle [\mathbf{v}(\mathbf{r}') - \mathbf{v}(\mathbf{r}' + \mathbf{r})]^2 \rangle,$$

is related to the autocorrelation function through

$$D(r) = 2[C(0) - C(r)].$$

The spectral energy density introduced above is the sine-Fourier transform of the structure function,

$$E(k) = \frac{2}{\pi} \int_0^\infty D(r)\,(kr) \sin(kr)\,\mathrm{d}r.$$

Similar functions can be constructed for any random field — pressure, density, polarization degree, brightness, etc. However, in contrast to these quantities, the velocity and magnetic fields are vectors and their adequate description (in the random phase approximation) is given by the correlation tensor. Autocorrelation and structure functions introduced above are traces of corresponding matrices and characterize the kinetic and magnetic energies.

Dominant contribution to the structure function at the scale r comes from motions of the same scale: velocities at scales greater than r drop out since

$\mathbf{v}(\mathbf{r}') \simeq \mathbf{v}(\mathbf{r}' + \mathbf{r})$ for them, while velocities at smaller scales nearly cancel after averaging. Thus, a rough estimate is

$$D(r) \simeq 2v_0^2(r/l_0)^{2/3} \qquad \text{for} \qquad r < l_0 \tag{VI.7}$$

for the Kolmogorov spectrum and

$$D(r) \simeq 2v_0^2(r/l_0)^{1/2} \qquad \text{for} \qquad r < l_0 \tag{VI.8}$$

for the Kraichnan spectrum. Note that spectral exponents differ only in $\frac{1}{6}$ and it is very difficult to distinguish observationally these two spectra, which are very different physically.

At scales exceeding the energy-range one, $r > l_0$, velocities are uncorrelated and

$$D(r) \simeq 2v_0^2 \qquad \text{for} \qquad r > l_0.$$

Thus, the energy-range scale can be observationally determined as that at which structure and autocorrelation functions become constant.

First observational structure functions were obtained for the Orion Nebula by von Hoerner (1951) who used the line-of-sight gas velocities. A similar procedure was carried out for line-of-sight velocities of interstellar gas clouds by Kaplan and Pronik (1953). The structure (and autocorrelation) function of line-of-sight cloud velocities is shown in Figure VI.12 (Kaplan, 1955). The result is compatible, within observational errors, with the laws of both Kolmogorov (VI.7) and Kraichnan (VI.8). From these data the energy-range scale is estimated as $l_0 \simeq$ 80 pc.

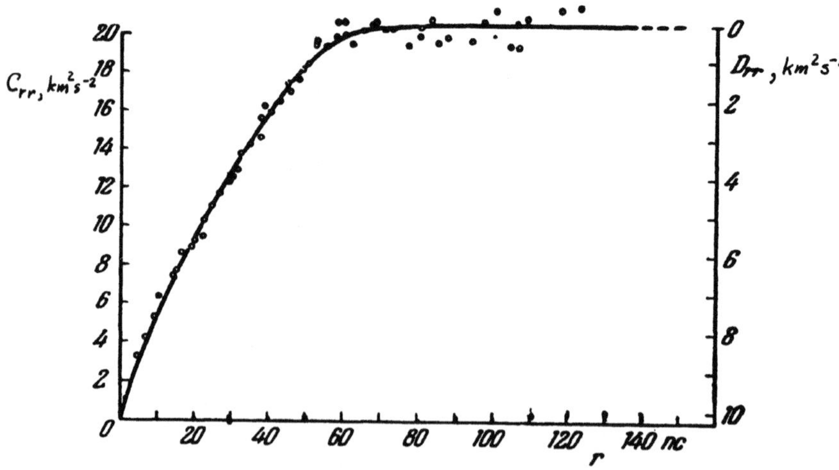

Fig. VI.12. Structure $D(r)$ and autocorrelation $C(r)$ functions of the line-of-sight velocities of interstellar gas clouds (Kaplan, 1955, 1966, §18).

Turbulence in the Orion Nebula was also studied by Pikelner and Shajn (1954) using the brightness fluctuations autocorrelation function. For magnetic field investigations, the structure functions of interstellar starlight polarization fluctuations and of intensity fluctuations of the non-thermal radio background are especially interesting. Such studies have been pioneered by Klimishin and Bazilevich (1963). Later, such investigations were carried out by Jokipii *et al.* (1969) (see Section III.4). The autocorrelation function of intensity fluctuations of the radio background was discussed at length in Section IV.6 (see Figure IV.7). From these observations, $l_0 \simeq 150$ pc.

Correlation analysis of brightness fluctutations of λ 21 cm emission was undertaken by Perry and Helfer (1972) for the scales 50—100 pc in the Perseus arm and by Baker (1973) for small clouds less than 7 pc across in the solar vicinity of 200 pc radius (see also Mebold *et al.*, 1974). Baker has found an anisotropy of velocities at these scales: the HI clouds in the region $l = 87.5°$, $b = 24.5°$ move toward the Sun. Spectra of small-scale (< 5 pc) λ 21 cm brightness fluctuations have been obtained by Crovisier and Dickey (1983).

Angular broadening of radiosources and scintillation of pulsars (see Rickett, 1977) reveal fluctuations in the density of interstellar free electrons. These fluctuations are interpreted as being the result of turbulent motions of interstellar gas at scales 10^8—10^{13} cm. Figure VI.13 reproduces the spectrum of density fluctuations compiled by Armstrong *et al.* (1981). Data for larger scales $l > 10^{18}$ cm are based on observations of velocity fluctuations, and density fluctuations are estimated through the relation

$$\frac{\delta n_e}{\langle n_e \rangle} \propto \frac{\delta v}{a_0}$$

(Neugebauer *et al.*, 1978), where $\langle n_e \rangle = 0.03$ cm^{-3} is the average electron density and $a_0 \simeq 10$ km s^{-1} is the Alfvén velocity. This relation is valid for the solar-wind plasma and should be applied to the interstellar medium with caution. For example, in molecular clouds the relation $\delta v \propto \rho^{-0.38}$ seems to hold (see Larson, 1981 and the discussion below). In addition, observations are lacking in the range $10^{14} \lesssim l \lesssim 10^{18}$ cm. Nevertheless, it seems that interstellar gas participates in turbulent motions of an enormously wide range of scales,

$$10^8 \text{ cm} \lesssim l \lesssim 3 \times 10^{20} \text{ cm} \simeq 100 \text{ pc!}$$

The power spectrum of electron density $P(k)$ is approximated by

$$P(k) \propto k^{-3.7 \pm 0.6}$$

for $10^{-13} \lesssim k \lesssim 10^{-8}$ cm^{-1}. Unfortunately, the statistical error is so large that not only both Kolmogorov's $P \propto k^{-11/3}$ and Kraichnan's $P \propto k^{-7/2}$ spectra fit the observations, but even the shock wave spectrum $P \propto k^{-4}$ does.

The statistical error reduces when motions of larger scales are taken into account:

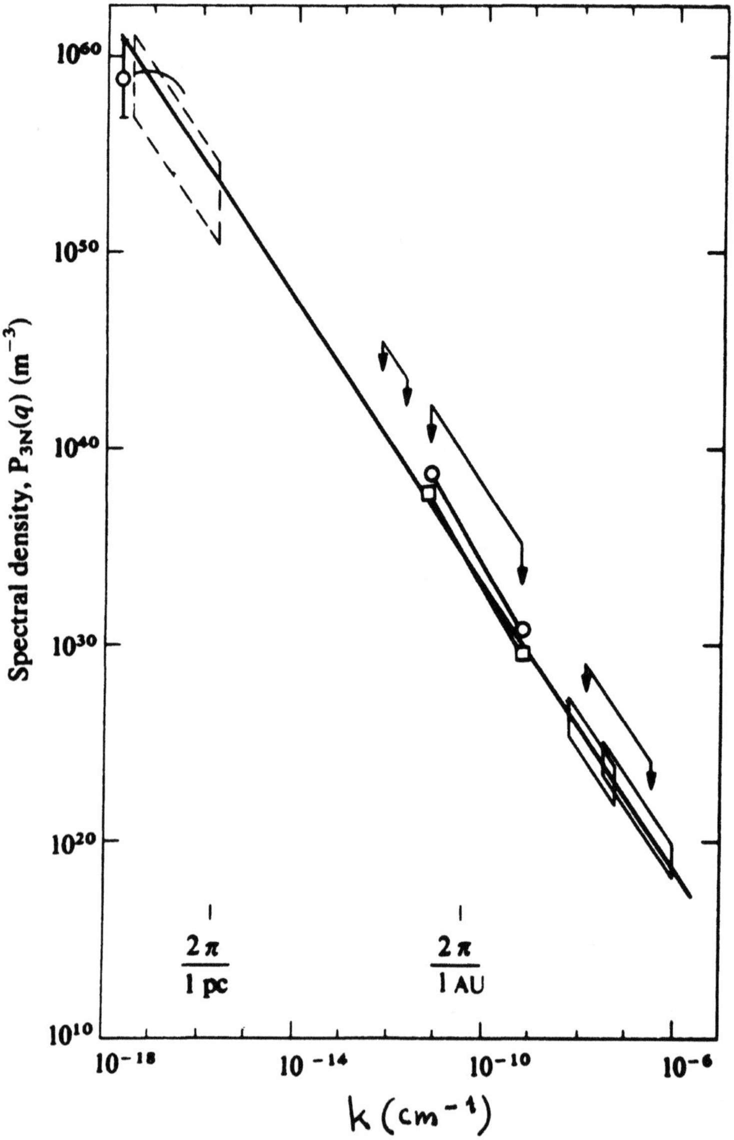

Fig. VI.13. The power spectrum of density fluctuations of thermal electrons in the interstellar medium (Armstrong *et al.*, 1981).

$$P(k) \propto k^{-3.6 \pm 0.2}$$

for $10^{-20} \lesssim k \lesssim 10^{-8}$ cm^{-1}, but both Kolmogorov's and Kraichnan's spectra still fit the result.

Turbulence can be hardly maintained over such a wide range of scales by one general energy source (Armstrong *et al.*, 1981), and continuity of the spectrum should be viewed as accidental in this event. However, the continuity may turn out to be an artifact of improper transformation from velocity fluctuations to density ones. Moreover, even at smaller scales, inhomogeneities at scales 10^{13} cm and 10^{10} cm may be of different origins (Chashei and Shishov, 1980). A possibility for extending the turbulence toward smaller scales due to anisotropy induced by magnetic field is discussed by Higdon (1984).

Apart from the determination of autocorrelation and structure functions, observational investigations of interstellar turbulence often use one more method, much less rigorous from a mathematical point of view. It is based on the observed correlation between velocity dispersion and the size of the sampled region (Larson, 1979). It is assumed that such a dependence can provide an estimate of the structure function. Velocity dispersion is deduced from the broadening of absorption lines λ 21 cm of neutral hydrogen and emission lines λ 2.6 mm of the CO molecule taking account of the contribution from thermal broadening, which is generally relatively small (see, however, Myers, 1983). Larson (1979) also uses velocity dispersions of young stars suggesting that their motions bear the memory of motions of progenitor interstellar gas.

Figure VI.14 shows the dependence of velocity dispersion on the region size

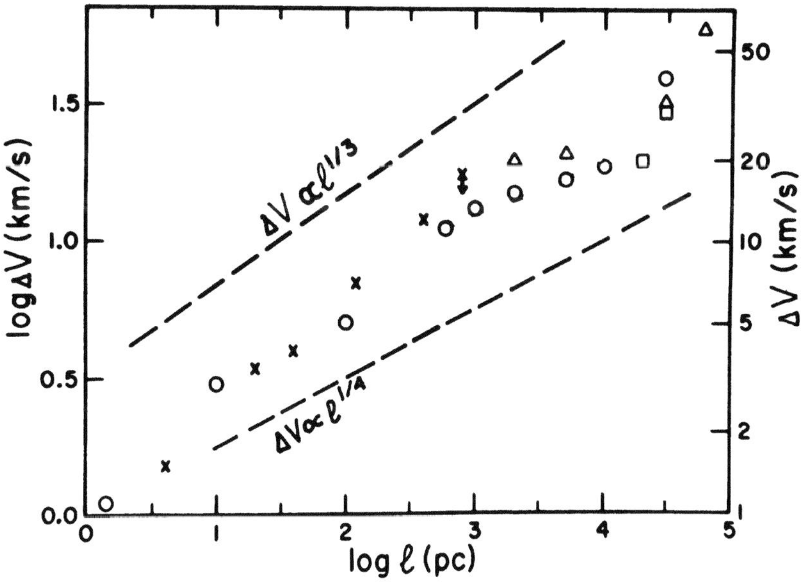

Fig. VI.14. Correlation between velocity dispersion and the size according to Larson (1979). Data are used on young stars (crosses) and interstellar gas (circles) in the Milky Way galaxy, and gas dispersions in M31 (triangles) and M81 (squares).

compiled by Larson (1979), who also uses data from M31 and M81 galaxies. The dispersion Δv monotonously grows with scale l in a wide range of scales. He approxiamtes this dependence at $l \lesssim 1.1$ kpc by

$$\Delta v\ (\mathrm{km\ s^{-1}}) \simeq 1.1\ (l/1\ \mathrm{pc})^{0.37},$$

which is somewhat steeper than Kolmogorov's law $\Delta v \propto l^{0.33}$. However, the data can be fitted even better by Kraichnan's dependence $\Delta v \propto l^{0.25}$ for $l < 100$ pc. In principle, we may assert a rough agreement between the observed correlation and the spectrum of random Alfvén waves. Comparison with the Kolmogorov's spectrum seems to be less justified since magnetic fields undoubtedly play an essential role in the interstellar turbulence.

Larson (1979) has mentioned the possibility of extending the correlation between velocity dispersion and scale up to scales of 10 kpc, comparable with overall galactic dimensions. It seems, however, questionable that these large-scale motions (warpings and corrugations of gas layers, asymmetries in rotation curves, etc.) are of the same origin as turbulent motions at scales below 100 pc. Gravitation and galactic rotation are undoubtedly important at large scales. At the scales 100—1000 pc and larger, there probably exists turbulence of two-dimensional Jeans waves whose inertial-range spectrum is $E(k) \propto k^{-7/4}$ (Churilov and Shukhman, 1982).

Substantial efforts have been devoted to the study of the correlation between velocity dispersion and scale for molecular clouds (see, e.g., Larson, 1981; Leung *et al.*, 1982; Myers, 1983 and references therein; Scalo, 1984). The general result is an establishment of a systematic growth of velocity dispersion with scale. According to Larson (1981),

$$\Delta v\ (\mathrm{km\ s^{-1}}) = 1.1\ l^{0.38},$$

where l is the cloud diameter measured in pc and majority of clouds have $l \lesssim 20$ pc. The slope is significantly larger for giant molecular clouds $10 \lesssim l \lesssim 100$ pc:

$$\Delta v\ (\mathrm{km\ s^{-1}}) = (0.88 \pm 0.16)\ l^{0.62 \pm 0.05}$$

(Sanders *et al.*, 1985).

Interpretation of these correlations is far from being clear. Dense molecular clouds are possibly gravitationally bound (Solomon *et al.*, 1979; Fleck, 1980; Larson, 1981; Leung *et al.*, 1982; Myers, 1983) and internal chaotic velocities follow from the virial equilibrium condition as

$$\Delta v \simeq \left(\frac{6GM}{5l}\right)^{1/2} \tag{VI.9}$$

where M and l are the mass and diameter of the cloud, respectively. With the observationally found relation between the density of molecular clouds and their size,

$$\rho \propto l^{-1} \tag{VI.10}$$

(Larson, 1981; Myers, 1983), virial equilibrium implies the power-law dependence of velocity dispersion on the cloud size:

$$\Delta v \propto l^{1/2}.$$

This relation fits observations of many dense clouds within observational errors. There are, however, indications that for larger clouds $l \gtrsim 0.1$ pc

$$\rho \propto l^{-2}$$

(Quiroga, 1983).

Under the assumption that the mass of molecular clouds is conserved in their formation, the relation (VI.10) would imply a one-dimensional character of their collapse. This may be due to shock-wave compression (Larson, 1981) or a strong magnetic field (Baker, 1979).

The lifetime of molecular clouds seems to be two orders of magnitude greater than a free-fall time. Equilibrium of molecular clouds can be maintained by the observed internal chaotic motions (Fleck, 1980; Larson, 1981) and/or strong (10^{-4}—10^{-5} G) internal magnetic fields (Baker, 1979).

Thus, there is an additional source of chaotic motions in molecular clouds — gravitational energy (cf. Scalo and Pumphrey, 1982). It should be noted also that molecular clouds are a site of active star formation and young stars contribute to the generation of turbulent motions (Leung *et al.*, 1982; Silk, 1985).

Turbulent broadening of spectral lines in molecular clouds greatly exceeds thermal broadening. This implies that the turbulence is supersonic. This brings up the problem of maintenance of supersonic motions which seem to be bound up with rapid dissipation. Nevertheless, the observations of Myers (1983) show that properties of turbulence in molecular clouds are similar both in subsonic and supersonic regimes. Slow dissipation of the turbulent energy may be attributed to a low effective cross section for collisions of condensations (turbulent cells) because these collisions are predominantly oblique (Scalo and Pumphrey, 1982). Another possibility of reconciling supersonic motions with prolonged maintenance of turbulence is based on the assumption that the observed velocity dispersions refer to dense cold condensations moving in warmer gas for which the observed motions are subsonic (Larson, 1981).

A theory of turbulence in molecular clouds which takes into account effects of self-gravitation and explains correlations $\rho \propto l^{-1}$ and $\Delta v \propto l^{1/2}$ is proposed by Henriksen and Turner (1984).

One more obstacle in interpretation of velocity dispersion — scale correlation is due to a possible role of compressibility (von Weizsäcker, 1951; Fleck, 1983). The observed correlation between cloud densities and sizes directly indicates the effects of compressibility. On the other hand, dependences $\Delta v \propto l^{1/3}$ and $\Delta v \propto l^{1/4}$ correspond to the Kolmogorov's and Kraichnan's spectra, respectively, for an incompressible medium where the energy flux is conserved per unit mass. Fleck (1983) proposes to account for compressibility phenomenologically through the energy cascade with energy flux conserved per unit volume,

$$\varepsilon_V = \varepsilon\rho = \text{const},$$

rather than per unit mass. This gives, for the Kolmogorov's spectrum,

$$\Delta v \propto (l/\rho)^{1/3},$$

and for

$$\rho \propto l^{-1},$$
$$\Delta v \propto l^{2/3}.$$

This dependence is considerably steeper than

$$\Delta v \propto (l/\rho)^{0.21}$$

observed in molecular clouds (Fleck, 1983). This contradiction is interpreted as an indication of energy injection at small scales.

However, we can propose an alternative interpretation presuming that the basic spectrum is of Kraichnan's rather than of Kolmogorov's type. Indeed, similar considerations give for the spectrum of magnetohydrodynamic waves (see (VI.5)) with $\varepsilon_V = \text{const}$

$$\Delta v \simeq \left(a_0 \varepsilon_V \frac{l}{\rho} \right)^{1/4} \tag{VI.11}$$

which is close to the dependence inferred from observations by Fleck (1983). Important role of magnetohydrodynamic waves in molecular clouds is also emphasized by Myers (1986).

To summarize, observational data on turbulent motions in molecular clouds do not contradict the concept of turbulent energy cascade but indicate Kraichnan's rather than Kolmogorov's form of spectrum. Interpretation of velocity dispersion — size correlation is burdened with many uncertainties and lack of rigour (see Scalo, 1984). For example, measurements of spectral line broadening, which are the basis of this method, provide information on scatter of line-of-sight velocities. The sampled line-of-sight length grows with the growth of the size of the observed region. Accordingly, the number of fluctuations which contribute to the observed line broadening grows. Even for completely random fluctuations, which have nothing in common with turbulent energy cascades, the velocity dispersion would thus grow as $l^{1/2}$, which is very close to the observed dependence. Isolation of regular velocity gradients present another problem. Direct construction of auto-correlation and structure functions or other mathematically rigorous methods are undoubtedly more reliable, though more labour-consuming, tools for investigation of interstellar turbulence. Useful discussion of the methods of observational study of interstellar turbulence can be found in Spicker and Feitzinger (1987a and references therein).

Direct comparison of these methods was performed by Roy and Joncas (1985) who have made detailed measurements of a velocity field in the HII region Sharpless 142 with 41 000 values of the line-of-sight velocity measured with spatial resolution 0.05 pc (the diameter of this HII region is about 28 pc). The authors have performed a comparison of the dependence of velocity dispersion on the scale with the structure function of velocity fluctuations. Velocity dispersion — scale correlation is very clearly pronounced and has the form

$$\Delta v \; (\mathrm{km\,s^{-1}}) \simeq 6.46 \; l^{0.30 \pm 0.02}$$

with l in pc. The structure function for $1.6 \lesssim r \lesssim 9$ pc has a regular form

$$D(r) \propto r.$$

Thus, the observed structure function would correspond to the correlation $\Delta v(l) \propto l^{1/2}$, very different from the observed one.

An estimate of autocorrelation function for velocity fluctuations in the Taurus molecular complex based on 1200 velocity measurements taken at 0.6 pc spacing is presented by Kleiner and Dickman (1985). Their results show no correlation of velocities at scales below 0.6 pc, while velocity dispersion in this molecular complex seems to be consistent with the dispersion — size correlations inferred for molecular clouds. Again we see a contradiction of results based on correlation studies and on velocity dispersion — size dependence. Data on interstellar turbulence basing solely on dispersion — size correlations should be considered with proper caution.

Chaotic gas motions are also observed in external spiral galaxies, and their velocities are comparable with those in the Milky Way. For example, the root mean square velocity of neutral hydrogen clouds in the Andromeda Nebula (M31) is 12 km s^{-1} in the inner parts (Emerson, 1976) and 10 km s^{-1} in the outer regions (Newton and Emerson, 1977). According to the observations of Warner *et al.* (1973), chaotic velocities in M33 are 15 km s^{-1} near the nucleus and fall down to 9 km s^{-1} at 6 kpc radius. However, Newton (1980a) found that the velocity dispersion of HI in M33 does not vary with radius and is estimated as 9 $\pm$ 3 km s^{-1} for $r \lesssim 6$ kpc. In IC 342 HI velocity dispersion is 9.8 km s^{-1} and is constant at all radii (Newton, 1980b). A close value 10 $\pm$ 1 km s^{-1} is observed in NGC 3938 (van der Kruit and Shostak, 1982). In NGC 628, HI velocity dispersion falls down from 9—10 km s^{-1} in the central region to 7 km s^{-1} at the edge of the optical disc, while it is constant at 7—8 km s^{-1} in NGC 1058 (van der Kruit and Shostak, 1983). Decrease in the intensity of turbulent motions is observed toward larger radii in M51: from 30 km s^{-1} in the central part to 17—20 km s^{-1} in the main part of optical disc (Tully, 1974).* First results of the study of the turbulence in the Large Magellanic Cloud are presented by Spicker and Feitzinger (1987b).

When observations are carried out at small spatial resolution, the regular velocity gradient contributes to the observed dispersion. Growth of observed dispersions in the central parts of some galaxies may be due to this bias. Data obtained at low resolution frequently show the growth of dispersion toward the center. However, higher resolution studies give a constant value (cf. Newton, 1980b). Observations of neutral hydrogen in spiral galaxies are generally consistent with the constant value 7—10 km s^{-1} for gas velocity dispersion, and are nearly the same for all galaxies (van der Kruit, 1981). Analysis of a large sample (200) of spiral galaxies by Lewis (1984) indicates isotropy of HI velocity dispersion, $\Delta v_z \simeq \Delta v_r \simeq \Delta v_\varphi$. Moreover, one-dimensional dispersion of 8 km s^{-1} seems to be typical for all spiral galaxies independently of morphological type, luminosity and position within the galaxy.

VI.4. Mean Helicity

Rotation of interstellar gas and non-uniform gas density distribution provide interstellar turbulence properties rarely met in laboratories. One such property which is of paramount importance for the generation of large-scale magnetic fields is violation of mirror symmetry. This peculiarity follows from symmetry arguments. Indeed, a rotating stratified medium can be characterized by the scalar product of pseudo-vector of angular velocity and the vector of density gradient $\Omega\nabla\rho \simeq \Omega\,\partial\rho/\partial z$, that is by the pseudo-scalar quantity which changes the sign under reflection in the central plane of the galactic disc, $z \to -z$.

From the physical point of view, violation of the mirror (reflection) symmetry is associated with predominance of, say, right-handed turbulent cells over left-handed ones. An extra screw rotation is produced by the Coriolis force. Consider a fluid element which moves away from the central plane of the disc at velocity v_z (Figure V.4). A decrease in the ambient gas density, e.g. according to the exponential law with the scale height $h = -\rho/\nabla\rho$, makes the element expand, i.e. produces the lateral velocity component v_r. This gives rise to Coriolis forces $\propto \Omega v_r$, which give the element a screw rotation.

The angular momentum per unit mass of the turbulent cell grows by the value $r^2|\mathrm{rot}\,\mathbf{v}| \sim rv_r\Omega\tau$ in the correlation time $\tau \sim l_0/v_0$. From continuity of the flow, $\mathrm{div}\,\rho\mathbf{v} = 0$, we obtain $v_r/r \sim v_z/h$. Thus, the fluid element acquires the vorticity

* In all cases, dispersions of line-of-sight velocities are given. The total, three-dimensional dispersions are $3^{1/2}$ times larger, provided chaotic velocites are isotropic.

$|\text{rot}\, \mathbf{v}| \sim v\Omega\tau/h$. Those fluid elements that descend, and thus contract, acquire the curl of the same modulus, but of the opposite sign. However, the scalar product $\mathbf{v} \cdot \text{rot}\, \mathbf{v}$, which is called helicity, is the same (and negative) both for ascending and descending fluid elements in the upper half of the disc. The signs of this scalar product are opposite in the upper and lower halves of the disc.

The equation of generation of the mean magnetic field includes the pseudoscalar quantity* (see Chapter VII)

$$\alpha = -\frac{\tau}{3}\langle \mathbf{v}\, \text{rot}\, \mathbf{v}\rangle.$$

Angular brackets $\langle \ldots \rangle$ denote an ensemble average which coincides in favourable conditions with the average over all turbulent cells. This average can be conveniently understood as the time-average taken over the correlation time τ. This brings us to the estimate

$$\alpha(z) \simeq \begin{cases} l_0^2\Omega/h & \text{for} \quad 0 < z < h, \\ -l_0^2\Omega/h & \text{for} \quad -h < z < 0, \end{cases} \tag{VI.12}$$

where $l_0 \simeq \tau v_0$ is the correlation length.

The above qualitative arguments have implied that the correlation time is the shortest characteristic time of the system, that is $\tau \lesssim \Omega^{-1}$ or the Rossby's number exceeds unity, $Ro \equiv v_0/l\Omega \gtrsim 1$. In other words, we have assumed that displacement of the fluid element under the action of the Coriolis's force is small and this force produces only screw rotation. The displacement $\Omega v_r \tau^2$ reaches the correlation length l_0 in the time $(l_0/\Omega v_0)^{1/2} \simeq \tau Ro^{1/2}$. When $Ro \gtrsim 1$, the correlation time τ is the shortest characteristic time and we obtain (VI.12). However, for $Ro < 1$ the fluid element travels through many correlation cells in time τ and averaging proceeds over the residence time in a single cell, $\tau Ro^{1/2}$. In this event the mean helicity $\langle \mathbf{v} \cdot \text{rot}\, \mathbf{v}\rangle$ is $Ro^{1/2}$ times smaller. Accordingly, the value of α is Ro times smaller, since the factor τ should be also replaced by $\tau Ro^{1/2}$:

$$|\alpha(z)| \simeq \frac{l_0}{\text{h}} v_0. \tag{VI.13}$$

It should be noted that in the vortex turbulence $|\alpha|$ never exceeds the characteristic turbulent velocity v_0. Existence of this upper limit for the mean helicity follows from the obvious fact that the number of turbulent cells with uniform screwedness cannot exceed the total number of cells.

Existence of this upper limit is associated with the simplest picture of turbulence adopted above, where the correlation time is of the order of vortex revolution time, $\tau \sim l_0/v_0$, and characteristic scales of rot $\mathbf{v}$ and $\mathbf{v}$ are both equal

* We often also call this quantity, whose dimension is cm s^{-1}, helicity, hoping that this will not cause confusion.

to l_0. If at least one of these conditions is violated, $|\alpha|_{max}$ in principle may exceed v_0. Indeed, an order of magnitude estimate gives

$$|\alpha| = \left|\frac{\tau}{3}\langle \mathbf{v} \cdot \mathrm{rot}\, \mathbf{v}\rangle\right| \lesssim \frac{\tau}{3}\frac{v_0^2}{l_r},$$

where l_r is the characteristic scale of rot $\mathbf{v}$ distribution. In the vortex turbulence $\tau \sim l_0/v_0$ and the maximal mean helicity may be large,

$$|\alpha|_{max} \sim \tfrac{1}{3} v_0 \frac{l_0}{l_r} \gtrsim v_0,$$

provided the vorticity in turbulent cells is concentrated within long thin tubes of radius l_r (or in sheets of small thickness l_r) and curvature radius l_0 (the intermittency). In the linear approximation, vorticity behaves similarly to magnetic field (Batchelor, 1953), which indeed possesses the required intermittent properties in turbulent fluids (Zeldovich *et al.*, 1987). Note that the well known realisability condition $|F(k)| < 2kE(k)$ for the spectra of helicity and energy (Moffatt, 1978) implies the random phase approximation which is not valid in the case of intermittent flow (Frisch, 1985).

Another potential possibility is the turbulence with great correlation time $\tau > l_0/v_0$, for example, composed of long random wave trains.

Let us estimate the Rossby number in the Galaxy. For the solar vicinity $v_0 \simeq 10$ km s^{-1}, $l_0 \simeq 100$ pc, $\Omega \simeq 25$ km s^{-1} kpc^{-1} and $Ro \simeq 3$, that is Ro larger than, or of the order of, unity. In the central parts, $r < 2$–3 kpc, where $\Omega \gtrsim 70$ km s^{-1} kpc^{-1}, the Rossby number is below unity. A qualitative dependence of the mean helicity on galactocentric radius is shown in Figure VI.15.

The interstellar turbulence is considerably helical. According to our estimates, the value of α is a few tenths of the turbulent velocity.

We draw the reader's attention to the fact that until now the mean helicity of turbulence in dynamo systems, including discs of spiral galaxies, has been determined only by qualitative order-of-magnitude estimates. However while our knowledge of other hydrodynamic parameters of galaxies is approximate, a situation not uncommon in astrophysics, it is always based on direct observational data and is often verified by quantitative theory. In the laboratory, helical flows of liquid metals have been produced and observed (Steenbeck *et al.*, 1968; Kirko, 1985). Mean helicity is more typical not of laboratory flows, but rather of cosmical bodies where convection or turbulence in rotating stratified medium is a widespread property. Direct measurement of helicity of solar convection is hampered by the necessity of numerical calculation of the curl of the three-dimensional velocity field. Even more remote is the possibility of measuring the helicity in planetary cores isolated from us by opaque shells. The situation is much more promising in the case of the Galaxy, which is our primary concern here: this is the

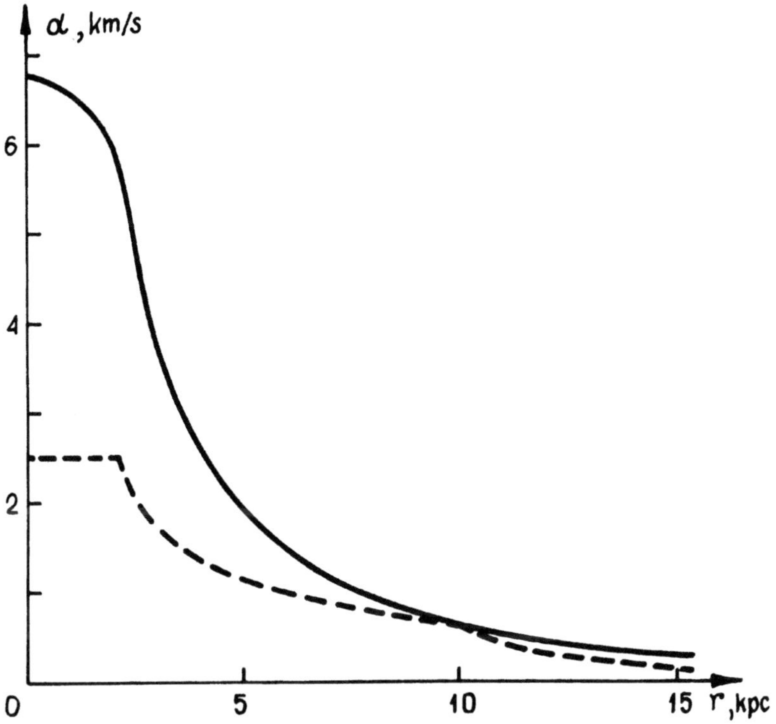

Fig. VI.15. Dependence of the mean helicity of interstellar turbulence on the radius in the Galaxy. The solid line is for the disc shape of the parted hyperboloid of rotation and the broken line for the disc given by (VI.1).

only dynamo system which can be seen from the inside! This gives us the opportunity to use an integral, rather than a differential, manifestation of the mean helicity as the measure of mirror asymmetry of velocity field.

The correlation tensor of homogeneous, isotropic velocity field **v** can be always represented as

$$\langle v_i(\mathbf{x})v_j(\mathbf{x}+\mathbf{r})\rangle = \delta_{ij}A(r) + \frac{r_i r_j}{r^2}B(r) + e_{ijk}C(r)r_k$$

(Batchelor, 1953), where A and B describe the mirror-symmetric part of the velocity field, e_{ijk} is the unit completely antisymmetric tensor, and the value of C at $r = 0$ is proportional to the mean helicity of the field **v**. Therefore, observational determination of the mean helicity can be accomplished through measurements of the correlation tensors in two regions whose positions are mirror symmetric, $\mathbf{r}_1 = -\mathbf{r}_2$, with respect to the observer (the Earth), i.e. which lie in directions $(l^{\mathrm{II}}, b^{\mathrm{II}})$ and $(l^{\mathrm{II}} + 180°, -b^{\mathrm{II}})$ in galactic coordinates. The difference between these tensors yields the mean helicity. Possibly, measurements of a cross-correlation, e.g. $\langle v_1 v_2\rangle$, would suffice. Observation of all components would allow us to verify the validity of the pseudo-scalar α approximation: in the general case of anisotropic, inhomogeneous turbulence the mean helicity is a tensor.

If only Doppler measurements are used, the derivation of cross correlations $\langle v_1(\mathbf{x})\, v_2(\mathbf{x} + \mathbf{r})\rangle$ requires that the points $\mathbf{x}$ and $\mathbf{x} + \mathbf{r}$ are separated by an angle close to $\pi/2$. Accuracy will decrease with smaller r and the value $C(0)$ probably should be determined through extrapolation.

It seems plausible that even existing surveys of Galactic neutral hydrogen contain information necessary for the determination of the mean helicity of interstellar turbulence. Our order-of-magnitude estimate shows that $\alpha \sim 0.1\, v_0$ and velocity measurements with an accuracy of the order of or less than 1 km s^{-1} would suffice. The sensitivity of a survey does not play a decisive role: it is sufficient to make measurements in the close vicinity of the Earth within one or two correlation cells, 100—200 pc, or even less. The principal requirement is the possibility of making a comparison between observations at points lying at opposite ends of the celestial sphere's diameter. A device ideally fit for this purpose is an orbital telescope.

VI.5. Magnetic Fields and Star Formation

The problem of star formation is closely related to problems of the large-scale galactic magnetohydrodynamics. Formation of a star is an evolutionary process spanning millions years, in which magnetohydrodynamic effects play diverse and remarkable roles. In this section we only briefly discuss modern views on the role of magnetic fields in star formation and refer to basic papers and reviews that consider this problem in detail and where elaborate references can be found.

Magnetic fields influence the process of star formation apparently at all its stages. At early stages, dense gas clouds condense in the interstellar environment with much help from the Rayleigh-Taylor instability (Hoyle and Harwit, 1958; Parker, 1966, 1982, Ch. 13). Owing to the stabilizing influence of cosmic rays (Nelson, 1985; Ptuskin, 1986), this instability cannot destroy completely the hydrostatic equilibrium of the galactic gaseous disc but rather causes isolation of interstellar cold, dense gas clouds. The physical mechanism of the instability is briefly discussed in Section 2 of this chapter. A typical horizontal wavelength of growing disturbances is of the order of 1 kpc. This instability leads to condensation of a cold interstellar gas in magnetic pockets that can be revealed obesrvationally, e.g. through optical polarization observations in vicinity of star formation regions (Appenzeller, 1974). Together with gravitational and thermal instabilities (see Kaplan and Pikelner, 1979), the magnetic Rayleigh-Taylor instability determines the cloudy structure of the interstellar medium.

Magnetic fields are no less important in the interstellar clouds themselves, where they determine the character of internal turbulent motions (Section 3 of this chapter, Section VII.12) and contribute to the cloud equilibrium (Section III.5). There exists a pronounced statistical relation between gas density of interstellar clouds and internal magnetic field strengths (Section III.5 and Fig. III.4). When the gas number density exceeds the value of order 10^2 cm^{-3}, magnetic field strengths are proportional to approximately the square root of the gas density. This relation is in a good agreement with theoretical predictions.

It is usually considered that magnetic fields of interstellar clouds and protostellar nubulae are the result of compression and subsequent passive evolution of the interstellar field trapped by an evolving cloud. However, magnetic Reynolds numbers and lifetimes of molecular clouds are large enough to make principally possible considerable amplification of internal magnetic fields by the small-scale turbulent dynamo that generates chaotic magnetic fields at the scales below the correlation scale of the internal turbulence. This dynamo operates when magnetic Reynolds number exceeds approximately 10^2; the characteristic growth time of the field is of the order of l/v, where l and v are the correlation length and velocity of the turbulence, respectively (Artamonova and Sokoloff, 1986; Kleeorin *et al.*, 1986). In molecular clouds $l/v \simeq 1$ pc/1 km s^{-1} $\simeq 10^6$ years which does not exceed the presumable lifetime of molecular clouds. Generation of the turbulent magnetic fields in giant molecular clouds is discussed by Dogiel *et al.* (1986).

It seems plausible that it is the magnetohydrodynamic turbulence in dense molecular clouds that is a progenitor of the rotation of stars and planetary systems (Ruzmaikina, 1986, 1987), although one more source of protostellar angular momentum is also traditionally considered, the overall galactic rotation (see, e.g., Hopper and Disney, 1974).

A curious and perplexing property of planetary systems, exemplified by the Solar system, is an extremely inhomogeneous distribution of the angular momentum of rotation per unit mass. The specific angular momentum of the Sun, $\simeq 10^{15}$ cm^2 s^{-1}, is 10^5 times smaller than the angular momentum associated with the orbital motion of the planets; meanwhile, the total mass of the planetary system is some 700 times smaller than the solar mass. Such inhomogeneous distribution of the mass and angular momentum is energetically preferable (Lynden-Bell and Pringle, 1976) and what remains to be understood is a sufficiently efficient mechanism of braking of a protostar, of outward transport of its angular momentum to the surrounding protoplanetary nebula. These mechanisms are the magnetic braking and the turbulent transfer of the angular momentum. Magnetic stresses can rapidly transfer the angular momentum of a collapsing protostar into the protoplanetary nebula thereby allowing further collapse (see Mestel and Paris, 1979, 1984; Mouschovias, 1978, 1985; Ruzmaikina, 1983, 1985; Mestel, 1985). The mechanism of magnetic braking can be described in simple terms as follows. A cloud rotating relative to the ambient interstellar gas with angular velocity $\mathbf{\Omega}$ is connected with the latter by the galactic magnetic field that here can be considered frozen both into the cloud gas and ambient medium. The azimuthal component of the magnetic field evolves according to the induction equation:

$$\frac{\partial H_\varphi}{\partial t} = r(\mathbf{H} \cdot \nabla)\Omega,$$

where r is the distance to the rotation axis and the cylindrical coordinates with z-axis parallel to $\mathbf{\Omega}$ is used. This component of the magentic field takes part in the redistribution of the specific angular momentum of the cloud, as described by the azimuthal component of the Navier-Stokes equation:

$$\rho \frac{\partial(\Omega r^2)}{\partial t} = \mathbf{H} \cdot \nabla(rH_\varphi/4\pi).$$

Combining these two equations we obtain the following wave type equation for the angular velocity:

$$4\pi\rho \frac{\partial^2\Omega}{\partial t^2} = r^{-2}\mathbf{H} \cdot \nabla(r^2\mathbf{H} \cdot \nabla\Omega),$$

which takes very simple form when the external field is parallel to the rotation axis:

$$\frac{\partial^2\Omega}{\partial t^2} = V_A^2 \frac{\partial^2\Omega}{\partial z^2},$$

where $V_A = H(4\pi\rho)^{-1/2}$. Thus, the angular momentum is transported from the cloud by torsional Alfvén waves. This process is effective if its characteristic time does not exceed the free fall time. By the order of magnitude, this inequality holds when the cloud mass is below $2\sqrt{3}\Phi/(3\pi\sqrt{G})$, where Φ is the magnetic flux through the cloud and G is the gravitation constant (see Mestel, 1985). For instance, magnetic fields produced the observed distribution of the angular momentum in the Solar system (Ruzmaikina, 1985). Note that the mean-field turbulent hydromagnetic dynamo possibly was active in the inner parts of the presolar nebula (Safronov and Ruzmaikina, 1985).

Imprints of strong magnetic fields of the presolar nebula can be found even in the present-day Solar system. A strong remanent magnetization was found in six meteorites of the primitive origin, carbonaceous chondrites. This fact can be interpreted as indication of magnetic field strength in the presolar nebula (Levi and Sonett, 1978). Remanent magnetic field strength in these meteorites is of order 1 G, and this very high strength is attributed to the magnetic field of the nebula. Such strong fields can be explained by compression of the interstellar magnetic field in the course of the nebula formation, with major portion of the magnetic flux lost due to ambipolar diffusion.

In recent years, new ideas have been developed concerning global aspects of star formation process. According to the point of view which has become classical in the seventies, star formation in spiral galaxies is triggered by passage of spiral shock waves that compress interstellar gas and thereby promote star formation within spiral arms (see Kaplan and Pikelner, 1974; Section 10.VII. below). Bright, rapidly evolving massive stars outline prominent, regular spiral arms that reproduce, in an amplified form, a spiral perturbation of the gravitational field produced by the density wave. Along with this regular way, star formation might proceed in a less ordered mode. Compression of interstellar gas, that triggers star formation, also occurs in the supernova shells that are more or less randomly distributed in space and time within a galaxy. In addition, extremely large number of factors that influence the star formation rate results in probabilistic character of star formation when only few basic parameters of interstellar environment can be actually specified. These ideas have been developed into the theory of stochastic self-propagating star formation (Elmegreen and Lada, 1977; Gerola and Seiden, 1978; Seiden and Gerola, 1982; Schulman and Seiden, 1986) according to which bright massive stars are being formed preferentially and with some finite probability within shells compressed by supernova explosions and stellar winds. Rapid evolution of massive young stars results in explosions of a new generation of supernovae that are distributed within the shells and the star formation propagates randomly in space. Differential rotation of the host galaxy stretches the regions populated by newborn bright stars into spiral segments. A combinatior of such segments can produce a picture that resembles spiral patterns observed in galaxies.

Although the scenario of stochastic star formation is very attractive, it seems improbable that it can explain all morphological properties of spiral galaxies. It is possible, however, that regular (triggered by spiral shock waves) and stochastic star formation processes coexist in spiral galaxies: the former determines a global structure of the spiral pattern while the latter produces its smaller scale peculiarities, branchings, breaks, etc. Stochastic star formation may dominate in irregular galaxies and in some spiral galaxies with patchy spiral patterns (e.g., M33).

Correspondence between large-scale galactic magnetic fields and galactic spiral patterns is an intriguing and intricate problem. Magnetic lines of the large-scale fields are spirals with typical inclination to the circumferential direction of 10°–20° (Baryshnikova *et al.*, 1987; see Chapter VII below). It is essential that these fields can be excited even in an *axisymmetric* gaseous disc, i.e. they may arise *before* the spiral pattern. Moreover, they are able to provoke formation of the spiral pattern even without density waves (or provoke excitation of the latter). First, the interstellar gas condences in spiral regions of higher or lower strength of magnetic field in non-axisymmetric magnetic structures (Fujimoto and Tosa, 1980). This may lead to formation of the two-armed spiral pattern in galaxies with bisymmetric magnetic fields. Note that in disc galaxies more complicated magnetic structures can be generated, e.g. the quadrisymmetric structures whose aximuthal wave number is $m = 2$ (Starchenko and Shukurov, 1987). These structures would induce four-armed spiral patterns. Since magnetic modes with higher azimuthal wave numbers are localized at greater galactocentric distances, thereby multi-level spiral patterns may also arise, with number of arms growing in the outer parts of galaxies. Thus, large-scale galactic magnetic fields may serve as a factor that maintains spiral density waves. Of course, these ideas deserve quantitative consideration. Second, large-scale magnetic lines (no matter, whether the field is axisymmetric or not in this case) can serve as guiding lines for the self-propagating star formation (Feitzinger and Spicker, 1987) and thereby regulate it producing more ordered and persistent spiral pattern.

A possible role of magnetic fields in formation of the spiral structure of galaxies was a subject of active research in 1950-ies before development of the density wave theory (see, e.g., Pikelner, 1966, Sect. 17). Modern theory of galactic magnetic fields calls for modern approaches to this problem.

CHAPTER VII

THE GALACTIC DYNAMO

VII.1. Introduction

In Section V.5 we have shown, at a qualitative level, how differential rotation and helical turbulence together generate the magnetic field. Now we intend to give a more substantiated, detailed picture of the generation process. A reasonable approximation to the galactic dynamo provides the problem of magnetic field growth in a prescribed flow of conductive fluid (the kinematic dynamo). First, this is an unescapable step in the study of the magnetic field origin. Second, the characteristic growth time of the field is only an order of magnitude shorter than the galactic lifetime. Therefore, it should be expected that the generation proceeds currently in the linear or in a weakly non-linear regime which inherits typical properties from the linear stage.

The theory of the kinematic dynamo is thoroughly developed from the fundamental point of view and has numerous applications to magnetic fields in the Milky Way and other spiral galaxies, in the Sun and stars, in the Earth and planets, and in the fast breeder nuclear reactors and tokamaks (Moffatt, 1978; Parker, 1979; Vainshtein *et al.*, 1980; Krause and Rädler, 1980; Zeldovich *et al.*, 1983; Kirko, 1985; Strauss, 1985). Applications still deal primarily with models of the mean, large-scale magnetic field generation. However, recent theoretical results shed light on fluctuative, small-scale fields and substantiate the possibility of the growth of an unaveraged, random magnetic field in a turbulent flow (Molchanov *et al.*, 1985). At high magnetic Reynolds numbers the generated magnetic field has an intermittent structure: the average magnetic energy is stored predominantly in sharp magnetic peaks which stand out against the background of relatively weak and uniform chaotic fields.

The growth and distribution of the mean magnetic field in a galactic disc are discussed in detail in §§3—9. Section 10 is devoted to the generation of mean magnetic fields in rigidly rotating objects. The role of spiral shock waves is discussed in §11. Sections 12 and 13 consider non-linear effects and the spectrum of small-scale fields. This chapter concludes with discussion of the seed magnetic fields in spiral galaxies.

The idea of the galactic dynamo was first considered independently by Parker (1971) and Vainshtein and Ruzmaikin (1971). Further developments are due to Stix (1975), Soward (1978), White (1978), Ruzmaikin *et al.* (1979, 1980b) and

Isakov *et al.* (1981). Salient features of two- and three-dimensional distributions of magnetic fields in spiral galaxies were first considered by Ruzmaikin and Shukurov (1981) and studied in detail by Ruzmaikin *et al.* (1985; 1986), Baryshnikova *et al.* (1987a, b), Sawa and Fujimoto (1986), Fujimoto and Sawa (1987), and Ruzmaikin (1987). Spectrum of the interstellar magnetic fields is discussed by McIvor (1977) and Ruzmaikin and Shukurov (1982), while Vainshtein and Ruzmaikin (1972) consider non-linear effects in the galactic dynamo. The inter-relation between galactic magnetic fields and the spiral pattern is discussed by Roberts and Yuan (1970) and Ruzmaikin (1976), the content of the latter paper being presented in §11.

VII.2. The Mean Magnetic Field

Magnetic field embedded in a turbulent conductive medium is random. Magnetic lines cannot be confined to harmonious, regular configurations, they are strongly tangled, especially at small scales. It is then quite natural to divide the total field into the large-scale, smooth component **B** and the small-scale, tangled field **b** and consider them separately. The following representation is widely used:

$$\mathbf{H} = \mathbf{B} + \mathbf{b}. \tag{VII.1}$$

Generally speaking, this equality has a purely symbolic meaning. In order to make it a working concept, one must define in which sense the average value is understood. Usually, **B** is considered as the large-scale component, i.e. the spatial average, or equivalently, a combination of modes with small wave numbers k, if we use the language of Fourier analysis. This approach is the basis of the scheme known as the two-scale method in electrodynamics of turbulent media (Moffatt, 1978; Krause and Rädler, 1980). Two problems are immediately evident. First, at what wave number k_0 the distinction is found between the long-wavelength part of the magnetic field spectrum and the short-wavelength part? The natural boundary is provided by the maximal scale of the turbulent motions. Assuming that the principal, or at least one of the most important sources of the interstellar turbulence are supernova explosions, we can identify $l_0 \equiv 2\pi/k_0$ with the typical size of supernova remnants, 50—150 pc (see Section VI.3). We should note that this scale is only a few times smaller than the half-thickness of the ionized Galactic disc, 400 pc (see Section VI.2), in which the large-scale field is generated. Therefore, the 'mean' magnetic field obtained from observations at high Galactic latitudes can be only a very crude approximation to the ensemble-averaged field.

Another problem is associated with intermittency in the spatial distribution of the field **H**, i.e. with the presence of concentrated random structures that may introduce typical scales exceeding l_0, for instance the mean distance between magnetic concentrations and the typical length for elongated configurations (ropes). Taking adequate account for the contribution of elongated structures into the mean field requires some additional procedures, e.g. averaging over their random orientations.

An obvious advantage of the spatial average is the possibility of direct comparison with observational data. Because of this, volume averages are widely used in Chapter IV where we discuss the observational data. From the formal point of view, however, of more relevance is the magnetic field averaged over an ensemble of turbulent pulsations, i.e. over all possible realizations of the velocity field.* In this case, both abovementioned problems are avoided. In order to compare theoretical results with observational data, it is sufficient to relate the ensemble and spatial averages to each other, although is may be not a simple task in practice.

We derive the equation governing evolution of the (ensemble) averaged magnetic field in Section 4. This equation follows from the general equation for magnetic field evolution in a moving medium, to which we turn now.

VII.3. Evolution of Magnetic Field in a Moving Medium

Hydrodynamic motions of interstellar gas in galaxies are non-relativistic, $v \ll c$ (see Chapter VI), therefore the displacement current may be neglected in the Maxwell equations. The motion of conductive medium appears exclusively in the Ohm law $\mathbf{j} = \sigma(\mathbf{E} + \mathbf{c}^{-1}\mathbf{v} \times \mathbf{H})$. With the help of the Ampère law $\mathbf{j} = (c/4\pi)$ rot $\mathbf{H}$ and using the solenoidality of magnetic field div $\mathbf{H} = 0$, we obtain from the Maxwell equation $\partial\mathbf{H}/\partial t = -c$ rot $\mathbf{E}$ the closed equation for a magnetic field,

$$\frac{\partial \mathbf{H}}{\partial t} = \text{rot}\, \mathbf{v} \times \mathbf{H} + \nu_m \Delta \mathbf{H}, \tag{VII.2}$$

where $\nu_m = c^2/4\pi\sigma$ is the magnetic diffusivity and σ is the electric conductivity of the medium. Equation (VII.2) is known as the *induction equation.*

A solution to the Equation (VII.2), which is linear in magnetic field, can be obtained in a general form for arbitrary velocity field (Molchanov *et al.*, 1985). To clarify the idea, let us begin with a simple and useful case of perfectly conducting ($\nu_m = 0$), incompressible (div $\mathbf{v} = 0$) fluid. In this case the magnetic field evolution is described through the well known concept of frozenness.

* Formally, the use of time-averages is also admissible. However, these possess neither "observability" of spatial averages (characteristic times of the interstellar turbulence far exceed the observer's lifetime), nor the rigour of the statistical averages.

In the perfectly conducting incompressible fluid, the induction equation

$$\frac{\mathrm{d}H_i}{\mathrm{d}t} = H_k \nabla_k v_i, \qquad \frac{\mathrm{d}}{\mathrm{d}t} = \frac{\partial}{\partial t} + v_k \nabla_k, \qquad i, k = 1, 2, 3$$

is equivalent to the field transfer along the Lagrangian path of a fluid element. Indeed, let us solve this equation by successive approximations (Parker, 1979). Let $\boldsymbol{\xi} = \mathbf{x}|_{t=0}$ be the initial position of a fluid element. After the interval Δt we have

$$H_i(\boldsymbol{\xi} + \Delta\mathbf{x}, \Delta t) = H_i(\boldsymbol{\xi}) + \frac{\mathrm{d}H_i}{\mathrm{d}t} \Delta t$$

$$= \frac{\partial(\xi_i + v_i(\boldsymbol{\xi})\,\Delta t)}{\partial \xi_j} H_j(\boldsymbol{\xi})$$

$$= \frac{\partial x_i(\boldsymbol{\xi}, \Delta t)}{\partial \xi_j} H_j(\boldsymbol{\xi}).$$

Applying this procedure consequently, we obtain

$$H_i(\mathbf{x}, t) = H_i(\mathbf{x}(\boldsymbol{\xi}, n\Delta t), n\Delta t)$$

$$= \frac{\partial x_i(\boldsymbol{\xi}, n\Delta t)}{\partial x_k(\boldsymbol{\xi}, (n-1)\Delta t)} \cdots \frac{\partial x_l(\boldsymbol{\xi}, 2\Delta t)}{\partial x_m(\boldsymbol{\xi}, \Delta t)} \frac{\partial x_m(\boldsymbol{\xi}, \Delta t)}{\partial \xi_j} H_j(\boldsymbol{\xi}) \tag{VII.3}$$

$$= \frac{\partial x_i(\boldsymbol{\xi}, t)}{\partial \xi_j} H_j(\boldsymbol{\xi}).$$

Therefore, in order to follow the evolution of a frozen-in magnetic field, one needs to know only the fluid element's trajectory

$$x_i(\boldsymbol{\xi}, t) = \xi_i + \int_0^t v_i(\mathbf{x}, s)\,\mathrm{d}s,$$

which is completely described by the velocity field. We have presented the trajectory in the form $\mathbf{x} = \mathbf{x}(\boldsymbol{\xi}, t)$. This expression can be inverted in time to give the Lagrangian (initial) coordinates in terms of the Eulerian ones:

$$\boldsymbol{\xi}_t = \mathbf{x} - \int_0^t \mathbf{v}(\boldsymbol{\xi}, t - s)\,\mathrm{d}s \tag{VII.4}$$

The second line from (VII.3) implies that the magnetic field transfer along the trajectory over a finite time interval can be represented as a product of transportations over infinitesimal pieces of the trajectory. In the limit $\Delta t \to 0$ we obtain an infinite product of matrices that are close to the unit matrix (mathematicians call this product the multiplicative integral):

$$G_{ij}(t, \boldsymbol{\xi}, \mathbf{x}) \equiv \prod_{s=0}^{t} \left(\delta_{ij} - \frac{\partial v_i(\boldsymbol{\xi}, t-s)}{\partial x_j} \, ds \right)$$

$$= \prod_{s=0}^{t} \left(\delta_{ij} + \frac{\partial v_i(\mathbf{x}, s)}{\partial x_j} \, ds \right) \tag{VII.5}$$

$$= \frac{\partial x_i}{\partial \xi_j},$$

which directly shows how the transfer function (the Green function) is connected with the velocity field.

The form

$$H_i(\mathbf{x}, t) = G_{ij} H_{0j}(\boldsymbol{\xi}) \tag{VII.6}$$

of the solution is preserved in a compressible flow as well but the Green function is no longer given by (VII.5). The differential equation for G_{ij} can be obtained by substitution of (VII.6) into equation (VII.2) and for $\nu_m = 0$ it is given by

$$\frac{d}{ds} G_{ij}(t-s, \mathbf{x}, \boldsymbol{\xi}_s) = -G_{lj} \frac{\partial v_i(t-s, \boldsymbol{\xi}_s)}{\partial x_l} + G_{ij} \frac{\partial v_l(t-s, \boldsymbol{\xi}_s)}{\partial x_l}. \tag{VII.7}$$

The solution to this equation also can be obtained as an infinite product convenient for application of successive approximations in the derivation of the magnetic field.

The remarkable property of the Lagrangian solution (VII.6) is the fact that it can be generalized to the case when magnetic diffusion affects the field transfer, that is when the field is described by the general induction equation (VII.2). It is known from the theory of Brownian motion that diffusion can be described as a result of averaging over motions along a bundle of random trajectories (see, e.g., McKean, 1969). The positions along a random trajectory $\mathbf{W}_t$ are, at any moment, Gaussian random variables with the root mean square deviations proportional to $t^{1/2}$ (for $\mathbf{W}_{t=0} = 0$) and with independent increments in time. Note that the Gaussian statistics is required only for the random walk, while the velocity field may have arbitrary statistical properties and even be regular. Thereby the random trajectories, that generalize the form (VII.4) by addition of random walk, can be defined through

$$\boldsymbol{\xi}_t = \mathbf{x} - \int_0^t \mathbf{v}(t-s, \boldsymbol{\xi}_s) \, ds + (2\nu_m)^{1/2} \mathbf{W}_t, \tag{VII.8}$$

where $\mathbf{W}_t$ is the stochastic (Wiener) process whose properties are

$$M\mathbf{W}_t = 0,$$
$$M(\mathbf{W}_{ti} \cdot \mathbf{W}_{tj}) = \delta_{ij} t. \qquad \text{(VII.9)}$$

The symbol M denotes an average over the distribution of $\mathbf{W}_t$. Equation (VII.8) describes a set of random trajectories which reach the point $\mathbf{x}$ under consideration at the moment t. We emphasize that, in contrast to $\mathbf{x}$, the Lagrangian coordinate $\boldsymbol{\xi}_t$ is a random variable. For this reason, (VII.8) does not allow an inversion $\mathbf{x} = \mathbf{x}(\boldsymbol{\xi}_t, t)$ similar to the Lagrangian one. Moreover, the trajectory (VII.8) is not differentiable since at $\Delta t \to 0$ the value of $|\Delta\boldsymbol{\xi}|$ is proportional to $(\Delta t)^{1/2}$ rather than to Δt. This situation is described by saying that the random trajectory has only one half derivative (a simple discussion of these concepts can be found in Zeldovich and Sokoloff, 1985).

The analysis of magnetic field transfer along a single random trajectory (VII.8) (defined by a certain realization of the stochastic process $\mathbf{W}_t$) is accomplished by the solution (VII.6) of the Equation (VII.2) without the Laplacian. The true solution is obtained by averaging over all random trajectories which intersect at the point $\mathbf{x}$ at the moment t:

$$H_i(\mathbf{x}, t) = M_x G_{ij}(t, \boldsymbol{\xi}_t, \mathbf{x}) H_{0j}(\boldsymbol{\xi}_t). \qquad \text{(VII.10)}$$

This is the general solution to the Cauchy problem for the induction equation (VII.2). Its validity can be verified by substituting it into the induction equation with help of (VII.7) and (VII.9).

The way of description of magnetic diffusion which we have discussed is mathematically rigorous (see Molchanov *et al.*, 1985 for details). It is similar to Feynman's method of path integration in quantum mechanics. Its validity and convenience in application to the hydromagnetic dynamo problem is due to the fact that it exploits the explicit Lagrangian solution to the induction equation. What remains to be done after the Lagrangian solution is obtained, is to average this solution over the set of random trajectories with known statistical weights. Of course, path integration generally also represents a complicated problem. Nevertheless, in this way one can derive, in a simple manner, equations for the mean field $\langle H_i \rangle$, the mean square $\langle H_i H_j \rangle$, and other higher moments of magnetic field $\langle H_i H_j \dots H_k \rangle$.

It is proper to note that the random trajectory (in contrast to the Lagrangian one) and the field transfer along it are fictitious. The physical meaning has only the field averaged over the whole set of trajectories. The averaging procedure makes the situation non-invariant with respect to the time inversion, as it should be when diffusion is present.

Averaging over fictitious random trajectories should not be mixed with physical averaging over a random velocity field. For example, the technique of random trajectories is applicable also to the case of molecular diffusion in a deterministic laminar flow.

VII.4. The Equation for the Mean Magnetic Field

The equation that describes evolution of the mean magnetic field in a turbulent flow was proposed by Steenbeck, Krause and Rädler (1966) and was further justified later by a number of authors who have adopted certain restrictive assumptions — either Gaussian statistics of the velocity field, or its two-scale structure, or low magnetic Reynolds numbers (see Moffatt, 1978; Parker, 1979; Krause and Rädler, 1980; Rädler, 1980). Existence of the growing solutions of this equation is shown to be associated with non-vanishing mean helicity of the flow.

Molchanov *et al.* (1983) have proposed a simple and rigorous derivation of governing equations for the mean field and higher field moments under the single assumption of short-correlated (δ-correlated in time) velocity field. Their results seem to remain unchanged when the correlation time of the velocity field is not very long, i.e. at worst of the order of turnover time of the turbulent eddy (Dittrich *et al.*, 1984).

The velocity field which we consider can be conveniently thought of as the limiting case $\Delta t \to 0$ of velocity fields $\mathbf{v}^{\Delta}(t, \mathbf{x})$ that are time-independent over periods $\Delta t : (0, \Delta t)$, $(\Delta t, 2\Delta t)$, . . . and are statistically independent at each time interval. In the limit $\Delta t \to 0$, this means that

$$\langle v_i(t, \mathbf{x}) v_k(t', \mathbf{y}) \rangle = 2 \, \frac{l_0}{v_0} \, \delta(t - t') \, V_{ik}(\mathbf{x}, \mathbf{y}),$$

were l_0 and v_0 are the characteristic scale and velocity, respectively, V_{ik} is the correlation tensor, the angular brackets denote averaging over the velocity field distribution, and δ denotes the Dirac's delta function. Thus, at small Δt the velocity $\mathbf{v}^{\Delta}$ has the magnitude of order $v_0(l_0/v_0 \, \Delta t)^{1/2} \sim (\Delta t)^{-1/2}$ (Molchanov *et al.*, 1985). For the sake of simplicity, we consider first the case when the average velocity is null, $\langle v_i(t, \mathbf{x}) \rangle = 0$, and omit the dimensional factor $2l_0/v_0$ (which means that the time is measured in units $2l_0/v_0$).

The idea behind the method of derivation of the mean-field equation is the following. Starting with a magnetic field at moment t we calculate the field after the short period Δt according to (VII.10):

$$H_i(t + \Delta t, \mathbf{x}) = M_x \{ G_{ij}(\Delta t, \mathbf{x}, \boldsymbol{\xi}_{\Delta t}) H_j(t, \boldsymbol{\xi}_{\Delta t}) \}. \qquad \text{(VII.11)}$$

Our aim is to average this relation over an ensemble of turbulent velocities. When averaging, we may take advantage of the smallness of Δt on the right-hand side of (VII.11). However, the direct expansion into Taylor series is hampered by the fact that the magnetic field $H_j(t, \boldsymbol{\xi}_{\Delta t})$ at the moment t is already tangled and rugged. Therefore, we apply the two-step procedure of averaging over the velocity field described below. At the first stage we average for the moments before the given moment t to obtain a smooth field at the moment t and only then we proceed to averaging the random process over the period from t to $t + \Delta t$ in order to obtain the solution, i.e. the average field at $t + \Delta t$ expressed in terms of the average field at t and the velocity field.

Average the expression (VII.11) over the velocity field with interchange in the sequence of two averages on the right-hand side of (VII.11), $\langle M\rangle \rightarrow M\langle\ldots\rangle$. As we described above, the correlator $\langle G_{ij}(\Delta t, \boldsymbol{\xi}_{\Delta t})H_j(t, \boldsymbol{\xi}_{\Delta t})\rangle$ can be evaluated by division of the velocity-averaging procedure into two stages. Begin with the averaging over the period from some initial moment to the moment t when solely the field H_j undergoes the averaging, that depends on the velocity field at moments earlier than t. The field thus averaged is a smooth function of position which can be expanded into a Taylor series keeping terms of the orders up to Δt, the term

$$\frac{1}{2}\,\frac{\partial^2\langle H_j\rangle}{\partial x_k\,\partial x_l}\,(\boldsymbol{\xi}_{\Delta t} - \mathbf{x})_k(\boldsymbol{\xi}_{\Delta t} - \mathbf{x})_l$$

being among them. With the same accuracy, expressions (VII.8) and (VII.7) yield (recall that $v^{\Delta} \sim (\Delta t)^{-1/2}$):

$$(\boldsymbol{\xi}_{\Delta t} - \mathbf{x})_k = \sqrt{2\nu_m}\,\mathbf{W}_k + v_k(t, \mathbf{x})\,\Delta t + \\ + \sqrt{2\nu_m}\,\frac{\partial v_k}{\partial x_l}\int_0^{\Delta t}\mathbf{W}_l\,\mathrm{d}t + \tfrac{1}{2}\,v_l\,\frac{\partial v_k}{\partial x_l}\,(\Delta t)^2,$$

$$\begin{aligned} G_{ij}(\Delta t, \boldsymbol{\xi}_{\Delta t}) = \delta_{ij} - \frac{\partial v_i(t,\mathbf{x})}{\partial x_j}\,\Delta t - \\ - \sqrt{2\nu_m}\,\frac{\partial^2 v_i}{\partial x_j\,\partial x_l}\int_0^{\Delta t}\mathbf{W}_l\,\mathrm{d}t + \\ + \frac{(\Delta t)^2}{2}\,\frac{\partial}{\partial x_k}\left(v_i\,\frac{\partial v_k}{\partial x_j} - v_k\,\frac{\partial v_i}{\partial x_j}\right) + \\ + \Delta t\,\delta_{ij}\,\frac{\partial v_l}{\partial x_l} + \\ + \delta_{ij}\,\frac{\partial^2 v_l}{\partial x_l\,\partial x_p}\,\sqrt{2\nu_m}\int_0^{\Delta t}\mathbf{W}_p\,\mathrm{d}t + \delta_{ij}\,\frac{\partial^2 v_l}{\partial x_l x_p}\,v_p\,\Delta t - \\ - 2\,\frac{\partial v_i}{\partial x_j}\,\mathrm{div}\,\mathbf{v}(\Delta t)^2 - \delta_{ij}(\mathrm{div}\,\mathbf{v})^2(\Delta t)^2. \end{aligned} \qquad \text{(VII.12)}$$

Now we perform the averaging of $G_{ij}(\Delta t, \xi_{\Delta t}) \langle H_j \rangle$ over the velocity field over the period from t to $t + \Delta t$. The correlations of the type

$$\left\langle v_k \frac{\partial v_i}{\partial x_j} \cdot \frac{\partial \langle H_j \rangle}{\partial x_k} \right\rangle$$

split into

$$\left\langle v_k \frac{\partial v_i}{\partial x_j} \right\rangle \frac{\partial \langle H_j \rangle}{\partial x_k}$$

due to the fact that the random field v_i is short-correlated. Finally, averaging over the Wiener random trajectories gives for the total mean field $B_i(t, \mathbf{x})$

$$\begin{aligned} \frac{\partial B_i(t, \mathbf{x})}{\partial t} = & - \left\langle v_k \frac{\partial v_i}{\partial x_j} \right\rangle \frac{\partial B_j}{\partial x_k} + \tfrac{1}{2} \frac{\partial}{\partial x_k} \left\langle v_i \frac{\partial v_k}{\partial x_j} - v_k \frac{\partial v_i}{\partial x_j} \right\rangle B_j + \\ & + \frac{\partial}{\partial x_j} \left[(\nu_m \delta_{kj} + \tfrac{1}{2} \langle v_k v_j \rangle) \frac{\partial B_i}{\partial x_k} \right] + \\ & + \frac{\partial B_i}{\partial x_k} \langle v_k \operatorname{div} \mathbf{v} \rangle + \left\langle v_p \frac{\partial v_l}{\partial x_l \partial x_p} \right\rangle B_i - \\ & -2 \left\langle \frac{\partial v_i}{\partial x_j} \operatorname{div} \mathbf{v} \right\rangle B_j + \langle (\operatorname{div} \mathbf{v})^2 \rangle B_i \end{aligned} \qquad \text{(VII.13)}$$

where the properties (VII.9) have been used and the limit $\Delta t \to 0$ has been taken. Here the symbol $\langle \ldots \rangle$ includes only the spatial part of the two-point velocity-field correlator taken at coincident points. Return to the dimensional units is accomplished through adding the factor $2l_0/v_0$ to the right-hand side of Equation (VII.13).

It is interesting to note that the coupling of different field components does not occur in the diffusion term of Equation (VII.13) in the considered approximation. Such coupling occurs, however, when finiteness of the correlation time is taken into account.

Equation (VII.13) is valid in the general, inhomogeneous and anisotropic case. Only the assumption of short-correlated velocity field is essential. When the turbulence is locally homogeneous,

$$\frac{\partial}{\partial x_l} \langle v_i v_k \rangle = 0,$$

Equation (VII.13) simplifies to

$$\frac{\partial B_i}{\partial t} = \left\langle v_i \frac{\partial v_k}{\partial x_l} \right\rangle \frac{\partial B_l}{\partial x_k} + (\nu_m \delta_{kl} + \tfrac{1}{2}\langle v_k v_l \rangle) \frac{\partial^2 B_i}{\partial x_k \partial x_l}. \qquad \text{(VII.14)}$$

The uniform tensor of the third rank that appears with the first derivative of the magnetic field is antisymmetric with respect to i and k. Hence, it can be represented as

$$\left\langle v_i \frac{\partial v_k}{\partial x_l} \right\rangle = e_{ikm} \alpha_{lm} - \delta_{il} q_k + \delta_{kl} q_i,$$

where e_{ikm} is the totally antisymmetric unit tensor, δ_{il} is the Kronecker delta, and α_{lm} and q_i are some tensor and vector. Multiplying this representation by e_{ikj} and taking advantage of the flow homogeneity, it can be verified that $\mathbf{q} = 0$ in incompressible flow, while α_{lm} is a symmetric tensor given by

$$\alpha_{ik} = \tfrac{1}{2}\langle v_i \omega_k + v_k \omega_i - \delta_{ik} v_l \omega_l \rangle,$$
$$\boldsymbol{\omega} = \text{rot}\,\mathbf{v},$$

whose trace,

$$\alpha_{ii} = -\tfrac{1}{2}\langle \mathbf{v}\,\text{rot}\,\mathbf{v} \rangle,$$

is the mean helicity of the flow. Owing to the symmetry, the tensor α_{ik} can always be transformed into the diagonal form. We should note that generation of the mean field is possible even for vanishing mean helicity $\langle \mathbf{v} \cdot \text{rot}\,\mathbf{v} \rangle$ when, however, the tensor $\langle v_i \omega_k \rangle$ is not identical to zero (Krause and Rädler, 1980; Molchanov *et al.*, 1983).

The polar vector q_i is not zero in a fluid with weakly inhomogeneous density, when the representation (VII.14) still can be applied. The corresponding term in the mean field equation has the form rot $(\mathbf{q} \times \mathbf{B})$ and describes the field transfer at the effective velocity $\mathbf{q}$ (diamagnetism of the turbulence: Zeldovich, 1956; Krause and Rädler, 1980; Vainshein *et al.*, 1980).

In astrophysical applications, the isotropic, weakly inhomogeneous approximation is usually applied, in which the equation for the mean field, when supplemented with the mean velocity $\mathbf{V}$, takes the form

$$\frac{\partial \mathbf{B}}{\partial t} = \text{rot}(\mathbf{V} \times \mathbf{B}) + \text{rot}\,\alpha \mathbf{B} - \text{rot}\,\beta\,\text{rot}\,\frac{\mathbf{B}}{\mu}, \qquad \text{(VII.15)}$$

where α is the mean helicity of the flow; by definition,

$$\alpha_{ij} = \delta_{ij} \alpha \qquad \text{or} \qquad \alpha = -\frac{\tau}{3}\langle \mathbf{v} \cdot \text{rot}\,\mathbf{v} \rangle$$

in dimensional units. The coefficient β includes both molecular and turbulent magnetic diffusivities, and μ is the turbulent magnetic permeability, $\beta\nabla\mu = \mathbf{q}$. In contrast to Equation (VII.14), the last equation is valid for a locally isotropic velocity field.

Before proceeding to analysis of Equation (VII.15) in application to spiral galaxies, let us recall the typical values of parameters that appear in (VII.15). First, the linear velocity of rotation $V = \Omega r$ is typically 200 km s^{-1}, which gives $\Omega \simeq 10^{-15}$ s^{-1} for the radius $r = 10$ kpc. The estimate of the turbulent magnetic diffusivity $\beta \simeq \frac{1}{3} l_0 v_0$ gives the value $\beta \simeq 10^{26}$ cm^2 s^{-1} for the correlation scale $l_0 \simeq 100$ pc and the velocity $v_0 \simeq 10$ km s^{-1} of interstellar turbulence. Finally, the estimate $\alpha \simeq l_0^2\Omega/h$ gives $\alpha \simeq 1$ km s^{-1}. The generation region in spiral galaxies is essentially confined to the ionized gaseous disc whose half-thickness is at least $h \simeq 400$ pc and the radius is 10—20 kpc. All relevant parameters of the gaseous discs of spiral galaxies were discussed in length in Chapter VI.

VII.5. Field Distribution Across the Disc

When solving the mean-field equation, to the first approximation the gaseous galactic disc may be considered as a thin turbulized slab of thickness $2h$ which rotates with angular velocity $\Omega(r)$. Let us neglect the effects of compressibility, diamagnetism and deviations from the axial symmetry; let us put $\mu = 1$ and $\beta = \text{const}$. With these simplifications, the equation for the mean *axisymmetric* magnetic field has the following components in polar cylindrical coordinates (r, φ, z) with the z-axis coincident with the rotation axis:

$$\frac{\partial B_r}{\partial t} = -\frac{\partial}{\partial z}(\alpha B_\varphi) + \beta(\Delta \mathbf{B})_r,$$

$$\frac{\partial B_\varphi}{\partial t} = G B_r + \beta(\Delta \mathbf{B})_\varphi, \qquad \text{(VII.16)}$$

$$\frac{\partial B_z}{\partial t} = \frac{1}{r}\frac{\partial}{\partial r}(r\alpha B_\varphi) + \beta(\Delta \mathbf{B})_z,$$

where B_r, B_φ and B_z are the components of the mean magnetic field, and $G \equiv r\,\mathrm{d}\Omega/\mathrm{d}r$ is the measure of differential rotation. In the second equation we have omitted the term $-(\partial/\partial z)(\alpha B_r)$. Such an omission is justified when $G \gg \alpha/h$ which inequality is valid in the main portions of galactic discs (see Chapter VI).

Axial symmetry, $\partial/\partial\varphi = 0$, allows to describe both field components B_r and B_z in terms of a single scalar function, the azimuthal component of the vector potential A_φ(rot $\mathbf{A} = \mathbf{B}$):

$$B_r = -\frac{\partial A_\varphi}{\partial z}, \qquad B_z = \frac{1}{r}\frac{\partial}{\partial r}(rA_\varphi),$$

This representation guarantees that div $\mathbf{B} = 0$.

Below we drop the subscript φ at A_φ and B_φ: $B \equiv B_\varphi, A \equiv A_\varphi$.

The first step is to solve our problem at fixed radius when it is reduced to a one-dimensional eigenvalue problem. In this case $(\Delta\mathbf{B})_r \approx \partial^2 B_r/\partial z^2$, $(\Delta\mathbf{B})_\varphi \approx \partial^2 B/\partial z^2$, $(\Delta\mathbf{B})_z \approx \partial^2 B_z/\partial z^2$ and the third equation from (VII.16) becomes trivial. Hence, we can restrict our attention to the first two equations from (VII.16), or to the equivalent equations

$$\frac{\partial A}{\partial t} = \alpha B + \beta\frac{\partial^2 A}{\partial z^2}, \qquad \frac{\partial B}{\partial t} = -G\frac{\partial A}{\partial z} + \beta\frac{\partial^2 B}{\partial z^2}. \tag{VII.17}$$

We turn now to the boundary conditions for our problem. In applications to spiral galaxies, although conductivity may be rather high at $|z| > h$, force-free conditions seem to be applicable outside galactic gaseous discs, because the gas is rarefied there and the pressure in galactic coronae is probably lower than within the discs. Moreover, the large-scale field outside the disc probably should be potential since force-free fields with rot $\mathbf{B} \neq 0$ do not decay at infinity. Now the axial symmetry immediately implies that $B = 0$ for $|z| > h$, while the potential A obeys the following equation:

$$\frac{\partial^2 A}{\partial z^2} + \frac{\partial}{\partial r}\left(\frac{1}{r}\frac{\partial}{\partial r}rA\right) = 0,$$

whose solution is a linear combination of the functions of the form $\exp(ikz)J_1(kr)$, where k is determined from the boundary condition rot $\mathbf{B} = 0$ at $r = R$ as $k \sim R^{-1} \ll h^{-1}$ with R the radius of the galactic disc. Therefore, the boundary condition at the surface of a thin disc $|z| = h \ll R$ reads $\partial A/\partial z \simeq kA \approx 0$. In other words, we suppose that the solutions under consideration behave outside the disc as the axisymmetric, independent of r, magnetic field outside an infinite slab. Such outer solution is $B = 0$, $B_r = 0$. The field of the actual disc of course decreases at infinity and this is due to the finite dimensions of the disc; the situation is quite similar to the electrostatic field of a charged plate. For an infinite slab, $\partial A/\partial z = 0$. Hence, our first-order assumption is: the edge of the disc makes a negligible contribution to the generation of the mean magnetic field.

The boundary conditions at $|z| = h$ follow from continuity of the azimuthal and meridional fields:

$$B(\pm h) = 0$$

$$B_r = -\frac{\partial A}{\partial z}(\approx h) \approx 0.$$

We note again that the first condition exactly follows from the axial symmetry and potential structure of the outer field. The second boundary condition is approximate and applicable to a thin disc. These boundary conditions are often referred to as *vacuum boundary conditions.*

In the limiting case of a slab surrounded by the medium with infinite conductivity, when the field diffusion across the boundaries is prohibited, the exponential growth of the magnetic field with $\partial B_r/\partial z = \partial B/\partial z = 0$ at $|z| = h$ is impossible (Parker, 1971). We should realize that suppression of magnetic diffusion across the boundaries of the generation region hinders not the field generation, understood as the growth of magnetic energy, but impedes the development of a magnetic mode that can grow with preserved shape (eigenmode). It seems plausible that in a finite disc with such boundary conditions the role of the edge is always considerable and the adequate analysis should be based on two-dimensional equations. It is thus expected that eigenfunctions of the corresponding eigenvalue problem cannot be factorized, i.e. presented as a product of single-variable functions of z and r. Choudhuri (1984) finds the growing solutions for the dynamo with impenetrable boundaries and estimates the lower bound on the radial scale of the generated field.

Adopting the vacuum boundary conditions, let us find out axisymmetric solutions to Equations (VII.17) of the form

$$B(t, z) = B(z)e^{\gamma t},$$
$$B_r(t, z) = B_r(z)e^{\gamma t}, A(t, z) = A(z)e^{\gamma t}.$$

To simplify notations we keep in this section the same letters B, B_r and A for time independent quantities. It is convenient to introduce dimensionless variables $z \to z/h$, $t \to t/(h^2/\beta)$, $\gamma \to \gamma/(\beta/h^2)$ and $\alpha \to \alpha_0\alpha(z)$ and dimensionless amplitudes of the field and potential whose ratio is $A_0/B_0 = \alpha_0 h^2/\beta$. In this section we denote dimensionless quantities by the same symbols as corresponding dimensional ones. In dimensionless form, the generation equations are

$$\left(\gamma - \frac{\partial^2}{\partial z^2}\right) B_r = -\frac{\partial}{\partial z}(\alpha B),$$

$$\left(\gamma - \frac{\partial^2}{\partial z^2}\right) B = DB_r, \tag{VII.18}$$

$$B(\pm 1) = B_r(\pm 1) = -\frac{\partial A}{\partial z}(\pm 1) = 0.$$

Here we have introduced the dimensionless combination

$$D = \frac{G\alpha_0 h^3}{\beta^2},$$

which is called the dynamo number and which plays an essential role in the problem under consideration. The first equation from (VII.18) may be replaced by the equivalent one,

$$\left(\gamma - \frac{\partial^2}{\partial z^2}\right) A = \alpha B. \tag{VII.19}$$

Thus, the problem is reduced to a one-dimensional eigenvalue problem.

Let us now discuss the general properties of the generation equations (VII.18, VII.19).

(i) *Symmetry.* The equations possess an important symmetry: since $\alpha(z) = -\alpha(-z)$, the system (VII.18, VII.19) is invariant with respect to transformation $z \to -z$, provided

$$A(-z) = A(z), B_r(-z) = -B_r(z), B(-z) = -B(z)$$

or, otherwise,

$$A(-z) = -A(z), B_r(-z) = B_r(z), B(-z) = B(z).$$

Therefore, solutions to the generation equations can be divided into two classes conveniently distinguished according to the symmetry of $B(z)$: the odd solutions (dipolar in B_r, B_z) and the even ones, quadrupolar in B_r, B_z.

(ii) *Decay modes.* The generation equation can be easily solved in the absence of the sources, $\alpha = G = 0$. Growth rates of the odd modes can be shown to be

$$\gamma_k^D = -k^2\pi^2,$$

while for the even modes

$$\gamma_k^Q = -(k + \tfrac{1}{2})^2\pi^2,$$

where $k = 0, 1, 2, \ldots$. The corresponding eigenfunctions are doubly degenerate. For example, the following pair corresponds to γ_1^D:

$$A = \cos \pi z, \qquad B = 0$$

$$A = 0, \qquad B = \sin \pi z.$$

Notice that the dipole mode with $\gamma_k^D = 0$ is trivial, the corresponding eigenfunction is identical zero, $\mathbf{B} \equiv 0$ even for $\alpha \neq 0$ and $G \neq 0$. Hence, the lowest quadrupole mode decays four times as weakly as the lowest dipole one. This fact is closely associated with the property of the lowest quadrupole mode to be generated preferentially at a larger growth rate than the dipole ones. This property is of paramount importance in application to spiral galaxies. (Note also that the preference of even, quadrupole modes is a specific feature of the disc geometry; in spherical bodies the dipole mode is preferred and exemplified by magnetic fields of the Sun and the Earth.) Evidently, at small α and G, i.e. small dynamo numbers, the eigenfunctions and eigenvalues are close to the freely decaying solution. Numerical calculations confirm this. Moreover, in spiral galaxies only a single, the lowest even (quadrupole) mode can be generated.

(iii) *Properties of the growing even solution.* Integrating Equations (VII.16) over the interval $0 < z < 1$ for the case of a thin disc and for a smooth function $\alpha(z)$ gives:

$$\frac{\partial}{\partial t}\int_0^1 B_r(t, z)\,\mathrm{d}z = \beta\left.\frac{\partial B_r}{\partial z}\right|_0^1,$$

$$\frac{\partial}{\partial t}\int_0^1 B(t, z)\,\mathrm{d}z = G\int_0^1 B_r\,\mathrm{d}z + \beta\left.\frac{\partial B}{\partial z}\right|_0^1. \tag{VII.20}$$

Note that the source $\alpha(z)$ drops out from the integral equations because $\alpha(0) = 0$ and $B(1) = 0$! Consider the function $B(z)$ that has no zeros and is, say, positive $B > 0$ over the interval (0, 1). It is expected that such a function corresponds to the lowest excited mode. An important conclusion can be obtained from equations (VII.20) for the growing even (quadrupole) solution. In this case

$$\frac{\partial B_{\varphi, r}}{\partial z}(0) = 0 \qquad \text{and} \qquad \frac{\partial B_\varphi}{\partial z}(1) < 0$$

and, therefore, at $G < 0$ we obtain the following inequalities:

$$\int_0^1 B_r\,\mathrm{d}z < 0, \qquad \beta\frac{\partial B_r}{\partial z}(1) < 0. \tag{VII.21}$$

Hence, the radial component of the magnetic field changes the sign near the disc surface.

Integral forms (VII.20) of the generation equations also help to elucidate the role of the viscosity in the dynamo mechanism. It would seem at first glance that magnetic viscosity can be safely neglected for the growing solutions. Put, however, $\beta = 0$ in (VII.20). It can be seen immediately that $\int_0^1 B_r\,\mathrm{d}z = \text{const}$, while $\int_0^1 B_\varphi\,\mathrm{d}z$ can grow only linearly in time. In other words, a non-stationary dynamo is impossible when $\beta = 0$. As follows from the second condition from (VII.21), the dynamo action requires non-vanishing 'viscous flux' of the field across the boundary. The role of viscosity can be also seen directly from the generation equations. For definiteness, assume again $G < 0$ and $B > 0$. The source $-\partial(\alpha B)/\partial z$ generates the radial field whose sign is the opposite to that of the radial field B_r near the symmetry plane (where $B_r < 0$). The positive B_r near the surface gives, through differential rotation $-|G|B_r$, the negative contribution to $\partial B/\partial t$, which can be compensated only by the viscous term, $\beta\,\partial^2 B/\partial z^2$. In order to provide such compensation, the latter term must be positive near the disc surface, i.e. the field must be transported outwards. It also has become clear that, in addition to conditions (VII.21), the function $B(z)$ must have the inflection point situated at somewhat smaller z than the zero of B_r.

We must bear in mind, on the other hand, that the viscosity should not be excessively large, otherwise the field would rapidly decay within the main part of the disc and would be carried out toward the disc surface.

The discussion above referred to the growing solutions. In the stationary case ($\partial/\partial t = 0$) we have

$$\beta \frac{\partial B_r}{\partial z} = 0 \qquad \text{and} \qquad \beta \frac{\partial^2 B}{\partial z^2} = 0 \qquad \text{at} \qquad z = 1,$$

i.e. both components, B_r and B, are sign-fixed (and opposite in signs) over the disc thickness. Figures VII.1a, b, and c show the solutions that differ from the stationary one to various extents. These solutions are computed by Ruzmaikin *et al.* (1980b) for $\alpha = \sin \pi z$ and $D = -100$, -20, and -8, respectively. The latter value of D is the critical dynamo number for the field generation (i.e. corresponds to $\gamma = 0$) for the chosen dependence of α on z.

When $\alpha(z)$ is discontinuous, the second condition from (VII.20) of course remains valid, while the product of B and the algebraic sum of α jumps at discontinuity points should be added to the right-hand side of the first equation from (VII.20). For example, for $\alpha = \theta(z) - \theta(-z)$ (Parker, 1971), where $\theta(z)$ is the step function, we obtain

$$\frac{\partial}{\partial t} \int_0^1 B_r \, dz = -2B(0) + \beta \frac{\partial B_r}{\partial z}(1),$$

and the second condition from (VII.21) takes the form

$$\beta \frac{\partial B_r}{\partial z}(1) < 2B(0).$$

Solutions for discontinuous α may qualitatively differ from those for continuous distributions of the mean helicity, although the dynamo is also impossible for $\beta = 0$.

A similar integral relation can be obtained for a δ-function form of $\alpha(z)$. In this case additional terms include the values of derivative of B at the points where $\alpha(z)$ is applied.

(iv) *Large dynamo numbers* (*Isakov et al., 1981*). Numerical solution of the eigenvalue problem (VII.18) shows (Ruzmaikin *et al.*, 1980b) that eigenfunctions concentrate more and more in the vicinity of the symmetry plane $z = 0$ with growth of the magnitude of the dynamo number. The order of magnitude estimate

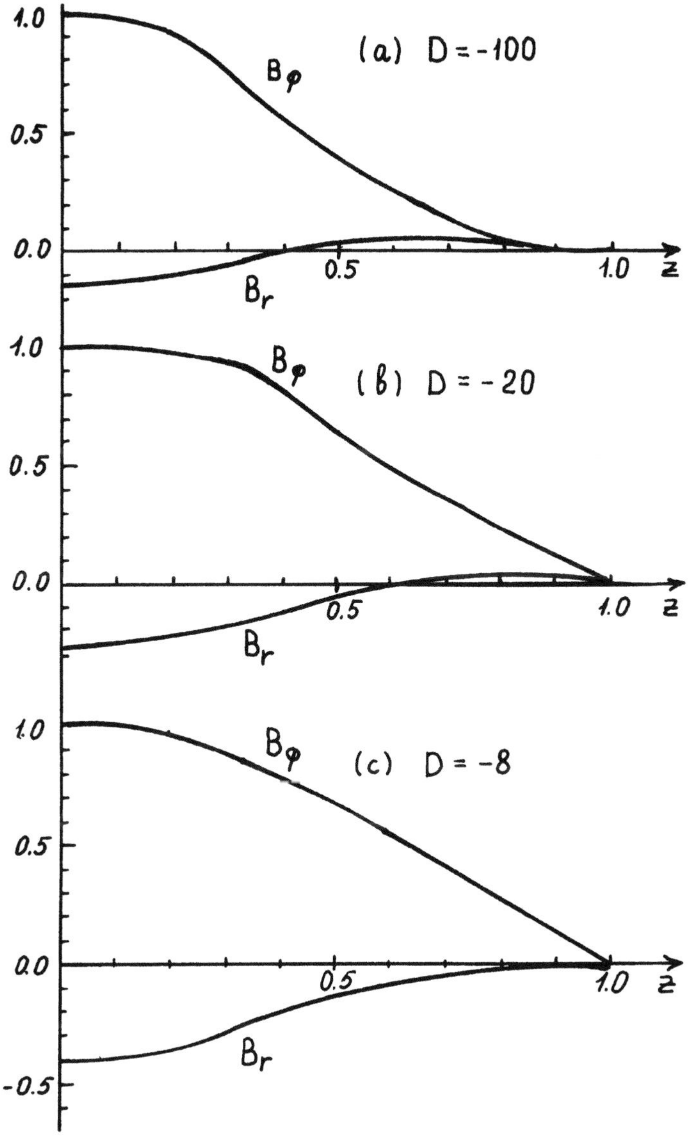

Fig. VII.1. The distribution of the even large-scale magnetic field across a thin disc. (a): $D = -100$; (b): $D = -20$; (c): $D = -8$. The last value of the dynamo number is the critical one for the adopted mean helicity $\alpha = \sin \pi z$.

in this vicinity gives from (VII.18)

$$B_r/B = O(|D|^{-1/2}), \qquad \gamma = O(|D|^{1/2})$$

and the condition $\gamma B = O(A)$ gives the spatial scale of the solution as $O(|D|^{-1/4})$.

Let us consider asymptotic forms of solutions for $|D| \gg 1$. These forms differ for even and odd solutions.

Discussing asymptotic solutions for $|D| \to \infty$, which are highly concentrated near $z = 0$, we can replace the boundary conditions (VII.18) by the simple requirement that eigenfunctions decay at infinity.

Consider first the even (quadrupole) solutions. Represent them as

$$\begin{pmatrix} B_r \\ B \end{pmatrix} = \begin{pmatrix} |D|^{-1/2}\mu \\ \nu \end{pmatrix} w(\xi), \qquad \gamma = \gamma_0 |D|^{1/2}, \tag{VII.22a}$$

with new variable

$$\xi = (z|D|^{1/4})^2 \tag{VII.22b}$$

introduced. The constants μ and ν are of order unity. At those positions where the generation proceeds most effectively, $z \simeq 0$, the mean helicity can be expanded as

$$\alpha = \alpha_1 z + \ldots \tag{VII.23}$$

and only the lowest-order term can be kept. As follows from physical considerations (see Section V.5), the mean helicity α is positive above the galactic disc and negative below. Thus, $\alpha_1 > 0$ with $\alpha_1 = \partial\alpha/\partial z(0)$.

Substituting (VII.22) and (VII.23) into (VII.18) we obtain a system of linear algebraic equations for μ and ν. This system has non-trivial solutions when $w(\xi)$ obeys the following equation:

$$(\gamma_0 w - w' - 2\xi w'')^2 \pm \alpha_1 w(w + 2\xi w') = 0, \tag{VII.24}$$

where the upper sign corresponds to $D > 0$ and the lower one corresponds to $D < 0$; and the prime denotes differentiating with respect to ξ. The boundary conditions

$$w(\infty) = 0, \qquad |w(0)| < \infty$$

supplement equation (VII.24).

The parameter α_1 can be removed from equation (VII.24) through the following transformation:

$$\xi = \tilde{\xi}\alpha_1^{-1/2}, \qquad \gamma_0 = \alpha_1^{1/2}\tilde{\gamma}_0.$$

Hence, we obtain the asymptotic form

$$\gamma_0 \simeq |D|^{1/2}[\alpha'(0)]^{1/2}\tilde{\gamma}_0, \qquad |D| \gg 1, \tag{VII.25}$$

where $\tilde{\gamma}_0$ is the constant that does *not* depend on either the form of $\alpha(z)$ or the dynamo number and whose value can be obtained from solutions of the non-linear equation (VII.24). However, its approximate value can be much easier derived from numerical solutions obtained for certain funciton $\alpha(z)$ whose specific form does not mean much. What matters, is the large magnitude of the dynamo number, $|D| \gg 1$. For example, Ruzmaikin *et al.* (1980b) use $\alpha = \sin \pi z$ and obtain $\gamma = 5.2$ for $D = -100$ for the even mode. This gives $\tilde{\gamma}_0 \approx 0.3$ for the negative D. For a positive value of the dynamo number, say $D = 300$, the same authors give $\gamma = 2.0 \pm 19i$. The corresponding estimate is $\tilde{\gamma}_0 \approx 0.07 \pm 0.6\,i$ for positive D.

Table 2 of Ruzmaikin *et al.* (1980b) for $D < 0$, which case is most interesting in applications to spiral galaxies, shows that the dependence of γ on D is indeed close to predictions (VII.25) for $|D| > 1$. The slope of this dependence allows us to obtain a somewhat better estimate of $\tilde{\gamma}_0 \approx 0.4$ for the case of negative dynamo numbers.

Dependence of the growth rate γ on the dynamo number for the quadrupole mode is shown in Fig. VII.2 for various distributions $\alpha(z)$. Dashed line shows the asymptotic dependence (VII.25).

Odd (dipole) solutions are obtained in similar fashion,

$$\begin{pmatrix} B_r \\ B \end{pmatrix} = z \begin{pmatrix} |D|^{1/2}\lambda \\ \eta \end{pmatrix} u, \qquad \text{(VII.26)}$$

where the factor z is singled out correspondingly to our previous knowledge of properties of the even solutions, $B(0) = B_r(0) = 0$ (see point (i) above). Again, the resolvability condition yields the equation for $u(\xi)$:

$$(\gamma_0 u - 3u' - 2\xi u'')^2 \pm 2\alpha_1 u(u + \xi u') = 0.$$

The eigenvalue preserves the functional form (VII.25) but with the other constant $\tilde{\gamma}_0$ estimated from the same numerical solution as $\tilde{\gamma}_0 \simeq 0.1 \pm 6\,i$ independently of the sign of D ($\gamma = 7.0 \pm 34\,i$ for $D = 10^3$).

We emphasize that the shape of solution and the growth rate γ depend asymptotically only on $\alpha'(0)$ and do not depend on other properties of $\alpha(z)$. In particular, the even (quadrupole) mode is always preferable, i.e. has the largest growth rate.

Note that in the case when $\alpha(z)$ has more zeros over the interval $0 < z < 1$, apart from that at $z = 0$, all the results obtained above are applicable in the vicinity of every zero point. It is interesting that the oddness of $\alpha(z)$ requires for the corresponding zero to be present in the interval $-1 < z < 0$. When quarupolar modes generated in the vicinity of two symmetric zero points are combined with proper signs (which is accomplished with the help of a proper configuration of the initial, seed field), the overall magnetic configuration could be dipolar (cf. Ruzmaikin *et al.*, 1979). There still remains some ambiguity with the observed configuration of the large-scale magnetic field in the Galaxy (see Chapter

IV). The most probable configuration has even parity. However, some authors insist on odd fields. The latter case, if confirmed, would indicate a rather complicated distribution of the mean helicity of interstellar turbulence across the Galactic gaseous disc with zeros at some intermediate heights.

(v) *Small dynamo numbers.* In this limiting case the right-hand sides of equations (VII.18) can be considered as small perturbations. Let us renormalize B_r in order to make this component comparable to B in order of magnitude, $B_r \to |D|^{1/2} B_r$. Preserving the old notation for the renormalized radial field, we recast the generation equations as

$$\left(\frac{d^2}{dz^2} - \gamma - V\right)\begin{pmatrix} B_r \\ B \end{pmatrix} = 0,$$

where

$$V = |D|^{1/2}\begin{pmatrix} 0 & \partial\alpha/\partial z \\ 1 & 0 \end{pmatrix}$$

is the perturbation potential for $D < 0$. The standard perturbation theory for a doubly degenerate energy level (see, e.g., Davydov, 1966) gives the following first-approximation correction to the eigenvalue of the decaying mode:

$$\Delta\gamma = \pm (V_{12} V_{21})^{1/2},$$

where V_{12} and V_{21} are the corresponding matrix elements calculated from undisturbed, free-decaying eigenfunctions (point (ii) above). Let us put $\alpha = z$ for the sake of simplicity. This gives

$$\gamma^D \simeq -\pi^2 \pm |D|^{1/2},$$

$$\gamma^Q \simeq -\frac{\pi^2}{4} \pm |D|^{1/2}$$

for the dipole and quadrupole modes, respectively. It is clear again that the even (quadrupole) model is excited as smaller dynamo numbers.

The condition $\gamma^Q(D) = 0$ gives, for $D < 0$, a crude estimate of the critical value D_{cr} of the dynamo number, at which generation commences:

$$D_{cr} \simeq -\frac{\pi^4}{16} \approx -6.$$

Of course, the critical dynamo number depends on the z-distribution of the helicity. Numerical calculations show that it lies between approximately -4 and -10. The rather low generation threshold, $D_{cr} = -4$, corresponds to the concentrated helicity applied at the middle points of the slab, $\alpha = \delta(z - \frac{1}{2}) - \delta(z + \frac{1}{2})$. Smooth distributions of the mean helicity give higher generation thresholds. For

$\alpha = \sin \pi z$ the critical dynamo number is $D_{cr} \approx -8$, while $D_{cr} \approx -11$ for $\alpha = z$. For the combination of step functions, $\alpha = \theta(z) - \theta(-z)$, $D_{cr} \approx -6$ (see Figure VII.2).

VII.6. Radial Field Distribution in Discs of Variable Thickness

In this section we discuss the radial structures of *axisymmetric* large-scale field

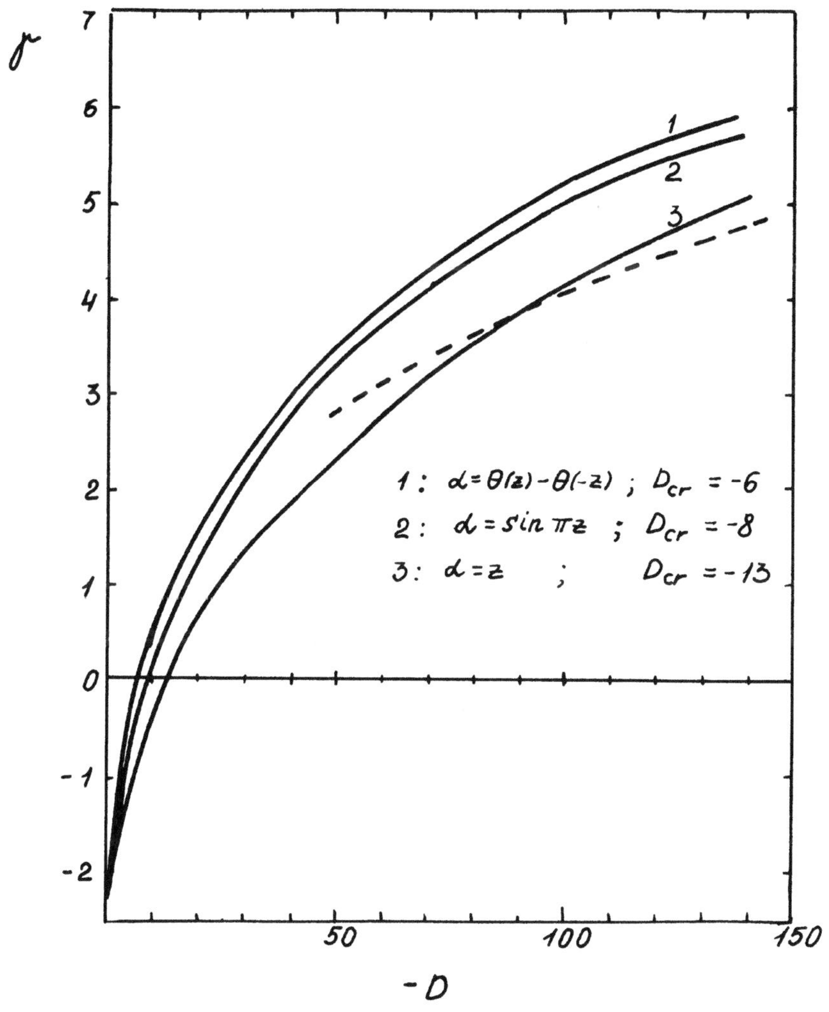

Fig. VII.2. Growth rate of the quadrupole mode in a thin disc versus dynamo number for different forms of the mean helicity distribution. The broken line gives the asymptotic dependence (VII.25) for $\alpha = z$.

generated by the turbulent dynamo in a radially non-uniform disc. When the radial dependence of both the generators and the field is included, the generation equations for the axisymmetric field take the form

$$\frac{\partial B}{\partial t} = -G\,\frac{\partial A}{\partial z} + \beta\,\frac{\partial^2 B}{\partial z^2} + \beta\,\frac{\partial}{\partial r}\left[\frac{1}{r}\,\frac{\partial}{\partial r}(rB)\right],$$

$$\frac{\partial A}{\partial t} = \alpha A + \beta\,\frac{\partial^2 A}{\partial z^2} + \beta\,\frac{\partial}{\partial r}\left[\frac{1}{r}\,\frac{\partial}{\partial r}(rA)\right] \tag{VII.27}$$

for $\partial/\partial\varphi = 0$ and the same assumptions as adopted in Equations (VII.17). We remember that B and A are the azimuthal components of the large-scale magnetic field and the vector potential respectively. When obtaining these equations, we have taken into account that the angular velocity of rotation only slightly varies with z within discs of spiral galaxies (see Section VI.1) and its vertical derivative is negligible as compared with the radial one.

Consider again exponentially growing solutions $(A, B) \propto \exp(\Gamma t)$ with the growth rate denoted by Γ and introduce dimensionless variables (R, Z, T) choosing different units across the disc and along the radius. This explicitly accounts for the thinness of galactic discs. Thus,

$$Z = z/h_0, \qquad R = r/r_0, \qquad T = t\beta/h_0^2.$$

we introduce also dimensionless angular velocity, its shear, and helicity:

$$\omega = \Omega/\Omega_0, \qquad g = G/\Omega_0, \qquad \alpha_* = \alpha/\alpha_0.$$

The units convenient for applications to spiral galaxies are $h_0 = 400$ pc, $r_0 =$ 10 kpc, $h_0^2/\beta \simeq 5 \times 10^8$ years, $\Omega_0 = 10^{-15}$ s^{-1}, $\alpha_0 = 1$ km s^{-1}. Further, we normalize the magnetic field by some arbitrary amplitude for instance the initial one B_0

$$A_* = A/(B_0 h_0), \qquad B_* = B\alpha_0 h_0/(B_0\beta),$$

and measure the growth rate Γ in units of the inverse diffusion time, β/h_0^2. The asterisks at dimensionless field, potential and helicity will be dropped below.

In terms of dimensionless variables equations (27) are

$$\Gamma B = -Dg\,\frac{\partial A}{\partial Z} + \frac{\partial^2 B}{\partial Z^2} + \lambda^2\,\frac{\partial}{\partial R}\left[\frac{1}{R}\,\frac{\partial}{\partial R}(RB)\right],$$

$$\Gamma A = \alpha B + \frac{\partial^2 A}{\partial Z^2} + \lambda^2\,\frac{\partial}{\partial R}\left[\frac{1}{R}\,\frac{\partial}{\partial R}(RA)\right], \tag{VII.28}$$

where the dynamo number is defined as

$$D = \frac{\alpha_0 \Omega_0 h_0^3}{\beta^2}$$

and is positive by definition (note that the sign of the rotational shear is contained in g), while

$$\lambda = h_0/r_0$$

is the small parameter that appears due to the great difference between vertical and horizontal dimensions of the galactic disc. (For chosen units, $\lambda = 0.04$). A simple consequence of this difference is a much more rapid magnetic diffusion across the disc (time scale $h_0^2/\beta \simeq 5 \times 10^8$ years) rather than along the radius (time scale $r_0^2/\beta \simeq 5 \times 10^{11}$ years).

The small value of λ which characterizes galactic discs allows us to obtain a simple asymptotic solution of the dynamo Equations (VII.28). Let us represent the solutions to Equation (VII.28) as

$$B(R, Z) = Q(R)\hat{B}(R, Z),$$
$$A(R, Z) = Q(R)\hat{A}(R, Z),$$

where $\hat{A}(R, Z)$ and $\hat{B}(R, Z)$ satisfy, by definition, the following one-dimensional equations:

$$\gamma(R)\hat{B} = -Dg \frac{\partial \hat{A}}{\partial Z} + \frac{\partial^2 \hat{B}}{\partial Z^2},$$
$$\gamma(R)\hat{A} = \alpha\hat{B} + \frac{\partial^2 \hat{A}}{\partial Z^2} \qquad \text{(VII.29)}$$

whose form coincides with the dynamo equations (VII.18) for an infinitesimally thin slab. As long as the coefficients g and α, the rotational shear and mean helicity, depend on R, solutions of the system (VII.29), $(\hat{A}, \hat{B})$ and γ also show the parametric dependence on the radius. The quantity $\gamma(R)$ may be considered as a local (for given R) growth rate of the field.

The half-thickness of the discs of spiral galaxies grows with the galactocentric distance,

$$H(R) = h(R)/h_0$$

(see Section VI.2). Here we introduce the dimensionless half-thickness, $H(R)$. Presuming that there are no electric currents flowing outside the disc (vacuum boundary conditions — see §4), the boundary conditions are

$$\frac{\partial \hat{A}}{\partial Z}(\pm H, R) = \hat{B}(\pm H, R) = 0. \qquad \text{(VII.30)}$$

Now it is quite natural to reduce the boundary-values problem (VII.29, VII.30) to the problem already solved, (VII.18), formulated for an infinitely thin slab. The only difference between them is that the boundary conditions are now formulated at the points whose positions depend on radius.

Decompose the coefficient $\alpha(R, Z)$ as

$$\alpha = \alpha_1(R) \cdot \alpha_2(Z)$$

and introduce new variables,

$$\tilde{Z} = Z/H(R), \qquad \tilde{B} = \hat{B}\alpha_1(R)H^2(R), \qquad \tilde{A} \equiv \hat{A}.$$

Equations (VII.29) and (VII.30) are recast as

$$\tilde{\gamma}(R)\tilde{B} = -D_{\text{eff}}(R)\frac{\partial \tilde{A}}{\partial \tilde{Z}} + \frac{\partial^2 \tilde{B}}{\partial \tilde{Z}^2},$$

$$\tilde{\gamma}(R)\tilde{A} = \alpha_2(\tilde{Z})\tilde{B} + \frac{\partial^2 \tilde{A}}{\partial \tilde{Z}^2}, \tag{VII.31}$$

$$\frac{\partial \tilde{A}}{\partial \tilde{Z}}(\pm 1) = \tilde{B}(\pm 1) = 0,$$

where

$$\tilde{\gamma}(R) = \gamma(R)H^2(R) \tag{VII.32}$$

and the effective dynamo number

$$D_{\text{eff}} = D\alpha_1(R)g(R)H^3(R) \tag{VII.33}$$

is defined exactly like the local dynamo number in the last section but includes radial variations of both the field generators, α and G, and the disc thickness.

Thus, the equations governing the field distribution across the disc are reduced to those whose properties are well understood and we may contentedly turn to the radial function, $Q(R)$.

Simultaneously with equations (VII.29), the following Schrödinger equation is derived for $Q(R)$:

$$\lambda^2 \frac{d}{dR}\left[\frac{1}{R}\frac{d}{dR}(RQ)\right] + [\gamma(R) - \Gamma]Q = 0, \tag{VII.34}$$

in which $-\gamma(R)$ plays the role of potential and the growth rate taken with opposite sign takes the place of total energy.

When obtaining the latter equation, we have neglected the radial derivatives of $\tilde{A}$ and $\tilde{B}$ in comparison with those of Q. Let us find out the range of validity of this approximation.

Upon observing that $\tilde{\gamma}(R)$ plays the role of the local growth rate in Equations (VII.31), the radial gradient of $\tilde{A}$ (and $\tilde{B}$) can be estimated as

$$\frac{1}{\tilde{A}} \frac{\partial \tilde{A}(R)}{\partial R} \simeq \frac{\partial \tilde{\gamma}}{\partial R}.$$

Now, estimating $\tilde{\gamma}$ according to the asymptotic expression (VII.25) reduces the original inequality

$$\frac{1}{\tilde{A}} \frac{\partial \tilde{A}}{\partial R} \ll \frac{1}{Q} \frac{\mathrm{d}}{\mathrm{d}R}(RQ)$$

to

$$|D_{\mathrm{eff}}| \ll \frac{4}{\gamma_0^2} \simeq 25.$$

The discussion below shows that this inequality is fulfilled in spiral galaxies where $|D_{\mathrm{eff}}|$ does not exceed 10–15 in the main parts of the discs, being larger only in the very central regions $r \lesssim 1$ kpc.

As a matter of fact, the approximation used in the derivation of the radial equation (VII.34) is quite similar to the adiabatic approximation of quantum mechanics: analysis of electron motion in an electric field produced by the nucleus neglects, to the first approximation, the motion of heavy nuclear particles. In galactic discs, where the diffusion rates across the disc and along it differ greatly ($\lambda \ll 1$), the vertical field distribution relaxes to its stationary form, dictated by Equations (VII.29, VII.30), much more rapidly than (and nearly independently of) the radial field distribution.* It is possible when $\Delta\Gamma \ll \Delta\gamma$, where $\Delta\Gamma$ and $\Delta\gamma$ are the differences between neighbouring eigenvalues Γ and γ, respectively (Davydov, 1966). This condition is, actually, just another form of the applicability condition obtained in the last paragraph. Below, we shall see that $\Delta\Gamma \lesssim 0.5$ and $\Delta\gamma \gtrsim 1$ in discs of spiral galaxies.

Thus, the radial distribution of the large-scale axisymmetric magnetic field in a spiral galaxy is governed (apart from the factor $\alpha(R)H^2(R) \propto \Omega(r)h(r)$] by the Schrödinger equation (VII.34) with the *real* potential $-\gamma(R)$, where γ can be understood as the 'local' growth rate. The actual, 'global' growth rate Γ is close to the maximal value of $\gamma(R)$ just like the lowest energy level in a potential well is close to the well depth.

The function γ monotonically depends on D_{eff}, i.e. on the product $\alpha_1(R)g(R)H^3(R)$. The functions $\alpha_1(R)$ and $H(R)$ are also monotonous, but the first is decreasing with R, while the second increases (see Figures VI.15 and VI.9

* We should note that the ratio of the relaxation times is of order λ rather than $r_0^2/h_0^2 = \lambda^2$ because the radial scale of the generated field is of order $\lambda^{1/2}$. This peculiarity is discussed at length in the next section.

respectively). In addition, the rotational shear

$$g(R) = \frac{r}{\Omega_0} \frac{d\Omega}{dr}$$

is a non-monotonous function of radius, as can be seen from Figures VI.2 and VI.5—VI.8 where typical cases are presented. As a result, the radial potential $-\gamma(R)$ has the form of an axisymmetric potential well whose bottom is situated at some distance from the galactic center. In the main part of the disc the form of $\gamma(R)$ can be estimated, with the help of asymptotic form (VII.25), as $\gamma \propto \Omega(r)/h(r)$.

To illustrate properties of the solutions for the magnetic field, consider a few model examples of $\gamma(R)$.

Let us begin with the simplest model in which generation sources are uniformly distributed across the ring $R_1 \leqslant R \leqslant R_2$ with impervious radial boundaries:

$$\gamma(R) = \begin{cases} \gamma_m & \text{for} \quad R_1 \leqslant R \leqslant R_2, \\ -\infty & \text{for} \quad R > R_2 \text{ and } R < R_1 \end{cases}$$

(Figure VII.3). Equation (VII.34) is reduced to the Bessel equation whose solution is given by

$$Q(R) = C_1 J_1 \left(\frac{(\gamma_m - \Gamma)^{1/2}}{\lambda} R \right) + C_2 Y_1 \left(\frac{(\gamma_m - \Gamma)^{1/2}}{\lambda} R \right),$$

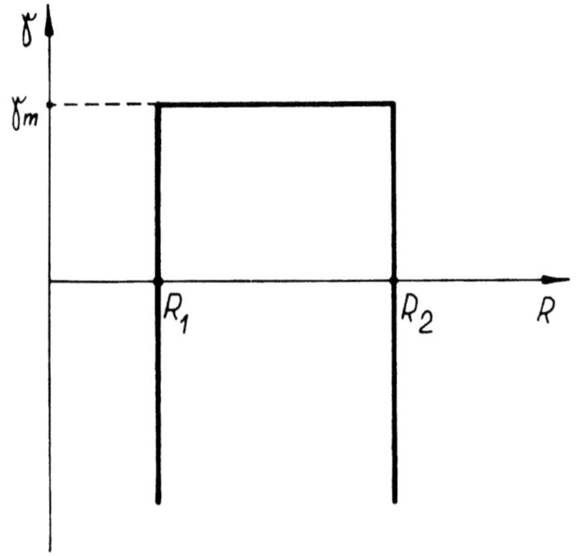

Fig. VII.3. A model radial potential.

where $J_1(x)$ and $Y_1(x)$ are the Bessel functions of the first and second kind, respectively. The boundary conditions

$$Q(R_1) = Q(R_2) = 0,$$

following from $\gamma(R) = -\infty$ outside the ring, yield the relation between C_1 and C_2 and the dispersion relation that determines the spectrum Γ_n of eigenvalues. The dispersion relation reads

$$\frac{Y_1(X_1)}{J_1(X_1)} = \frac{Y_1(X_2)}{J_1(X_2)},$$

where $X_{1,2} = R_{1,2}(\gamma_m - \Gamma)^{1/2}/\lambda$. The eigenvalues can be easily estimated with help of the asymptotic forms of the Bessel functions

$$J_1(x) \simeq \left(\frac{2}{\pi x}\right)^{1/2} \cos\left(x - \frac{3\pi}{4}\right),$$

$$Y_1(x) \simeq \left(\frac{2}{\pi x}\right)^{1/2} \sin\left(x - \frac{3\pi}{4}\right),$$

which are applicable for sufficiently high energy levels $\gamma_m - \Gamma \gg \lambda^2 R^{-2}$. This gives

$$\Gamma_n \simeq \gamma_m - \frac{\pi n \lambda^2}{R_2 - R_1},$$

where n is an integer. The difference between the neighbouring eigenvalues, $\Delta\Gamma = \lambda^2 n\pi/(R_2 - R_1) < 1$ for moderate $n < \lambda^{-2} \simeq 10^3$, while $\Delta\gamma \gtrsim \gamma_m = O(1)$ since we have seen in the last section that only *one* quadrupole mode can grow in spiral galaxies where the dynamo numbers generally do not exceed ten. Comparison of these estimates substantiates the applicability of the adiabatic approximation to the galactic discs.

Consider now another model, comparatively more realistic. In outer parts of galactic discs, say $R > R_1$, rotation curves become flat, $\omega \propto R^{-1}$, while the disc thickness grows nearly linearly with radius, $H \propto R$. Correspondingly,

$$\alpha_1(R) \propto \omega/H \propto R^{-2}, \qquad g(R) \propto R^{-1}$$

and

$$D_{\text{eff}} \approx \text{const.}$$

Hence, $\tilde{\gamma}$ is also approximately constant and (see (VII.32))

$$\gamma(R) \approx \frac{k^2}{R^2} \qquad \text{for} \qquad R > R_1$$

with certain constants k and R_1. The shape of this potential is given in Figure VII.4.

Now the radial equation (VII.34) takes the form

$$R^2 \frac{d^2Q}{dR^2} + R\frac{dQ}{dR} + \left[-\frac{\Gamma}{\lambda^2}R^2 + \left(\frac{k^2}{\lambda^2} - 1\right)\right]Q = 0. \qquad \text{(VII.35)}$$

Its exact solutions are cylindrical functions of imaginary argument and imaginary order:

$$Q(R) = Z_{i\nu}\left(i\frac{\Gamma^{1/2}}{\lambda}R\right),$$

where

$$\nu = (k^2/\lambda^2 - 1)^{1/2}.$$

For the qualitative analysis of the solution, it is sufficient to employ the asymptotic forms of cylindrical functions for large values of the argument. Using the theory of cylindrical functions (see Nikiforov and Uvarov, 1978), one can obtain (see Ruzmaikin *et al.*, 1985) the asymptotic solution to Equation (VII.35) which is valid at all points including the singular (turning) points where $\Gamma = \gamma(R)$. With boundary conditions $Q(\infty) = 0$ and $Q(R_1) = 0$ (impervious inner boundary) the asymptotic solution is

$$Q_n = C(p_n R)^{-1/2} S_n^{1/2} \begin{cases} \dfrac{\pi}{3^{1/2}} J_{-1/3}(S_n/\lambda) & \text{for} \quad R \leqslant R_{2n}, \\ K_{1/3}(S_n/\lambda) & \text{for} \quad R > R_{2n}, \end{cases}$$

where C is an arbitrary constant and

$$p_n = \Gamma_n^{1/2} \begin{cases} -(R_{2n}^2 - R^2)^{1/2} + R_{2n} \ln \left| \dfrac{R_{2n} + (R_{2n}^2 - R^2)^{1/2}}{R} \right| & \text{for} \quad R \leqslant R_{2n} \\ (R^2 - R_{2n}^2)^{1/2} - R_{2n} \arccos(R_{2n}/R) & \text{for} \quad R > R_{2n} \end{cases}$$

with $K_{1/3}(x)$ the McDonald function (the Bessel function of an imaginary argument). The radius R_{2n}, the outer boundary of the region where the nth eigenfunction is localized, is defined below. The eigenvalue spectrum is determined from the dispersion relation

$$(k^2 - \Gamma_n R_1^2)^{1/2} - k \ln \left| \frac{k + (k^2 - \Gamma_n R_1^2)^{1/2}}{R_1 \Gamma_n^{1/2}} \right| = -\lambda \mu_n \qquad \text{(VII.36)}$$

where μ_n are the zeros of the Bessel function $J_{-1/3}(x)$,

$$\mu_n \simeq -5\pi/12 + \pi n$$

where $n \gg 1$ is integer. It can be easily verified that the square root in (VII.36) can be neglected for $n \gg 1$, when $R_1\Gamma_n^{1/2} \ll k$, and

$$\Gamma_n \simeq \frac{k^2}{R_1^2 \cosh^2(\lambda\mu_n/k)} .$$

Expansion of the hyperbolic cosine for $\lambda\mu_n/k \ll 1$ yields a simple asymptotic estimate of the spectrum,

$$\Gamma_n \simeq \gamma_{\max} - \lambda^2 \frac{\mu_n^2}{2R_1^2} \qquad \text{(VII.37)}$$

where $\gamma_{\max} = k^2/R_1^2$.

The eigenfunction with number n is localized within the ring $R_1 < R < R_{2n}$, where the position of the right turning point is determined from the equation $\Gamma_n = \gamma(R_{2n})$ (see Figure VII.4) as

$$R_{2n} \simeq R_1 \cosh(\lambda\mu_n/k). \qquad \text{(VII.38)}$$

Thus, this example also gives a small difference between neighbouring eigenvalues, $\Delta\Gamma = \Gamma_n - \Gamma_{n-1} = O(\lambda^2)$. This model which is slightly more realistic than

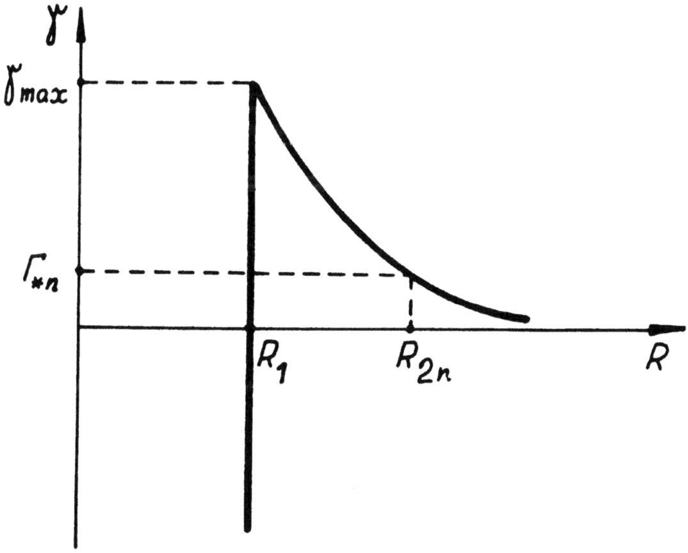

Fig. VII.4. A model radial potential.

the previous one, reveals a new important feature: the region where the nth eigenfunction is localized, $R_1 < R < R_{2n}$, becomes wider with n. The growth rate Γ_n, of course decreases with n. One more model example with Brandt rotation curve is considered in the next section.

Notice a general property of radial eigenfunctions $Q_n(R)$ that is valid for any form of the radial potential and follows from the oscillatory theorem for the Sturm—Liouville problem: the nth eigenfunction ($n = 1, 2, \ldots$) has $n - 1$ zeros (see, e.g., Courant and Hilbert, 1962). Thus, the first eigenfunction $Q_1(R)$ has a fixed sign, concentrates within the ring $R_1 < R \lesssim R_1 \cosh(1.87\,\lambda/k)$ and has the growth rate $\Gamma_1 \simeq \gamma_{\max} - 1.75\,\lambda^2 R_1^{-2}$. The next eigenfunction $Q_2(R)$ has one zero, its localization region is wider and the growth rate is smaller by approximately $10\,\lambda^2 R_1^{-2}$, etc.

We emphasize, however, that relations (VII.37) and (VII.38) are applicable only to high levels $n \gg 1$ and for lower eigenfunctions and eigenvalues they provide very crude, qualitative description. Numerical estimates from Equation (VII.36) for $R_1 = 1$, $k^2 = 2$ and $\lambda = 0.04$ show that $\Gamma_1 \simeq 1.48$ and $R_{21} \simeq 1.16$ (Ruzmaikin *et al.*, 1985). The lowest eigenfunction is localized within the interval $\Delta R = R_{21} - R_1 \simeq 0.16$ whose width is thus of order $\lambda^{1/2}$ rather than λ^2 as would follow from (VII.38). Correspondingly, the difference between neighbouring eigenvalues, while remaining small, is of order λ rather than λ^2. This fact is discussed in Section VII.8.

Mutual closeness of the eigenvalues Γ_n is a very important point of the galactic dynamo: $\Delta\Gamma_n \ll \gamma_m$ while $\Gamma_n \simeq \gamma_m = O(1)$ for moderate n. The cause of this peculiarity is the small value of the aspect ratio λ, i.e. the thinness of galactic discs. The consequence of a small difference between growth rates of modes with different radial structure is the fact that in spiral galaxies, where the growth times $\sim 1/\gamma_m \simeq 10^8$—$10^9$ years are comparable with the lifetime $\simeq 10^{10}$ years, the lowest mode could not have enough time to become dominant and the present radial distribution of the large-scale magnetic field would inherit some features of the initial, seed field. Note in this connection that some mechanisms of seed field formation prefer higher modes (see Section 14).

The models considered above give qualitative properties of radial distributions of the large-scale magnetic fields in spiral galaxies. Quantitative study necessitates the use of specific forms of the functions $\gamma(R)$ and $H(R)$ for every galaxy. Such applications are considered in the next section.

VII.7. Radial Distribution of Axisymmetric Fields in Spiral Galaxies

Now we turn to the estimation of radial distributions of the large-scale axisymmetric magnetic fields in specific spiral galaxies. In the previous Section we have seen how strongly the field distribution depends on the form of the radial potential, i.e., ultimately, on the form of the rotation curve and the shape of the gaseous disc. Therefore, here we consider one by one the most important or typical examples of spiral galaxies. It is only natural to begin with the Milky Way galaxy, although observational data on the global configuration of its magnetic field are far from being definite.

Rotation curves of spiral galaxies are obtained from observations with an accuracy quite sufficient for our purposes (Section VI.1).

Much worse is the situation concerning shapes of ionized gaseous discs in galaxies, i.e. with dependence of their thickness on galactocentric distance. Rather abundant observational estimates of the ionized disc half-thickness in the solar vicinity, which were discussed in Section VI.2, show large discrepancies although they seem to converge to the estimate $h_0 \simeq 400 \ldots 600$ pc. Direct evidence concerning the shape of the disc of free electrons and ions is absent. As indirect evidence, one may use data on the shape of the neutral gas disc, which thickens slowly towards the edge within the solar orbit and expands rapidly beyond the Sun. It is reasonable to believe that the electron disc behaves in the same fashion but only educated guesses are available concerning quantitative estimates. For external galaxies, even data on the neutral disc shape are scarce. We consider a model in which the ionized gaseous disc's surface is the parted hyperboloid of revolution,

$$h(r) = h_c[1 + (r/r_c)^2]^{1/2} \qquad \text{(VII.39)}$$

(Kaplan *et al.*, 1977). For $r \ll r_c$ the disc thickness grows slowly while for $r \gg r_c$ this growth is approximately linear. For our Galaxy, the values h_c and r_c should be adjusted so as to give, say, $h(10\text{ kpc}) = h_0 = 400$ pc and $h(0) = h_c = 150$ pc: the electron disc thickness in the Galactic central parts must be at least as thick as the neutral gas disc (see Section VI.2 for discussion of the origin of the free electron layer). This gives $r_c \approx 4$ kpc. For $r \gg r_c$ the disc (VII.39) thickens almost linearly giving $h(15\text{ kpc}) \approx 570\text{ pc} \approx 1.4\, h_0$.

Observations show that the neutral gas disc thickens very strongy towards the edge (see Figure VI.9) and $h_{\rm HI}$ (15 kpc) ≈ 375 pc $\approx 3\, h_{\rm HI}$ (10 kpc). If the ratio of thicknesses of the ionized and neutral discs does not depend on radius, the model (VII.39) underestimates the half-thickness. In this case the radius of the outer boundary of the magnetic field generation region is overestimated below.

In dimensionless variables, $H = h/h_0$ and $R = r/r_0$, the dependence (VII.39) takes the form

$$H(R) = H_c[1 + (2.5\, R)^2]^{1/2}. \qquad \text{(VII.40)}$$

The value 400 pc for the disc half-thickness in the solar vicinity corresponds to $H_c = 0.375$ (or $h_c = 150$ pc).

One more factor which influences the form of the radial potential, $\gamma(R)$, is the distribution of the mean helicity of turbulence across the disc, $\alpha_2(Z)$. The form of $\alpha_2(Z)$ determines the critical dynamo number (corresponding to $\gamma = 0$), i.e. the radial boundaries of the generation region. Dependence of γ on the dynamo number D and the form of $\alpha_2(Z)$ is discussed at length in §5 (see, e.g., (VII.25)).

Below we use the following approximation:

$$\alpha_2 = \begin{cases} 1 & \text{for} \quad z > 0, \\ 0 & \text{for} \quad z = 0, \\ -1 & \text{for} \quad z < 0, \end{cases}$$

which corresponds to the Gaussian density distribution along z (see Section VI.4). Strictly speaking, this discontinuous function should be smoothed because the mean helicity, an averaged quantity, cannot vary strongly over scales below the correlation scale, 100 pc. However, such smoothing gives only negligible corrections to the numerical results discussed below.

In the inner parts of spiral galaxies very large values of the angular velocity Ω and, correspondingly, of the rotational shear $r\,\mathrm{d}\Omega/\mathrm{d}r$ are observed. Owing to this, the computed local growth rate $\gamma(r)$ reaches very high values in the central parts of galaxies. However, the characteristic growth time of the large-scale magnetic field generated by the turbulent dynamo cannot be shorter than the revolution time of the turbulent cell, $\pi(l_0/v_0) \simeq 300\ \mathrm{pc}/10\ \mathrm{km\ s}^{-1} \simeq 3 \times 10^7$ years. In other words, there exists an upper limit $\gamma < 1/(3 \times 10^7\ \text{years}/5 \times 10^8\ \text{years}) \simeq 10$. At all points where the computed value of γ exceeded 10 we have put $\gamma = 10$. Note that this is only an order of magnitude limit. Actually, it may turn to be a few times lower which would be essential in some cases, e.g. in interpretation of our results for M51 (see below).

The procedure of solution for the problem of spatial distribution of the axisymmetric magnetic field within a galactic disc is as follows. Using the observed rotation curve and the adopted model of the disc shape, we calculate the radial shear $G = r\,\mathrm{d}\Omega/\mathrm{d}r$ and, with the use of (VI.12) and (VI.13), determine the radial distribution of the mean helicity, $\alpha_1(r)$. Now the effective dynamo number (VII.33) can be obtained as a function of radius. Then the solution (usually numerical) of the local boundary value problem (VII.31) with some specific choice of $\alpha_2(z)$ yields the distribution of the generated large-scale magnetic field across the disc, $\tilde{B}(z)$ and $\tilde{A}(z)$, and the "local growth rate" $\gamma(r)$ which is subsequently used in the radial Schrödinger equation (VII.34) for determination (numerical again) of the growth rates Γ_n of magnetic fields with different radial distributions $Q_n(r)$.

From the point of view of computer physics, we are dealing with a numerical solution of the two-dimensional, and even three-dimensional (in the next section), eigenvalue problem in, strictly speaking, an unbounded domain. Even with modern computers such a problem lies on the verge of possibilities. We solve this problem with quite common numerical opportunities due to a combination of numerical and analytical asymptotic methods based on thinness of galactic discs.

The procedure described above has been performed by us for the Milky Way galaxy and several nearby galaxies (Ruzmaikin *et al.*, 1985; Baryshnikova *et al.*, 1987a, b); preliminary results were obtained by Ruzmaikin and Shukurov (1981). Below we discuss the results of this analysis (Baryshnikova *et al.*, 1987b).

(i) *The Milky Way galaxy.* Figure VII.5 shows the local growth rate $\gamma(r)$ for our Galaxy. The calculations are based on the rotation curve of Burton and Gordon (1978) for $r < 10$ kpc (more precisely, the analytical approximation given in Table VI.1); at $r > 10$ kpc we used the flat rotation curve at $V = 250$ km s^{-1} (see Figure VI.1). The gaseous disc has been described by relation (VII.40) with $H_c = 0.375$ which corresponds to $h = 400$ pc near the Sun.

The broken lines in Figure VII.5. show the growth rates Γ_n for magnetic modes. The points where these lines intersect with the curve $\gamma(r)$ represent the boundaries of localization of the corresponding radial distributions $Q_n(r)$.

The given plot clearly shows that distribution of the large-scale magnetic field in the Galaxy is characterized by a gap within $4 \lesssim r \lesssim 7$ kpc (Ruzmaikin and Shukurov, 1981): the magnetic field is amplified and maintained only in the central region and in a wide outer ring. Within the outer ring, the field grows at maximal rate within $8 \lesssim r \lesssim 10$ kpc, where the lowest mode Q_1 is localized, and the growth rate is about 10^9 years.

Note now that the ionized galactic disc may be considerably thicker than was assumed in the derivation of the results presented in Figure VII.5. The half-

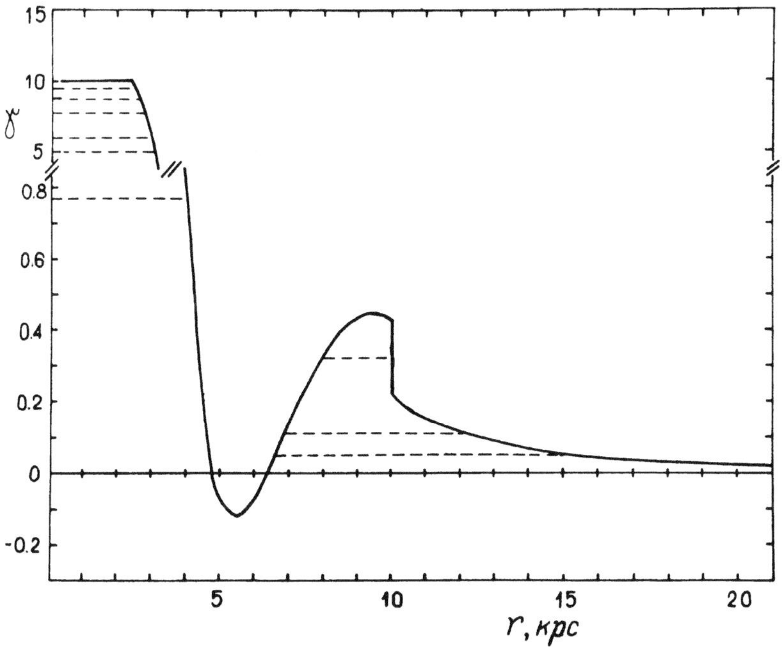

Fig. VII.5. The local growth rate for the Milky Way galaxy. Dashed lines show growth rates for axisymmetric modes and their localization regions. Two levels, $\Gamma = 3.3$ and $\Gamma = 1.9$ are omitted. The discontinuity at $r = 10$ kpc is an artefact of poor approximation in the rotation curve (cf. Figure VI.2) and does not affect the results.

thickness near the Sun, whose plausible value lies between 400 and 600 pc, may be as large as 1000 pc. Since in the Milky Way $d\Omega/dr < 0$ at all r (see Figure VI.2), D_{eff} is also negative at all radii. Absence of field generation in the intermediate part of the Galaxy is due to the fact that for $4 \lesssim r \lesssim 7$ kpc the effective dynamo number's value is below the generation threshold, $|D_{\text{eff}}| < |D_{\text{cr}}|$, which means $\gamma(r) < 0$ within this range. However, the effective dynamo number is proportional to h^2 (see (VII.33)) and $|D_{\text{eff}}|$ may be larger than the critical value if the disc is sufficiently thick. As Baryshnikova *et al.* (1987b) estimate, $\gamma(r) > 0$ at all radii for $h_c \gtrsim 160$ pc. Moreover, for $h_c \gtrsim 200$ pc the function $\gamma(r)$ becomes monotonous and eigenfunctions appear that are spread over the whole disc, $0 < r \lesssim 15$ kpc: the gap and the outer magnetic ring are washed out, the field distribution extends uninterrupted from the center of the Galaxy out to the outskirts (see Figure VII.6). Hence, the detailed form of the radial field distribution in the Galaxy essentially depends on the exact value of the gaseous disc thickness. As long as the half-thickness of the ionized disc near the Sun is below approximately 500 pc, the large-scale magnetic field concentrates in a ring around $r \simeq 10$ kpc and within the inner region $r \lesssim 4$ kpc. Otherwise the large-scale magnetic field is more or less uniformly distributed over the whole disc. Note that we may invert our arguments and infer the value of the disc thickness from detailed knowledge of the radial distribution of the large-scale magnetic field.

(ii) *The Andromeda Nebula.* Similarly to the Milky Way galaxy, the Andromeda Nebula (M31) has a double-peaked rotation curve (Figure VI.5). In this galaxy the minimum in rotation velocities is so prominent that the angular velocity even grows with r in the range $2 \leqslant r \leqslant 2.6$ kpc. Thus, in this ring the effective dynamo number is positive. When the dynamo number is positive, the magnetic field generation threshold in a disc is much higher than for negative dynamo numbers; quadrupole modes grow only for $D \gtrsim 300$ while dipole modes are excited for $D \gtrsim 500$ for $\alpha_2 = \sin \pi z$ (Ruzmaikin *et al.*, 1980b). These values far exceed those attained in the main parts of all spiral galaxies including M31. Therefore, in contrast to the Milky Way, the radial distribution of the large-scale magnetic field in the Andromeda Nebula is unavoidably ring-like, no matter how thick is its disc. The possibility of the ring-like field distribution in M31 was first proposed by Ruzmaikin and Shukurov (1981).

In Figure VII.7 we present the radial dependence of the local growth rate of the large-scale magnetic field in M31 computed by Baryshnikova *et al.* (1987b) with use of the rotation curve of Deharveng and Pellet (1975) and Haud (1981) and with the same disc (VII.40) with $H_c = 0.375$ which was used for the Galaxy. It is clearly seen that magnetic fields are generated for $r \lesssim 2$ kpc and $6 \lesssim r \lesssim 20$ kpc with the most effective generation near $r \simeq 10$ kpc.

The Andromeda Nebula is the closest spiral galaxy and its magnetic field is studied observationally in considerable detail (Beck *et al.*, 1980; Sofue and Takano, 1981; Beck, 1982). The remarkable property of the observed large-scale

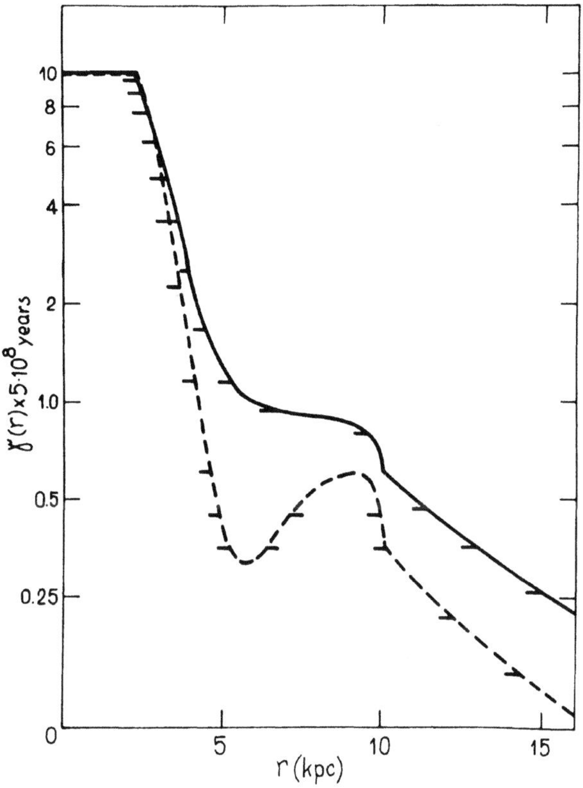

Fig. VII.6. Local growth rate in the Galaxy for the disc central half-thickness $h_c = 200$ pc (solid line) and $h_c = 160$ pc (broken line). Tick marks show positions of the growth rates, Γ_n, of axisymmetric modes.

magnetic field is its concentration within the ring around $r = 10$ kpc (Sofue and Takano, 1981, Beck, 1982), which exactly agrees with the prediction of the dynamo theory. The histogram in Figure VII.7 shows the radial distribution of the brightness temperature of polarized synchrotron emission in M31 as derived by Beck (1982). This distribution reflects the distribution of the large-scale magnetic field. We can satisfy ourselves that magnetic fields are observed at those radii where the eigenfunctions of the dynamo equations are localized.

From the discussion above we see that in a disc of variable thickness, e.g. in discs of spiral galaxies, the condition for large-scale magnetic field generation looks more complicated in comparison with a plane-parallel slab where it reads simply as $|D| > |D_{cr}|$ (see §5). Now it is required that the local growth rate $\gamma(r)$ considerably exceeds zero over a sufficiently wide range of galactocentric

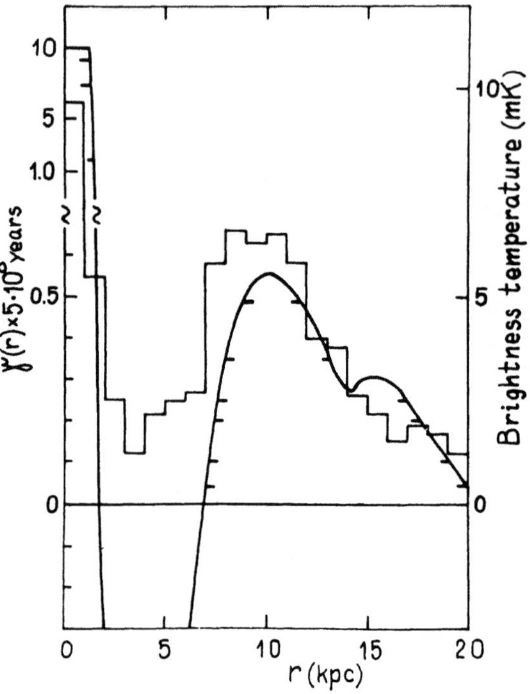

Fig. VII.7. The local growth rate in M31 (solid line) with tick marks showing positions of the growth rates of axisymmetric modes and respective localization regions of eigenfunctions. The histogram gives the observed distribution of polarized synchrotron emission from M31 (Beck, 1982). Note a remarkable coincidence of maxima and minima of both distributions. The local maximum of $\gamma(r)$ at $r = 14$ kpc is an artefact of poor approximation in the rotation curve (cf. Figure VI.5).

distances. Indeed, the Bohr-Sommerfeld quantum condition applied to the equation (VII.34) gives the following equation for the number n of magnetic modes excited in a galaxy:

$$\int_{R_1}^{R_2} (\gamma(R))^{1/2}\,\mathrm{d}R = \pi(n - \tfrac{1}{2})\lambda \tag{VII.41}$$

where $n = 1, 2, 3, \ldots$ and R_1 and R_2 are the radial boundaries of the region where $\gamma(R) > 0$: $\gamma(R_1) = \gamma(R_2) = 0$. Correspondingly, the generation condition is $n \geqslant 1$, or

$$\int_{R_1}^{R_2} (\gamma(R))^{1/2}\,\mathrm{d}R \geqslant \pi\lambda/2.$$

This condition can be expressed in terms of more convenient variables for the description of galaxies after specification of $\gamma(R)$, i.e. of the rotation curve and the shape of the gaseous disc.

Consider a simple model based on the expansion of $\gamma(R)$ near the maximum point,

$$\gamma(R) = \gamma_m - C^2(R - R_0)^2 + \ldots .$$

With this expansion, $R_{1,2} = R_0 \pm \gamma_m^{1/2}/C$ and condition (VII.41) takes the form

$$\gamma_m = \pi(n - \tfrac{1}{2})\lambda C. \tag{VII.42}$$

Consider the rotation curve

$$V = \Omega r = \frac{V_m(r/r_m)}{(1 + r/r_m)^{3/2}},$$

which is the Brandt rotation curve with the parameter $n = 1$. Assume that in those regions where $\gamma(r) > 0$ the disc has a constant thickness. In this case the asymptotic relation (VII.25) with D_{eff} from (VII.33) substituted for D gives

$$\gamma(R) \simeq 3(\tfrac{3}{2}\alpha_2'(0))^{1/2}\tilde{\gamma}_0 \frac{V_m h_0}{v r_m}\left(\frac{R}{R_m}\right)^{1/2}\left(1 + \frac{R}{R_m}\right)^{-2}, \tag{VII.43}$$

where $R_m = r_m/R_0$.

Note that the obtained expression always gives positive $\gamma(R)$. This shortcoming is due to our adoption of the asymptotic relation (VII.25) which is valid for $|D| \gg |D_{\text{cr}}|$ and is inapplicable for $\gamma(R) \simeq 0$. Therefore, we use the asymptotic form (VII.43) only for estimation of the shape of dependence γ on R only near the maximum of this function, i.e. for estimation of γ_m and C. Simple algebra now reduces (VII.42) to

$$\tilde{\gamma}_0(\alpha_2'(0)/2)^{1/2}\frac{V_m r_m}{h_0 v_0} = \pi^2(n - \tfrac{1}{2})^2,$$

where we recall that $\tilde{\gamma}_0 \approx 0.4$, $\alpha_2'(0)$ is the vertical derivative of dimensionless mean helicity at $z = 0$, h_0 is the disc half-thickness and v_0 is the turbulent velocity. Adopting typical values $\alpha_2'(0) = 1$, $v = 10$ km s^{-1}, $h_0 = 0.4$ kpc and $r_m = 5$ kpc, $V_m = 520$ km s^{-1} (parameters for the rotation curve of spiral galaxy IC342 – Rogstad *et al.*, 1973; see also Newton, 1980b), we obtain that in our model galaxy four magnetic modes are excited with different radial distributions. The maximal local growth rate is attained at $r = \frac{1}{3} r_m$ and the generated magnetic field is localized in a region of width $2\gamma_m^{1/2}/C \approx 1.5\, r_m$, that is between $r \approx 0$ and $r \approx 9$ kpc for $r_m = 5$ kpc. Note the difference between these results and those obtained for IC 342 by Ruzmaikin *et al.* (1985) where the same rotation curve was used but another form of $\alpha_2(z)$ was adopted. Such a strong difference underlines the importance of detailed knowledge of the helicity distributions for the dynamo models.

We should emphasize the following important point. Although the axisymmetric magnetic field can be concentrated within annular regions, this by no means implies that the magnetic lines are closed circles in this case! Enhancement of the magnetic field strength in certain region means that magnetic lines approach each other there, but the magnetic lines themselves are spirals. Indeed, when the field is generated, both azimuthal and radial (not to mention the vertical) field components grow simultaneously, at the same rate. By order of magnitude, the ratio of the radial and azimuthal components is estimated as

$$\frac{B_r}{B_\varphi} = O(|D_{\mathrm{eff}}|^{1/2})$$

(see point (iv) in Section VII.5). Therefore, the pitch angle between magnetic lines and circumference direction is typically arctan $B_r/B_\varphi \simeq 17°$ corresponding to $|D_{\mathrm{eff}}| = 10$. The larger $|D_{\mathrm{eff}}|$, the more tightly wound are the magnetic lines. In other words, in those places where the field is generated at the largest rate and where its strength is maximal at the linear stage, the magnetic lines wind tighter and approach each other. Hence the magnetic lines (to be precise, their projection on the galactic plane) of the ring-like axisymmetric structure, exemplified by a magnetic field in M31, have the form schematically shown in Figure VII.8.

The theory described above is linear in magnetic field and does not take into account the back-action of magnetic field on gas motions. Therefore, the developed theory can give us a picture of the early stages of the field evolution. The detailed forms of eigenfunctions do not correspond directly to the observed fields: non-linear effects distort linear distributions, equalizing field strength over the localization regions of eigenfunctions with different n (Ruzmaikin *et al.*, 1985). Considerable roles are played in the establishment of the stationary strength and configuration of the field by the distribution of the gas density and the value of the helicity of turbulence (see Section VII.12 below). However, some properties of linear solutions undoubtedly survive in the stationary field configuration.

First of all, in the regions where the field did not grow at the linear stage, it can never appear in the present configuration (the radial diffusion proceeds very slowly over scales comparable with the disc radius). Hence, the ring-like structures associated with negative local growth rates over certain ranges of radii are inherited by the stationary configuration.

Further, non-linear effects are probably quadratic in magnetic field **B** since the magnetic force rot **B** × **B** is quadratic in the Navier–Stokes equation. Therefore, non-linearity can hardly change the *sign* of the field generated at the linear stage. Note that higher ($n > 1$) radial modes $Q_n(r)$ are oscillatory along the radius and, therefore, in the regions where they dominate the field reversals along the radius can occur. Formation of the seed magnetic field by superposition of the fields ejected from stars (see §14 below) gives a pronounced preference to higher modes.

Fig. VII.8. Schematic picture of the projection onto the galactic plane of magnetic lines of the axisymmetric dynamo-mode of the large-scale magnetic field.

Note in this connection that the large-scale field reversals are probably observed along the radius in our Galaxy (see Chapter IV). We stress, however, that the field uniformly directed at all radii by no means contradict the dynamo theory but rather indicate a special phasing of modes in the seed field (see Ruzmaikin *et al.*, 1985).

It seems plausible that the ratio B_r/B_φ is only weakly affected by non-linear effects. This presumption is confirmed by the observation that the pitch angle of magnetic lines in spiral galaxies (see Beck and Gräve, 1987) approximately agrees with the estimate of the dynamo theory ($\sim 17°$) given above. Inside spiral arms the field lines are additionally aligned by refraction at spiral shock waves (see §11 below).

Finally, the growth time of the large-scale magnetic field in the Milky Way galaxy and other galaxies is not much shorter than 10^9 years, i.e. only ten times as short as galactic lifetimes, $\sim 10^{10}$ years. Therefore, the generation of *large-scale* magnetic field seems to proceed currently in only weakly non-linear regime.

VII.8. Generation of Non-Axisymmetric Magnetic Fields in an Axisymmetric Disc

Up to this point, we have discussed the generation only of axisymmetric large-scale magnetic fields. Let us turn now to a more complicated three-dimensional problem: whether or not non-axisymmetric magnetic fields can be generated and maintained in galactic discs.

It is well known that excitation of non-axisymmetric large-scale magnetic fields in differentially rotating spherical bodies (stars and planets) is strongly suppressed, if it is possible at all (Stix, 1971; Krause, 1971; Ivanova and Ruzmaikin, 1985; Rädler, 1986a, b). The cause of this discrimination lies in the fact that differential rotation warps non-axisymmetric fields into spirals with the fields directed oppositely in neighbouring turns (see Figure V.1). In the course of time the turns approach one another which catastrophically enhances the field dissipation (see Chapter V). In *spherical* bodies, such enhancement is usually fatal for non-axisymmetric fields. The situation is, however, very different in *thin* discs. In them, the dominant contribution to the field decay comes from the diffusion *across* the disc, along a small thickness. Therefore, as long as the radial distance between neighbouring turns of non-axisymmetric spiral field exceeds the disc half-thickness, the catastrophic enhancement of magnetic dissipation does not occur. Hence, amplification and maintenance of non-axisymemtric fields may be possible in thin discs provided the radial field scale considerably exceeds the disc thickness (Ruzmaikin *et al.*, 1986; Sawa and Fujimoto, 1986; Baryshnikova *et al.*, 1987a, b).

In order to substantiate this assumption, let us estimate the number of non-axisymmetric modes of magnetic field which can be excited in an *axisymmetric* disc by the turbulent dynamo. The azimuthal dependence of the field is $\mathbf{B} \propto \exp(im\,\varphi)$. The larger the azimuthal wave number, m, the closer are the turns with oppositely directed fields and the stronger is the dissipation. Consider an initially ($t = 0$) uniform field lying in the disc plane. At the moment t the radial distance between oppositely directed magnetic lines is given by

$$\Delta r_m(t) \simeq \frac{r_0}{m\omega_0 t},$$

where ω_0 and r_0 are the characteristic values of the angular velocity of rotation and the radial scale of its variations, respectively, $|\mathrm{d}\omega/\mathrm{d}r| \simeq \omega_0/r_0$ (cf. Section V.2). Accordingly, the characteristic time of the magnetic diffusion along the radius is

$$\tau_d(t; m) \simeq \frac{[\Delta r_m(t)]^2}{\beta},$$

where β is the magnetic turbulent diffusivity.

The field generators — the mean helicity and the very same differential rotation — must first overcome (for moderate m) the diffusion across the disc. The resulting generation time is determined by the disc thickness and to the lowest approximation does not depend on the azimuthal wave number because we assume that the radial and the azimuthal scales of the field far exceed the vertical one. Therefore, we can apply the asymptotic expression (VII.25) for the characteristic generation time:

$$\tau_g \simeq \frac{h^2}{\beta}\,\frac{1}{\tilde{\gamma}_0}\,D^{-1/2}.$$

Non-axisymmetric modes are discriminated only in the next approximation in h/r_0, when the radial diffusion is taken into account. Their generation is possible when

$$\tau_g \lesssim \tau_d(\tau_g; m).$$

Since $R_\alpha \simeq 1$ in spiral galaxies and variations in the dynamo number $D = R_\alpha R_\omega$ come predominantly from $R_\omega = \omega_0 h^2/\beta$, the last inequality brings us to the following estimate of the number of non-axisymmetric modes that can be generated by the turbulent dynamo in an axisymmetric disc:

$$m \lesssim \frac{r_0}{h}\,\tilde{\gamma}_0^{2/3} R_\omega^{-1/4} \approx 2 \qquad \text{(VII.44)}$$

for $r_0 = 3$ kpc, $h = 400$ pc, and $R_\omega = 10$, which values are typical for the discs of spiral galaxies.

A few lower non-axisymmetric modes can indeed be excited in discs of spiral galaxies. It is evident that the lower the azimuthal wave number m, the greater is the growth rate and the mode $m = 1$, which corresponds to bisymmetric structures observed in many galaxies (see Chapter IV), is the preferred one among non-axisymmetric modes. Note, however, that we have considered the axisymmetric underlying disc and thereby the absolute record in growth rate always belongs to the axisymmetric field $m = 0$.

Notice a weak dependence of the estimate (VII.44) on the value of R_ω which characterizes the differential rotation and considerably varies from one galaxy to another.

We emphasize that for $r_0 \simeq h$ the right-hand side of inequality (VII.44) surely does not exceed unity because $\tilde{\gamma}_0 \approx 0.4 < 1$ and $R_\omega > 1$. Thus, our estimate implies that excitation of non-axisymmetric magnetic fields is strongly suppressed in spherical bodies. This result agrees well with results of numerical calculations mentioned at the beginning of this section.

Let us proceed now to construction of three-dimensional solutions to the dynamo equations that describe the generation of non-axisymmetric fields in an axisymmetric disc. The solution to the dynamo equation (VII.15) with $\mathbf{V} = \mathbf{\Omega} \times \mathbf{r}$, $\mu = 1$ and $\beta = \text{const}$ is sought in the form

$$\mathbf{B} = \tilde{\mathbf{B}}(r, z)\exp(\Gamma t + im\varphi).$$

We introduce dimensionless variables in the same manner as in §6. Now, in dimensionless cylindrical polar coordinates (R, φ, Z) with the Z-axis coincident with the rotation axis,* the components of the dynamo equation (VII.15) are given by

$$\left(\Gamma + im\, R_\omega\, \omega + \frac{\lambda^2 m^2}{R^2}\right) \tilde{B}_r$$

$$= -R_\alpha \frac{\partial}{\partial Z} (\alpha \tilde{B}_\varphi) + \frac{\partial^2 \tilde{B}_r}{\partial Z^2} + \lambda^2 \frac{\partial}{\partial R}\left[\frac{1}{R}\frac{\partial}{\partial R}(R\tilde{B}_r)\right] +$$

$$+ im\,\lambda R_\alpha \frac{\alpha}{R} \tilde{B}_z - \frac{2\, im\, \lambda^2}{R^2} \tilde{B}_\varphi,$$

$$\left(\Gamma + im\, R_\omega\, \omega + \frac{\lambda^2 m^2}{R^2}\right) \tilde{B}_\varphi$$

$$= R_\omega R \frac{\mathrm{d}\omega}{\mathrm{d}R} \tilde{B}_r + \frac{\partial^2 \tilde{B}_\varphi}{\partial Z^2} + \qquad \text{(VII.45)}$$

$$+ \lambda^2 \frac{\partial}{\partial R}\left[\frac{1}{R}\frac{\partial}{\partial R}(R\tilde{B}_\varphi)\right] + \frac{2\, im\, \lambda^2}{R^2} \tilde{B}_r,$$

$$\left(\Gamma + im\, R_\omega\, \omega + \frac{\lambda^2 m^2}{R^2}\right) \tilde{B}_z$$

$$= \lambda R_\alpha \frac{1}{R}\frac{\partial}{\partial R}(R\alpha \tilde{B}_\varphi) - im\,\lambda R_\alpha \frac{\alpha}{R} \tilde{B}_r + \frac{\lambda^2}{R^2} \tilde{B}_z +$$

$$+ \frac{\partial^2 \tilde{B}_z}{\partial Z^2} + \lambda^2 \frac{\partial}{\partial R}\left[\frac{1}{R}\frac{\partial}{\partial R}(R\tilde{B}_z)\right],$$

* In the local Cartesian coordinates, generation equations do not include m, which is quite natural because axisymmetric and non-axisymmetric fields have similar local properties. Sawa and Fujimoto (1986) use the local Cartesian coordinates but apply a witty expedient using the differences of equations written for nearby spatial positions. In such form, the dependence of the generation equations on m is not excluded. However, the price is the higher order of equations to be solved and the necessity of exclusion of the spurious solutions.

where

$$R_\alpha = \frac{\alpha_0 h_0}{\beta}, \qquad R_\omega = \frac{\Omega_0 h_0^2}{\beta}$$

are again dimensionless numbers that characterize the strength of field generators, the mean helicity and differential rotation respectively. With the chosen units, $\lambda = 0.04$.

We have omitted the terms proportional to R_α in the second Equation (VII.45) for $\tilde{B}_\varphi$. These terms are generally small as compared to those containing R_ω in the same equation, since $R_\omega \simeq 10$ and $R_\alpha \simeq 1$ in galactic discs. As was pointed out by Rädler (1986, private communication), omitted terms can only facilitate the generation of non-axisymmetric fields.

Dynamo equations (VII.28) for axisymmetric magnetic fields contain only one dimensionless parameter, characterizing the magnetic field generators, which is called the dynamo number $D = R_\alpha R_\omega$. However, three-dimensional dynamo equations inevitably include R_α and R_ω separately. The physical reason for this difference is the fact that differential rotation (R_ω) and the mean helicity (R_α) act in different ways on non-axisymmetric magnetic fields: the differential rotation strongly distinguishes the axisymmetric magnetic modes while generation action of the mean helicity does not show a pronounced preference, being principally a local process (Rädler, 1986a).

The asymptotic solution to Equations (VII.45) is again constructed in the small parameter λ.

Before estimating the orders of magnitude of various terms in Equations (VII.45), we should emphasize the following important point. Although the dimensionless radius R and amplitudes $\tilde{B}_\varphi$ and $\tilde{B}_r$ are of order unity, the radial derivatives $\partial\tilde{B}_\varphi/\partial R$ and $\partial\tilde{B}_r/\partial R$ have another order of magnitude in λ, namely $\lambda^{-1/2}$. In other words, the radial scale of the solutions we are interested in is $\lambda^{1/2}$ times the disc half-thickness.

To make this point clear, we use a well known example from quantum mechanics. The basic wave function of a harmonic oscillator, $\psi = \exp(-x^2/2\hbar)$, has unit magnitude (we drop a normalization factor unessential here). However, its derivative,

$$\mathrm{d}\psi/\mathrm{d}x = -(x/\hbar)\exp(-x^2/2\hbar)$$

is of the order of $\hbar^{-1/2}$, not of $\hbar^{-1}$, since its maximal value $O(\hbar^{-1/2})$ is attained at the point $x = O(\hbar^{1/2})$.

An analogy between this example and our problem is much deeper than it would seem at first sight. The radial distribution of the large-scale magnetic field in a thin disc is described by the Schrödinger equation with λ taking place of the Planck's constant $\hbar$ (see also §6).

Let us find a solution to Equations (VII.45) in which similarly to the example considered above, radial derivatives are of the order $\lambda^{-1/2}$, that is $\partial/\partial R = O(\lambda^{-1/2})$, while z-derivatives are of the order λ^0, that is of order unity. Of course, we shall check this supposition in the solutions obtained.

With such ordering in Equations (VII.45), keeping only terms of two lowest orders in λ, we derive the following equations:

$$(\Gamma + im R_\omega \omega)\tilde{B}_r = -R_\alpha \frac{\partial}{\partial Z}(\alpha \tilde{B}_\varphi) + \frac{\partial^2 \tilde{B}_r}{\partial Z^2} + \lambda^2 \frac{\partial}{\partial R}\left[\frac{1}{R}\frac{\partial}{\partial R}(r\tilde{B}_r)\right], \tag{VII.46}$$

$$(\Gamma + im R_\omega \omega)\tilde{B}_\varphi = R_\omega g \tilde{B}_r + \frac{\partial^2 \tilde{B}_\varphi}{\partial Z^2} + \lambda^2 \frac{\partial}{\partial R}\left[\frac{1}{R}\frac{\partial}{\partial R}(R\tilde{B}_\varphi)\right], \tag{VII.47}$$

$$(\Gamma + im R_\omega \omega)\tilde{B}_z = \lambda R_\alpha \frac{1}{R}\frac{\partial}{\partial R}(R\alpha \tilde{B}_\varphi) - im\lambda R_\alpha \frac{\alpha}{R}\tilde{B}_r + \frac{\partial^2 \tilde{B}_z}{\partial Z^2}. \tag{VII.48}$$

We note immediately that, to the lowest orders, Equation (VII.48) for the vertical field component $\tilde{B}_z$ has split from the system and Equations (VII.46, VII.47) for $(\tilde{B}_r, \tilde{B}_\varphi)$ can be solved autonomously. Note also that $\tilde{B}_r$ and $\tilde{B}_\varphi$ are of order unity in λ† whereas $\tilde{B}_z = O(\lambda^{1/2})$. This can be seen from the fact that only in this case the source $\lambda R_\alpha (1/R)(\partial/\partial R)(R\alpha\tilde{B}_\varphi)$ and dissipation $\partial^2 \tilde{B}_z/\partial Z^2$ in Equation (VII.48) are of the same order in λ.

The system of Equations (VII.46) and (VII.47) for $\tilde{B}_r$ and $\tilde{B}_\varphi$ is rather close to similar generation equations (VII.28) for axisymmetric fields, which can be derived from (VII.46, VII.47) for $m = 0$. Recall that ω depends on R and those terms in Equations (VII.46) and (VII.47) that include m qualitatively affect solutions, since the substitution $\Gamma + im R_\omega \omega \to \Gamma'$ is not possible at all radii simultaneously.

Now we apply the same asymptotic procedure as used in §6 for axisymmetric fields. Represent again the solution in the form

$$\begin{pmatrix} \tilde{B}_r \\ \tilde{B}_\varphi \end{pmatrix} = Q(R) \begin{pmatrix} b_r(R, Z) \\ b_\varphi(R, Z) \end{pmatrix}$$

† Although $\tilde{B}_r$ and $\tilde{B}_\varphi$ are of the same order in λ, they have different orders of magnitude in another dimensionless parameter of the problem $D = R_\alpha R_\omega$ since $\tilde{B}_r/\tilde{B}_\varphi \simeq D^{-1/2}$ (see §5).

where b_r and b_φ satisfy the following one-dimensional equations:

$$\gamma(R)b_r = -R_\alpha \frac{\partial}{\partial Z}(\alpha b_\varphi) + \frac{\partial^2 b_r}{\partial Z^2}$$

$$\gamma(R)b_\varphi = R_\omega R \frac{\mathrm{d}\omega}{\mathrm{d}R} b_r + \frac{\partial^2 b_\varphi}{\partial Z^2} \qquad \text{(VII.49)}$$

equivalent to Equations (VII.29). For the radial function $Q(R)$ we obtain the Schrödinger equation with a complex potential:

$$\lambda^2 \frac{\mathrm{d}}{\mathrm{d}R}\left[\frac{1}{R}\frac{\mathrm{d}}{\mathrm{d}R}(RQ)\right] - U(R)Q = \Gamma Q, \qquad \text{(VII.50)}$$

where $U(R) = -\gamma(R) + imR_\omega \omega$.

One-dimensional equations (VII.49) (or (VII.29) which is essentially the same) describe field generation in the lowest order in λ. Notice that these equations do not include m: both axisymmetric and non-axisymmetric fields are identiclaly distributed across galactic discs and have, to the lowest order in λ, the same growth rate! Clearly, this peculiarity (already used in the beginning of this section) is a consequence of the disc thinness.

Solution of the system (VII.49) was thoroughly discussed above. Here we concentrate on an analysis of Equation (VII.50) which determines generation conditions and properties of non-axisymmetric fields.

Below we again use solutions to Equations (VII.49) with boundary conditions that correspond to an infinitely thin disc surrounded by vacuum:

$$b_\varphi(\pm h) = 0, \qquad b_r(\pm h) = 0.$$

Radial boundary conditions

$$Q(0) = Q(r_{\max}) = 0$$

with $r_{\max}(>1)$ chosen according to dimensions of galaxies, imply that the magnetic field is concentrated mainly within a galaxy. Adoption of the other outer radial boundary condition $(\mathrm{d}/\mathrm{d}r)(rQ) = 0$ at $r = r_{\max}$ has a negligible effect on the results.

Let us turn to analysis of the radial Schrödinger equation (VII.50). The radial potential U is a complex function of r. As was the case for axisymmetric fields, its real part $-\gamma(r)$ is derived through a solution of Equations (VII.49) or (VII.29) for the vertical field distribution with radial distributions of generation sources α and $r\,\mathrm{d}\Omega/\mathrm{d}r$, specific for every galaxy. The main difference in generation condition between different galaxies is due to different rotation laws (rotation curves). The angular velocity Ω enters both generators, the mean helicity α and rotational shear, $G = r\,\mathrm{d}\Omega/\mathrm{d}r$. Using observed rotation curves, one also can easily derive the imaginary part of the radial potential, $mR_\omega\omega(r)$.

When the radial potential $U(r)$ is determined from the local generation problem, solution of the boundary value problem for Equation (VII.50) gives the growth rates Γ (eigenvalues) and radial field distributions $Q(r)$ (eigenfunctions). We solve both the radial boundary value problem with boundary conditions $Q(0) = Q(r_{\max}) = 0$ and Equations (VII.49) with vacuum boundary conditions numerically by QR algorithm (Wilkinson and Reinsch, 1971; Ruzmaikin *et al.*, 1980b).

As an illustrative example, let us study properties of solutions of Equation (VII.50), taking as an example a polynomial approximation of the rotation curve of the spiral galaxy M51 (Goad *et al.*, 1979). This choice is not accidental. We shall see below that non-axisymmetric $m = 1$ mode is efficiently generated in this galaxy. The model studies in this section differs from the actual galaxy M51 in that here we investigate a wide range of R_ω values. This allows us, particularly, to analyse properties of the $m = 2$ mode, which cannot be generated in actual galaxy M51. Variation of the value of R_ω is equivalent to self-similar change in the rotation curve: $\tilde{V}(r) = (R_\omega / \tilde{R}_\omega) V(r)$.

The larger is R_ω, that is the farther from solid-body is the rotation, the more difficult is the generation of non-axisymmetric modes. In general form, this fact was stressed many times by Rädler (1986b). Note that below we vary R_ω and, thus, the imaginary part of the radial potential, without any changes in the real part, $\gamma(r)$. This means that we keep constant the product $R_\alpha R_\omega$, which, effectively, determines the value of $\gamma(r)$. Diminishing of R_ω means, thus, the growth of the ratio R_α / R_ω and, hence, facilitates generation of non-axisymmetric modes.

Dependence of the real part of the eigenvalue, Γ, on R_ω is shown in Figure VII.9 for modes $m = 0, 1, 2$. For the typical values of parameters we have $R_\omega = 12$. An important fact is that the mode $m = 2$ decays for $R_\omega > 8$.

The form of eigenfunctions $Q(r)$ for axisymmetric and non-axisymmetric modes is illustrated by Figure VII.10. The lines are shown where the field strength $\mathrm{Re}(Q\,e^{im\varphi})$ is maximal. It can be easily seen that this maximal value is $|Q|$ and, thus, the lines shown are determined by the equation $\mathrm{Re}(Q\,e^{im\varphi}) = |Q|$. The mode $m = 1$ is a two-armed spiral, while the spiral $m = 2$ has four arms. Generally speaking, variations of the radial function $Q(r)$ do not comprise the total dependence of the generated magnetic field on galactocentric distance. There is an additional factor $\alpha(r)h^2(r)$ in the azimuthal field which arises as a parametric radial dependence of the local solutions. However, $\alpha h^2 \propto l^2 \Omega h$ and for the flat rotation curve and linearly expanding disc this correction seems to be negligible.

We emphasize an important point that the lines shown in Figure VII.10 and further similar plots are not magnetic lines. The arrows on them show only a general field direction. At the spiral lines shown in these figures the field strength is maximal and, therefore, they can be visible on synchrotron radio-maps. (When comparing these results with observations, one should bear in mind that additional enhancement of magnetic fields occurs within spiral density waves. Thus, observa-

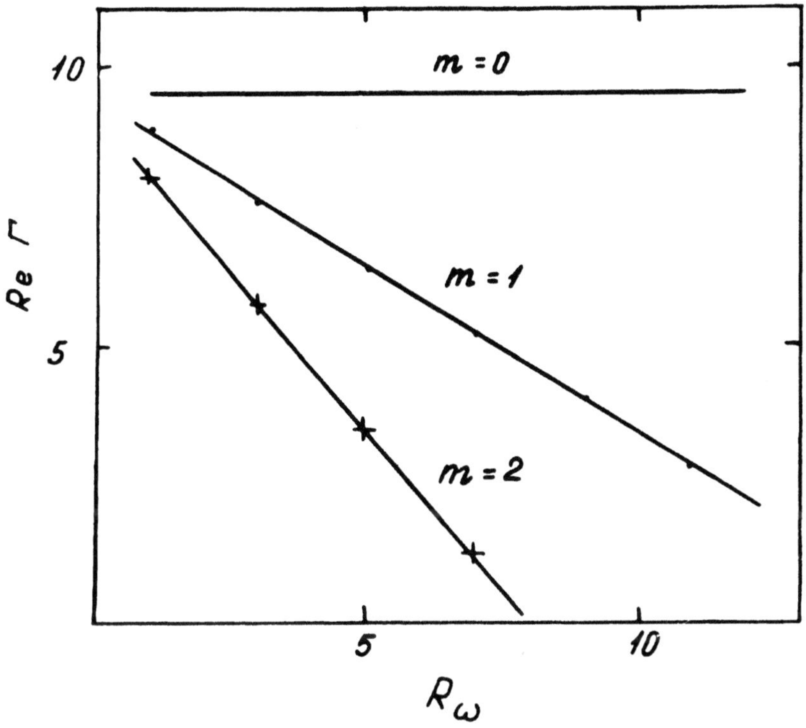

Fig. VII.9. Growth rates, Re Γ, of axisymmetric ($m = 0$) and non-axisymmetric ($m = 1, 2$) modes in the illustrative example versus the differential rotation measure, R_ω. The higher the deviation from solid-body rotation, the smaller are the growth rates of non-axisymmetric modes.

tions of magnetic fields in the interarm space seem to be most relevant.) The pitch angle of these spirals does not necessarily coincide with the pitch angle of magnetic lines. The angle between a magnetic line and the circumferential direction is of the order of

$$i = \arctan \tilde{B}_r/\tilde{B}_\varphi \simeq \arctan(R_\alpha R_\omega)^{-1/2} \simeq 17°$$

both for axisymmetric and non-axisymmetric fields, while the maximal-field spiral may be more tightly wound with typical pitch angle $i = 5°$.* However, in axisymmetric modes spiral form of magnetic lines comes together with a ring-like form of the region where the field is excited. Meanwhile, non-axisymmetric magnetic modes, also having spiral magnetic lines, have spirals as well as the lines

* Note that effective dynamo number D_{eff} generally decreases with radius in spiral galaxies. Hence, we expect that pitch of magnetic lines correspondingly grows toward the edges of galactic discs.

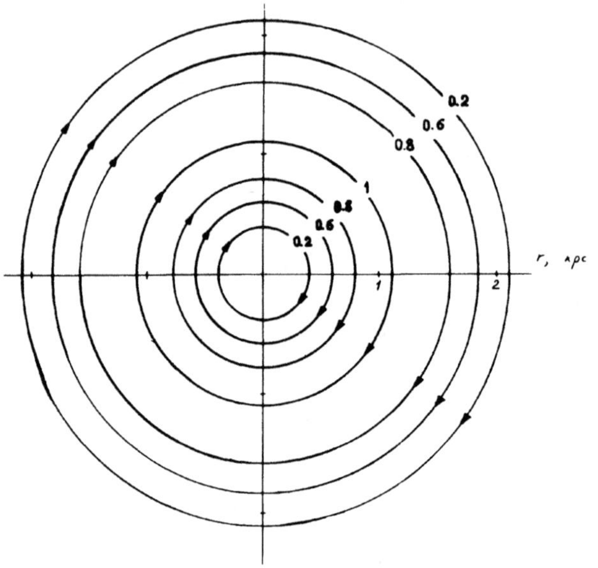

Fig. VII.10. Structure of the radial eigenfunction $Q(r)$ in the illustrative example. Numbers on the lines give the value of $|Q|$. The lines shown are not magnetic lines! (a). Axisymmetric mode $(m, n) = (0, 1)$; Re $\Gamma_{01} \approx 9.5$. The level lines of $|Q|$.

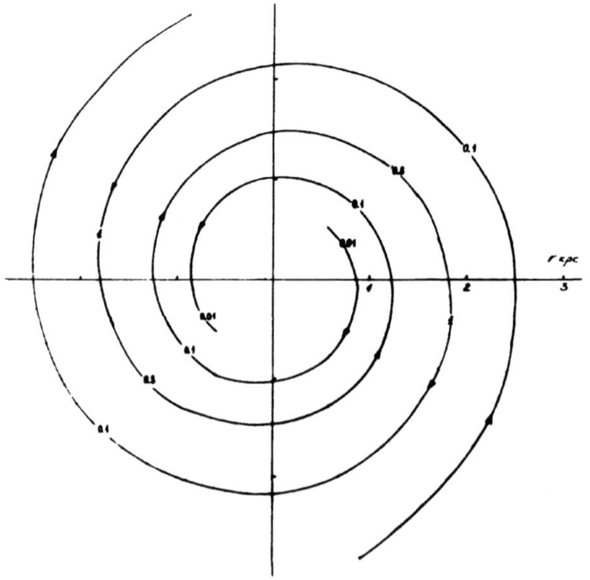

Fig. VII.10(b). Non-axisymmetric mode $(m, n) = (1, 1)$; Re $\Gamma_{11} \approx 5.1$; Im $\Gamma_{11} \approx 31.5$ (≈ 63 km s^{-1} kpc^{-1}).

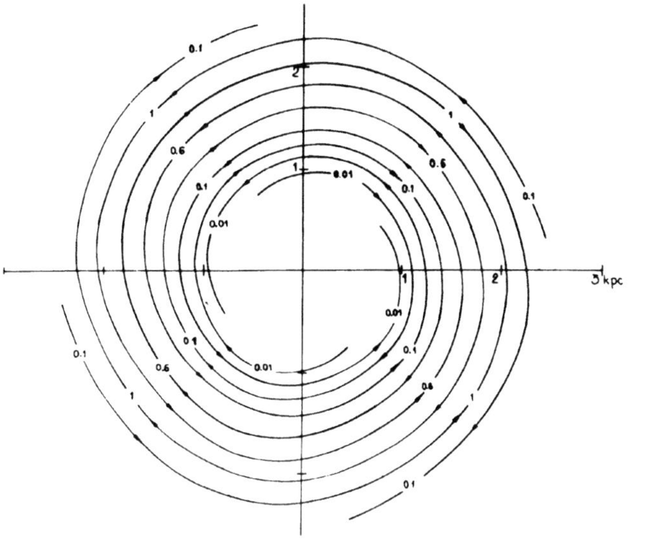

Fig. VII.10(c). Non-axisymmetric mode $(m, n) = (2, 1)$; Re $\Gamma_{21} \approx 1.2$; Im $\Gamma_{21} \approx 61.6$ (≈ 120 km s^{-1} kpc^{-1}). The level lines Re$[Q \exp(im\ \varphi)] = |Q|$ are shown in (b) and (c).

where the field is maximal. Projection of magnetic lines in the $m = 1$ mode onto the galactic plane are schematically shown in Figure VII.11.*

In a realistic situation a combination of axisymmetric and non-axisymmetric modes is excited. Instantaneous amplitudes of different modes are determined by their growth rates and by the configuration of the initial (seed) field. A characteristic feature of all reasonable combinations of modes is an axisymmetric structure in the inner parts of the disc which transforms into a bisymmetric configuration at greater distances from the center. This is due to the fact that the localization regions of the models with different azimuthal wave numbers m shifts outwards in r with growth of m. This property is clearly visible in Figure VII.12.

Equation (VII.50) was obtained under a natural assumption that the radial derivative of the eigenfunction is of the order of $\lambda^{-1/2}$. Now we should check whether the derived eigenfunctions behave properly as functions of λ. Figure VII.13 gives the dependence on λ of the widths ΔR of the $m = 0$ and $m = 1$ eigenfunctions for $R_\omega = 12$ and $R_\omega = 6$. Here ΔR is the radial width of the greatest region where an eigenfunction is of constant sign. Figure VII.13 assures

* We draw the reader's attention to the following delicate point. Magnetic field strength can be reproduced from the configuration of magnetic lines through calculation of their density. However, Figure VII.11 generally cannot serve for such a purpose since it represents the two-dimensional projection of magnetic lines (B_r, B_φ) and, for instance, these two components alone do not form together a divergency-free vector. In other words, the vertical field component not shown in Figure VII.11 may be essential at some points, especially in the central parts of the disc.

Fig. VII.11. Schematic picture of the projection onto galactic plane of magnetic lines of the $m = 1$ non-axisymmetric mode of the large-scale magnetic field.

that the dependence $\Delta R(\lambda)$ is comfortably close to $\lambda^{1/2}$ for a sufficiently wide range of λ that covers the range of λ which could be met in real galaxies. For small λ ($\lambda < 1.7 \times 10^{-2}$) our study is restricted by the lower accuracy of computed eigenfunctions.

When the disc becomes thicker, non-axisymmetric fields decay (cf. the estimate at the beginning of this section). When $R_\omega = 12$, the $m = 1$ mode decays for $\lambda > 5.5 \times 10^{-2}$. For $R_\omega = 6$ the decay occurs for correspondingly larger values $\lambda \gtrsim 0.12$. However, for $\lambda \gtrsim 9 \times 10^{-2}$ we observe deviations from the law $\partial/\partial R \propto \lambda^{-1/2}$: the disc becomes thick in this sense and the inequality $\lambda \lesssim 0.1$ thus gives a condition for applicability of our asymptotic solution. Broken lines in Figure VII.13 contain a range of possible values of λ given by observations of the Galactic gaseous disc (see Section VI.2). The applicability of Equation (VII.50) in this range seems to be well established.

Figure VII.14 gives the dependence of the growth rate, Re Γ, on λ for the mode $m = 0$ and for the non-decaying mode $m = 1$. We see that Re Γ linearly varies with λ which also confirms applicability of our presumption $\partial/\partial R \propto \lambda^{-1/2}$. Indeed, energy levels of the linear quantum oscillator are linear functions of the Planck's

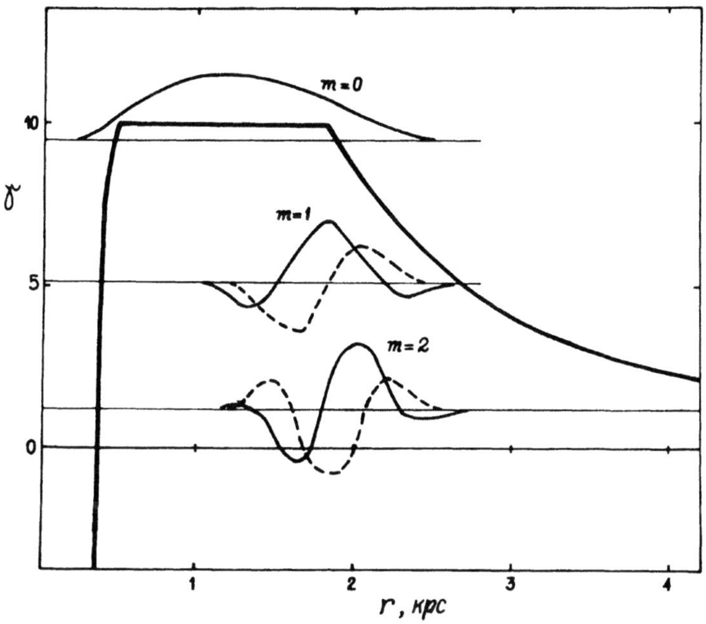

Fig. VII.12. The local growth rate (the real part of the radial potential) in the illustrative example (thick solid line). This distribution is identical with $\gamma(r)$ for the actual galaxy M51. Thin solid lines give Re $Q(r)$ and thin broken lines shown Im $Q(r)$ for indicated modes. Thin horizontal lines show growth rates of respective modes. The $m = 0$ results also coincide with those for M51.

constant, $\hbar$. Thus, in the working range of λ solutions behave like wavefunctions of the linear oscillator. To the first approximation, the solutions are determined by the maximal value max $[\gamma(R)] = \gamma_{\max}$ (depth of the potential well) and by the second derivative $\mathrm{d}^2\gamma/\mathrm{d}R^2$ at the bottom of the well (width of the potential well). Order of magnitude estimates $\lambda^2(\mathrm{d}/\mathrm{d}R)\,[(1/R)\,(\mathrm{d}/\mathrm{d}R)\,(RQ)] = O(\lambda)$ and $\gamma(R) = O(\gamma_{\max})$ in Equation (VII.50) give the following estimate of the eigenvalue:

$$\Gamma = O(\gamma_{\max} - C\lambda + im\,R_\omega\,\omega_{\min})$$

where $\omega_{\min}$ is the characteristic value of dimensionless angular velocity (see the next paragraph), and C is a constant of order unity.

For non-axisymmetric modes, eigenvalues Γ are complex. Non-axisymmetric magnetic structures rotate rigidly with angular velocity approximately $R_\omega\,\omega$ in a non-rotating frame, similar to spiral density waves. Analysis of numerical solutions gives

$$\mathrm{Im}\,\Gamma \simeq mR_\omega\,\omega_{\min},$$

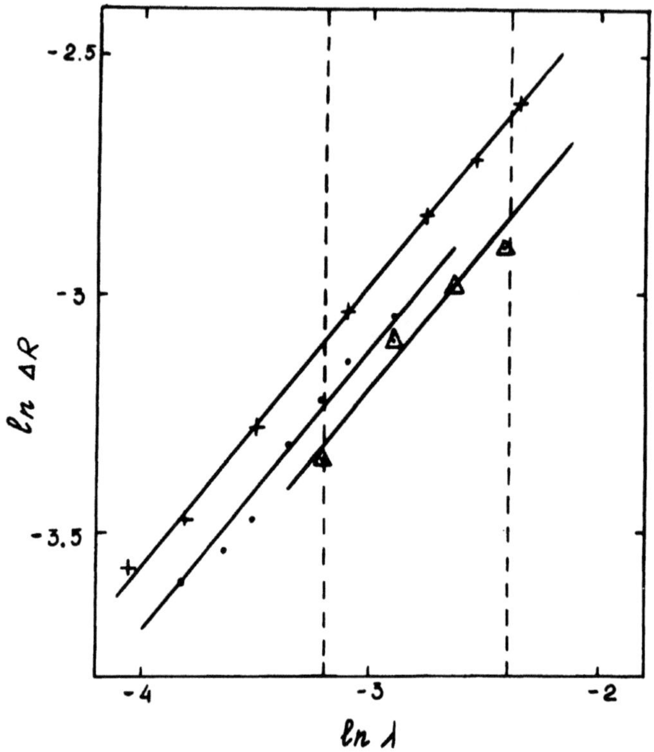

Fig. VII.13. The 'width' of eigenfunction versus λ in the illustrative example. Crosses: $(m, n) = (1, 1)$, $R_\omega = 6$. Dots: $(m, n) = (1, 1)$, $R_\omega = 12$. Triangles: $(m, n) = (0, 1)$, $R_\omega = 12$ and the widths are multiplied by e^{-2}. Solid lines show the asymptotic dependences $\Delta R \propto \lambda^{1/2}$.

where $\omega_{\min}$ is the value of a galaxy's angular velocity at the outer boundary of the localization region of the eigenfunction. This estimate also follows, in order of magnitude, from the arguments of the previous paragraph. Note also that the complex value of Γ that follows from the complex value of the radial potential in the Schrödinger equation, comprises the difference of the solutions described here from those for a 'classical' quantum oscillator.

Above we have discussed properties of growing solutions. Decaying (i.e. not maintained by a sufficiently strong dynamo action) non-axisymmetric modes deserve special attention. The rate of their decay is determined by the value of rotational shear: the greater is the shear $g = R\, \mathrm{d}\omega/\mathrm{d}R$, the more rapid is the decay. It means that what occurs is a rapid, after a few disc rotations, expulsion of non-axisymmetric field components to the region with the smallest possible differential rotation (see also Moffatt, 1978, §§3.8, 3.9). The rate of decay is therefore determined by the minimal possible value of g. As far as g decreases with

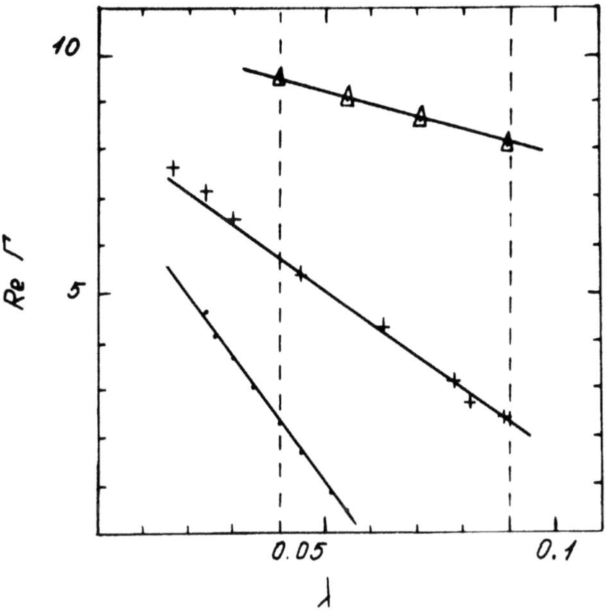

Fig. VII.14. The growth rate, Re Γ, versus λ in the illustrative example. Symbols are the same as used in Figure VII.13.

r in spiral galaxies (as $1/r$ for a flat rotation curve $\Omega \propto 1/r$), decaying non-axisymmetric modes are concentrated near the outer boundary of the galactic disc in a skin-layer of width $(\beta/\Omega_{\min})^{1/2}$. Since one cannot give a reasonable definition of the outer boundary of a galaxy, a realistic estimate of the expulsion time of non-axisymmetric fields which are not maintained by the dynamo is of the order of one rotation period in the main body of a galaxy.

Notice that these arguments make clear why the growing non-axisymmetric modes concentrate at greater radii than do axisymmtric ones and why the region of localization shifts to greater galactocentric distances with growth of m. While the mode $m = 0$ concentrates around the radius where $\gamma(r)$ is maximal (bottom of the radial potential well), the maximum of the $m = 1$ mode is shifted toward places where the shear $R\,\mathrm{d}\omega/\mathrm{d}R$ is smaller. Since Re Γ decreases with m, the region of localization becomes wider with m, i.e. higher modes can settle themselves down at even greater R (see Figure VII.12). Also due to this reason the angular velocity of rotation of magnetic spirals is close to $R_\omega \omega_{\min}$, where $\omega_{\min}$ is the disc's angular velocity at the outer boundary of the mth eigenfunction's localization region.

We have discussed at length properties of principal modes that have the largest growth rates for a given m. Evidently, there exists one more 'quantum number', n, which characterizes modes with different radial behaviour at given m. Modes with different n for $m = 0$ have been discussed in the previous two sections. Modes with $n > 1$ have lower growth rates, and, correspondingly, wider localization regions (compared to $n = 1$).* Therefore modes with $m > 0$, $n > 1$ may dominate at large distances from centers of galaxies. Note also that composition of the seed magnetic field from fields expelled by supernovae (see §14 below) gives a pronounced preference to higher modes which have smaller spatial scale. This preference can be preserved for time intervals comparable to the galactic lifetime.

VII.9. The Origin of Large-Scale Bisymmetric Magentic Structures

Non-axisymmetric magnetic structures, known as bisymmetric, are observed in a number of spiral galaxies (see Chapter IV). An outward resemblance between the bisymmetric configuration and that formed from a uniform frozen-in initial field under the action of differential rotation has provoked a suggestion of the primordial origin of the bisymmetric magnetic fields (Tosa and Fujimoto, 1978; Sofue *et al.*, 1986). It is, however, well known that the turbulent magnetic diffusion in a spiral galaxy destroys an external magnetic field, which lies in the galactic plane, in a very short time not exceeding one per cent of the galactic lifetime (see Chapter V). Even if one would not take into account the turbulent diffusion, it is clear that spiral galaxies have made several tens of rotations during their lifetimes and a primordial field should be wound up in a tight spiral, which is quite different from observed mildly wound configurations. Sawa and Fujimoto (1980) have attempted to pass by this difficulty hypothesizing that a relic magnetic field diffuses into the galactic disc from a gaseous corona where it could be preserved for a long time provided coronae are free from turbulence and rotate rigidly. However both these conditions are not fulfilled: both intense turbulent motions and differential rotation are observed e.g. in the corona of the Milky Way galaxy (York, 1982; Chapter VI here). An attempt to overcome the difficulty with the winding-up of the primordial field was also made by Kulsrud (1986).

Baryshnikova *et al.* (1987a) and Ruzmaikin *et al.* (1986) have proposed an interpretation of bisymmetric magnetic structures as a result of excitation of the first non-axisymmetric mode of the large-scale magnetic field by the dynamo action in the galactic disc. Similar ideas are independently considered by Sawa and Fujimoto (1986), who, however, adopted another approach to the solution of the dynamo equation.

Let us now turn to the application of the theory developed in the last section to specific galaxies. When comparing the results with observational data, one must

* The azimuthal 'quantum number', or wavenumber, m, is defined in a quite evident way. However, definition of the radial number, n, is less straightforward. For axisymmtric modes, when the oscillatory theorem is applicable, $n - 1$ is the number of zeros of the radial wave function $Q(r)$. For non-axisymmetric modes we enumerate wave functions according to their growth rates: the mode with the greatest growth rate, for a given m, is numbered by $n = 1$, the next is $n = 2$, etc.

bear in mind the following. We have constructed an asymptotic solution of the dynamo equations in a thin disc, $\lambda \ll 1$. The nature of an asymptotic solution differs markedly from, say, properties of perturbation series. In the perturbation theory, the accuracy of solution can always be improved by use of additional terms in the series, while the accuracy of an asymptotic solution is determined by the actual smallness of the expansion parameter. In other words, asymptotic series often diverge for any fixed value of the expansion parameter (λ in our case) and inclusion of further terms into the asymptotic expansion only deteriorates the result. Hence, an attempt to improve estimates given here would seem to be unwise: their accuracy is determined by the actual thickness of galactic discs. The decisive role in our asymptotic forms is played by the value of $\lambda^{1/2}$ which is of the order of 0.1 in galaxies.

The general property of asymptotic methods (well known from their special case: finite-difference schemes) is that they give good estimates of eigenvalues and reasonable estimates of eigenfunctions near the maxima of the latter. However, the accuracy of calculated eigenfuctions rapidly decreases both with decrease in their magnitude, i.e. with distance from the localization region, and with the order of the eigenfunction. Therefore, we cannot hope to obtain more than an order-of-magnitude agreement of spatial distributions discussed below with observed ones, especially in the outer parts of galaxies.

Our aim here is to discuss the origin of ring-like and bisymmetric magnetic structures in spiral galaxies and give a qualitative description of the mechanism of their maintenance. The detailed comparison of the theory with observations requires, first, numerical solution of three-dimensional (and non-linear) dynamo equations, and, second, observational data with high resolution and sensitivity. In addition, the lack of observational information on the hydrodynamic properties of spiral galaxies, especially on the mean helicity of interstellar turbulence and shapes of galactic ionized discs, strongly hampers further development of the theory. We should note optimistically, however, that general characteristics of the magnetic field generation process in specific galaxies are well within reach of modern theory. In particular, we shall see below that, in accordance with observations, our theory predicts that only axisymmetric fields can be generated in the Andromeda Nebula, while axisymmetric and non-axisymmetric fields are generated with almost equal ease in M51.

Note also that spiral galaxies are not very close to the axial symmetry. In particular, the spiral patterns cause departures from the axial symmetry but tightly wound patterns can be more successfully described by axisymmetric approximations. Some galaxies (e.g., IC 342) show departures from the axial symmetry induced probably by external sources. It can be expected *a priori* that account for violations of the axial symmetry would facilitate emergence of bisymmetric structures (cf. Vallée, 1984, 1986).

(i) *The Milky Way Galaxy.* Figure VII.15 shows the dependence of the growth

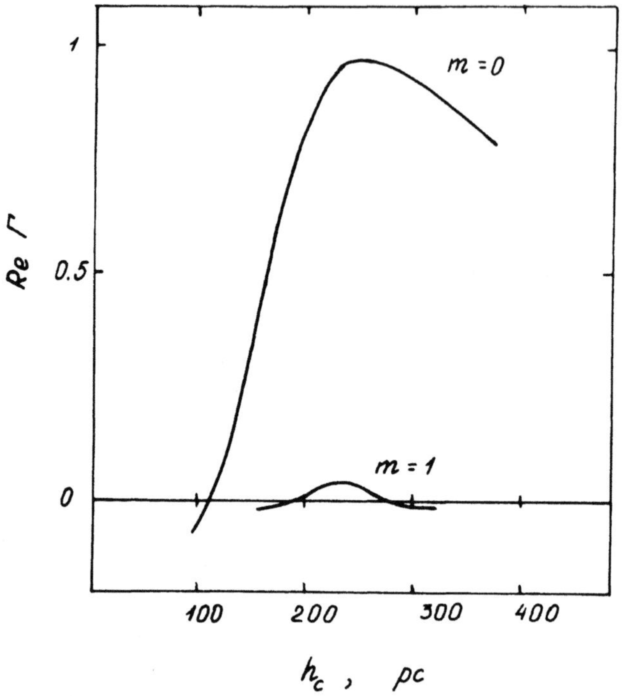

Fig. VII.15. The growth rate, Re Γ, of the lowest axisymmetric ($m = 0$) and non-axisymmetric ($m = 1$) modes localized in the outer part of the Galaxy, $r > 5$ kpc, in dependence of the disc central half-thickness.

rate Re Γ of the large-scale magnetic field in the Galaxy on the disc's central half-thickness for $m = 0$ and $m = 1$. The value $h_c = 150$ pc corresponds to half-thickness $h_\odot = 400$ pc in the solar vicinity; $h_c = 370$ pc gives $h_\odot = 1000$ pc. The mode $m = 2$ decays in our Galaxy for any value of the disc thickness. The computations were made with the outer radial boundary condition posed at $r_{\max} = 30$ kpc.

Note that there exists a rather narrow range of values of the disc half-thickness, $200 \lesssim h_c \lesssim 260$ pc ($540 \lesssim h_\odot \lesssim 720$ pc), for which generation of the $m = 1$ bisymmetric mode is possible.

The level lines $\mathrm{Re}(Q(r)\, e^{im\varphi}) = |Q|$ for $m = 1$ and $h_c = 200$ pc are shown in Figure VII.16. The field strength is maximal at this lines. Arrows indicate a general field direction. These lines should not be confused with magnetic lines!

The local growth rate $\gamma(r)$ in the Galaxy has two local maxima corresponding to a double-peaked form of the rotation curve. Hence, there exists two systems of eigenfunctions (Figure VII.16). The first is characterized by large growth rates

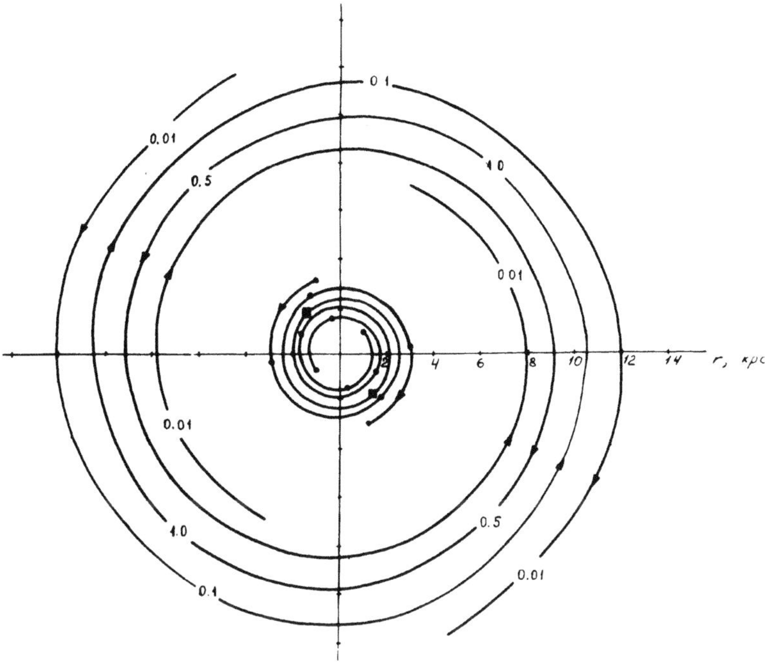

Fig. VII.16. The two lowest $m = 1$ eigenfunctions $Q(r)$ in the Galaxy localized near the center (Re $\Gamma \approx 1.4$) and in the outer part (Re $\Gamma \approx 0$). Shown are the maximal field lines. The filled squares on the inner spirals mark the points where $|Q| = 1$; dots mark the points where $|Q| = 0.25, 0.05, 0.01$ subsequently. Values of $|Q|$ on the outer spirals are given by numbers.

Re $\Gamma \gtrsim 1$ and these eigenfunctions are concentrated in the inner part, $r \lesssim 5$ kpc. Growth rates of another system are smaller, Re $\Gamma \lesssim 1$, but they are concentrated in the outer ring $5 \lesssim r \lesssim 15$ kpc. This means the existence of the minimum in radial distribution of the large-scale field in the Galaxy (Ruzmaikin and Shukurov, 1981). Figure VII.16 shows both the eigenfunction $(m, n) = (1, 1)$, which is concentrated at $r \lesssim 3$ kpc, and the eigenfunction $(m, n) = (1, 2)$ distributed in the outer ring.

Notice that with increase of the central disc thickness the radial potential $\gamma(r)$ in our Galaxy acquires only one minimum (see Figure VII.6). For $h_c > 160$ pc we have $\gamma(r) > 0$ everywhere. When h_c exceeds 200 pc, the minimum in the radial field distribution is washed out. Thus, observational verification of existence of the minimum in the strength of the large-scale magnetic field in our Galaxy at $r \simeq 5$ kpc would imply that the central disc half-thickness does not exceed 200 pc (for the functional form (VII.40)).

Thus, the global structure of the magnetic field in our Galaxy could be either axisymmetric or bisymmetric depending upon the thickness of the ionized gaseous disc. Generation of non-axisymmetric fields is admissible in the rather narrow range of values of the disc thickness. The ring-like distribution of magnetic field around $r \simeq 10$ kpc is possible if the disc thickness is not very large; otherwise the field is distributed uninterrupted over the whole galactic disc, $0 \lesssim r \lesssim 15$ kpc. We note again that observations of magnetic field distribution in our Galaxy can help to estimate the shape and thickness of the ionized gas disc.

(ii) *The Andromeda Nebula.* Dependence of Re Γ on h_c for the $m = 0$ mode, which is localized in the outer part of M31, $r \sim 10$ kpc, is plotted in Figure VII.17. *According to the model considered, all non-axisymmetric modes of the large-scale magnetic field decay in M31.*

Our result agrees with observations that show that the large-scale magnetic field in M31 is concentrated in an axisymmetric ring around $r = 10$ kpc (Sofue and Takano, 1981, Beck, 1982).

(iii) *The Whirlpool Galaxy (M51).* The large-scale magnetic field observed in the spiral galaxy M51 is probably bisymmetric (Tosa and Fujimoto, 1978; see,

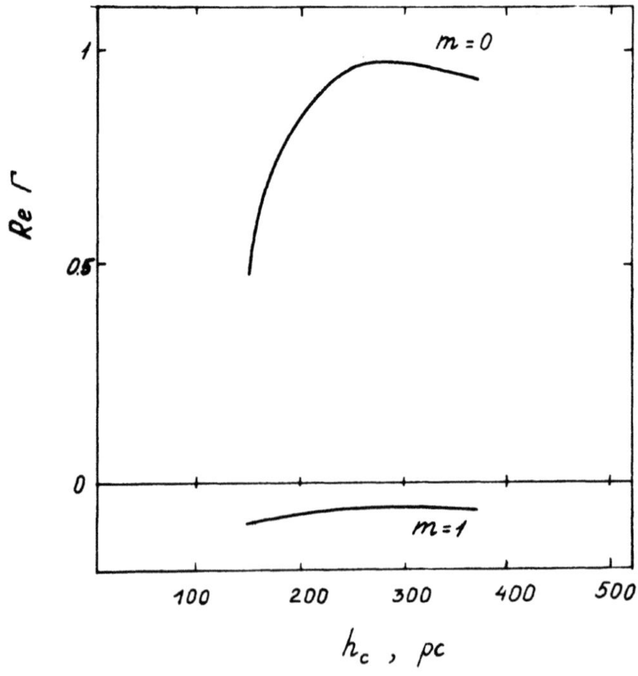

Fig. VII.17. The growth rate, Re Γ, versus the disc central half-thickness for the Andromeda Nebula. Results for the lowest axisymmetric ($m = 0$) and non-axisymmetric ($m = 1$) modes are given. All modes with $m > 0$ decay in M31.

however, Beck, 1986). This means that the growth rate of the non-axisymmetric mode $m = 1$ must be high in this galaxy. Indeed, our computations with use of the rotation curve of Goad *et al.* (1979) and the form (VII.40) for disc half-thickness show that *the growth time of the bisymmetric large-scale magnetic field* $m = 1$ *is as short as* 2×10^8 *years* ($\mathrm{Re}\,\Gamma \simeq 2.5$) *in M51.* The dependence of $\mathrm{Re}\,\Gamma$ on the central disc half-thickness is shown in Figure VII.18. Modes with $m = 1$, $n > 1$ and $m = 2$ decay.

We should stress that our model uses the axisymmetric underlying disc and distribution of generation sources does not depend on azimuthal angle φ. Therefore, axisymmetric fields in our model grow always faster than non-axisymmetric ones. The observed dominance of non-axisymmetric structures in some galaxies can be explained by the following effects.

First, non-axisymmetric fields are concentrated at larger r than are axisymmetric ones. Thus, bisymmetric structure in the main body of a galaxy may coexist with axisymmetric fields in the inner part. To verify this possibility, high-resolution radio observations of magnetic fields in central parts of galaxies are in need.

Second, violation of the axial symmetry of the disc by the spiral pattern promotes, through azimuthal variations in α, generation of non-axisymmetric

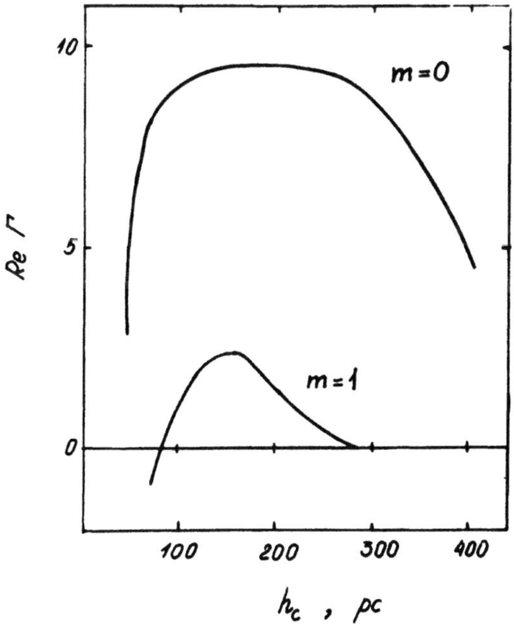

Fig. VII.18. The same as in Figure VII.17 but for the Whirlpool galaxy. The non-axisymmetric mode $m = 1$ has a large positive growth rate in M51.

modes. This effect, which can make non-axisymmetric $m = 1$ mode the dominant one, deserves further theoretical investigation.

Third, the initial (seed) magnetic field could be highly non-axisymmetric, with $m = 1$ mode distinguished to such an extent that this predominance has survived until now.

Finally, the growth rate of axisymmetric field in M51 is close to the upper limit $\mathrm{Re}\,\Gamma = 10$. Therefore, some modifications of the dynamo equations may be needed (e.g. replacement of the simplest form by a general integro-differential equation), which may make closer the growth rates of both modes.

We should also underline that large-scale magnetic fields in M51 have been detected up to distances $r \simeq 10$ kpc from the center, while the mode $(m, n) = (1, 1)$ has the magnitude as low as 0.01 even at $r = 2$ kpc. However, the growth time of this mode is $(\mathrm{Re}\,\Gamma_{11})^{-1} \simeq 2 \times 10^8$ years and in 10^{10} years the initial field in this mode can be amplified e^{50} times! Therefore, even a weak tail of the eigenfunction can easily grow up to the observed strength of 5 μG. In this event the generation regime could be highly non-linear in inner parts $r \lesssim 2$ kpc, while non-linearity is rather weak in outer regions.

Figure VII.19 shows the spatial structure of the mode $(m, n) = (1, 1)$ in M51. The angular velocity of rotation of this pattern is $\mathrm{Im}\,\Gamma = 51.5$, or $\mathrm{Im}\,\Gamma = 100$ km s^{-1} kpc^{-1} in physical units. The pitch angle of the shown spiral is nearly constant for $1 < r < 2$ kpc which indicates a logarithmic spiral. The typical values of the inclination angle are 4—5°. For comparison note that the angular velocity of the spiral pattern in M51 is, presumably, about 40 km s^{-1} kpc^{-1} (Goad *et al.*, 1979) and pitch angle of spiral arms is close to 18° (Tully, 1974).

We see now that in some galaxies the turbulent dynamo generates non-axisymmetric large-scale magnetic fields that can be identified with the observed bisymmetric magnetic structures. The possibility of generation of non-axisymmetric modes is associated with thinness of galactic gaseous discs. The generation process is determined primarily by local distribution of the generation sources and field diffusion across the disc. A rather weak difference in generation conditions for modes with different global, radial and azimuthal structure appears only in higher approximations in the ratio of the disc thickness to its diameter. The structure of generated fields depends on the form of the rotation curve of a given galaxy (the stronger differential is rotation, the weaker is generation of non-axisymmetric fields) and by the thickness of the ionized gaseous disc. The Andromeda Nebula (M31) has such a rotation curve that only axisymmetric fields can be generated there. In contrast, M51 rotates in such manner that a rapid growth (with characteristic time 2×10^8 years) of the magnetic mode $m = 1$ occurs, that corresponds to the bisymmetric magnetic structure observed in this galaxy. Less definite is the situation in our Galaxy. In a narrow, though compatible with observations, range of the disc thickness values generation of the bisymmetric

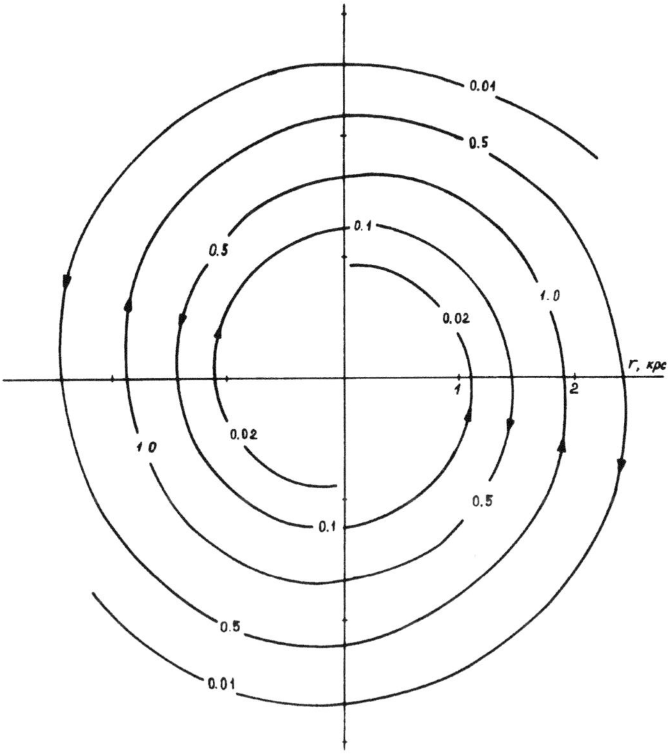

Fig. VII.19. Structure of the radial eigenfunction $Q(r)$ for $(m, n) = (1, 1)$ mode in M51. The disc central half-thickness is $h_c = 150$ pc. Meaning of the curves and notations are the same as in Figure VII.10b.

$m = 1$ magnetic mode is possible in the Milky Way. Therefore, theoretical prediction of the global structure of magnetic field in the Galaxy requires more precise determination of the parameters of the ionized gaseous disc. Observations still cannot determine definitively the large-scale magnetic field configuration in our Galaxy: only a small part of the Galaxy is investigated within 3—4 kpc from the Sun and both axisymmetric (Inoue and Tabara, 1981; Vallée, 1983c, 1984) and bisymmetric (Simard-Normandin and Kronberg, 1979; Sofue and Fujimoto, 1983) configurations have been proposed.

In reality, large-scale magnetic field configurations hardly can be described with a simple division into rings and bisymmetric structures. A wide range of modes is inevitably self-excited with various distributions along the radius (different n's) and the azimuth (different m's). Therefore, various combinations of axisymmetric and bisymmetric structures should be expected that are concentrated in different parts of galactic discs and even overlap. In addition, generation of higher non-axisymmetric magnetic modes $m = 2$ (and even higher) can be expected in principle, that correspond to spiral magnetic structures with four (and more) arms. Higher modes may prevail at greater galactocentric distances. The combination of axisymmetric field in the inner part of a galaxy with bisymmetric structure at its outskirts being even split into a four-armed spiral at still greater radii, seems to be quite natural. Note in this connection that multi-level spiral patterns are often observed in galaxies with a number of arms growing at outskirts.

The galactic dynamo model developed here explains the generation of bisymmetric magnetic structure in an axially symmetric disc. In this case axisymmetric modes are always preferential, they have greater growth rates than non-axisymmetric ones. Influence of the spiral arms undoubtedly facilitates the generation of non-axisymmetric fields and can, presumably, even make them dominant in such galaxies as M51 and M33. Since deviations from axial symmetry are most pronounced in weakly wound spirals, we can expect that bisymmetric structures are more frequent in galaxies with loosely wound spiral patterns (e.g. M33 and M51). Another effect that favours the generation of non-axisymmetric fields is a tidal interaction with galactic companions that obviously violates the axial symmetry (Vallée, 1983c, 1984, 1986). On the other hand, non-axisymmetric large-scale magnetic fields can be generated in disc galaxies even before the spiral pattern and promote the formation of the latter (cf. Fujimoto and Tosa, 1980).

Coexistence of axisymmetric fields in the inner parts of (some) galaxies with bisymmetric structures at greater radii can be an observational test of the theory proposed here. Observations of optical polarization in M51 by Scarrott *et al.* (1987) seem to confirm this point. However, even a failure in the search for axisymmetric fields accompanying bisymmetric structures by no means contradicts the dynamo theory. One more possibility should be investigated, generation of non-axisymmetric magnetic fields with account for violation of the disc's axial symmetry by spiral arms.

The asymptotic solution of the thin-disc dynamo equations constructed here, in which the radial field component is $\lambda^{-1/2}$ times smaller than the azimuthal one, describes the generation of tightly wound ($\lambda \ll 1$) magnetic spirals in an axisymmetric disc. When a disc is axially highly asymmetric or the spiral is weakly wound, other configurations may have greater growth rate, in which the radial and vertical field scales are comparable, $\Delta r \simeq h \simeq O(\lambda r)$ (Baryshnikova *et al.*, 1987b). Such fields can be generated, for instance, in barred spirals or under the influence of spiral arms in normal spiral galaxies.

In conclusion, we note a typical general feature of large-scale magnetic fields (both axisymmetric and non-axisymmetric) generated by the dynamo in a thin disc. In the outer parts of galaxies, where the gaseous disc may be considered thin, the ratio of the vertical field component to the aximuthal one is very small, $|\tilde{B}_z/\tilde{B}_\varphi| \simeq (h/r)(R_\alpha R_\omega)^{-1/2} \simeq 0.01$ (Ruzmaikin *et al.*, 1979, 1980b; see also discussion after

Equation (VII.48). However, B_z grows sharply in the central parts of a galaxy at galactocentric distances comparable to the disc thickness, $r \lesssim 0.5$—1 kpc, where all the field components become comparable. In other words, magnetic lines that run almost parallel to the disc in the main part of a galaxy bend vertically in the inner region (due to solenoidality of the field). This characteristic behaviour of the vertical component of the large-scale field can serve as another observational test for the galactic dynamo theory in general. Note that the field of primordial origin would show much less pronounced and less regular radial variations of B_z: such variations might be due only to spatial variations in compression of the gas with frozen-in magnetic field at galaxy formation stage (cf. Sofue, 1987).

VII.10. Large-Scale Magnetic Fields in Rigidly Rotating Objects

In a few sections above we have discussed self-excitation of the large-scale magnetic field in turbulent non-uniformly rotating discs. The general result is: the magnetic field can be intensified provided the mean helicity of the turbulence is sufficiently large while the rotation sufficiently deviates from the rigid-body law. But may the large-scale field be generated in those regions (or objects) where the rotation is uniform? The answer is affirmative: helical turbulence alone can generate the large-scale magnetic field. In differentially rotating media the meridional field component is produced from the azimuthal one by the helical turbulence, while the inverse process of amplification of the azimuthal component at the expense of the meridional one is due to differential rotation. Meanwhile, in a rigidly rotating medium both elements of the dynamo process are due to the helical turbulence. Dynamo models employing the mean helicity alone (known as α^2-dynamos) have been investigated, mainly numerically, by, e.g., Steenbeck and Krause (1966), Roberts (1972), Krause and Rädler (1980), Zeldovich *et al.* (1983), and Rädler (1986a).

The α^2-dynamo may possibly operate in such cosmic objects as peculiar magnetic stars and planetary cores. In the galactic realm, the α^2-dynamo could be operative in the regions of (near) solid-body rotation that are present in the central parts of many spiral galaxies (see Section VI.1), and in dwarf galaxies which seem to rotate rigidly and seem to host large-scale magnetic fields (Klein and Gräve, 1987). As we have already mentioned, the properties of dynamo-generated fields are strongly dependent on the geometry of the generation region. Available numerical models consider α^2-dynamo in spherical bodies (with applications to stars and planets). However, what we need here is the α^2-dynamo in the disc. Analytical models for particular cases of the delta-function and step function spatial distributions of the mean helicity across the disc have been considered by Ruzmaikin *et al.* (1980a); numerical investigation of the disc α^2-dynamo has been performed by Baryshnikova and Shukurov (1986). Below we discuss the results of analytical asymptotic solution of the α^2-dynamo problem for arbitrary distribution of the mean helicity (Isakov *et al.*, 1981; Sokoloff *et al.*, 1983; Shukurov *et al.*, 1983, 1985).

When the region occupied by helical turbulence of conducting fluid rotates rigidly, $\mathbf{V} = \mathbf{\Omega} \times \mathbf{r}$ with $\mathbf{\Omega} = \text{const}$, Equation (VII.15) reduces to

$$\frac{\partial \mathbf{B}}{\partial t} = R_\alpha \operatorname{rot} \alpha \mathbf{B} + \Delta \mathbf{B} \tag{VII.51}$$

in the reference frame rotating with the angular velocity $\mathbf{\Omega}$. We put $\beta = \text{const}$, $\mu = 1$ and use dimensionless coordinates measured in units of the spatial variations scale of the mean helicity, L, and dimensionless time measured in units L^2/β. We note that R_α is the turbulent magnetic Reynolds number based on the maximal value of the mean helicity (thus, $\max_{(\mathbf{r})}(\alpha) = 1$ in (VII.51)) and the scale L.

We look for asymptotic solutions to Equation (VII.51) for $R_\alpha \gg 1$ whose spatial scale, which turns to be of order R_α^{-1}, is small as compared with the helicity scale L of order of the generation region size but large in comparison with the turbulent scale. Therefore, we can neglect the boundaries of the generation region and consider solutions for an unbounded domain which decay at infinity as least as r^{-3}. In this case the only relevant spatial scale is the scale of spatial variations of the mean helicity.

We use the WKB asymptotic method, known from quantum mechanics, in its modern form developed by V. P. Maslov and his co-workers (see, e.g., Maslov, 1965; Maslov and Fedorjuk, 1981).

Consider Equation (VII.51) with the mean helicity of the general form, $\alpha = \alpha(\mathbf{r})$. The helicity $\alpha(\mathbf{r})$ is a smooth bounded function vanishing at infinity. Hence, at some points, $\alpha(\mathbf{r})$ attains extremal values (positive or negative). We shall begin the consideration with the case when the extrema are reached at some isolated points. The cases when $\alpha(r)$ reaches extremal values on some lines or surfaces are simpler and are considered later.

We shall look for solutions of the form

$$\mathbf{B}(t, \mathbf{r}) = \boldsymbol{\varphi}(\mathbf{r}) \exp[\gamma t + iR_\alpha^q S(\mathbf{r})]. \tag{VII.52}$$

Substitution of (VII.52) into Equation (VII.51) yields terms of order R_α^{2q}, R_α^{q+1}, R_α^q, R_α^1 and R_α^0. If $q > 1$, the single term of the highest order on the right-hand side of (51) comes from the Laplace operator and the corresponding field therefore decays. If $q < 1$, the Laplace operator does not make any contribution to the leading order terms and the corresponding field does not vanish at infinity. Relevant solutions arise therefore only for $q = 1$. Thus, the characteristic wavelength of the generated field is of order R_α^{-1}.

It is quite natural to expand the slowly varying amplitude $\boldsymbol{\varphi}(\mathbf{r})$ and the growth rate γ in power series in the parameter R_α:

$$\begin{aligned} \boldsymbol{\varphi}(\mathbf{r}) &= \boldsymbol{\varphi}_0(\mathbf{r}) + R_\alpha^{-1} \boldsymbol{\varphi}_1(\mathbf{r}) + \dots, \\ \gamma &= R_\alpha^2 \gamma_2 + R_\alpha \gamma_1 + \gamma_0 + \dots. \end{aligned} \tag{VII.53}$$

The expansion of γ begins with the second-order term, γ_2, because the highest order terms on the right-hand side of (VII.51) are of this order.

In order to estimate the growth rate, γ, it is sufficient to calculate γ_2; a threshold value $(R_\alpha)_{cr}$ for field amplification can be estimated if the value of γ_1 is also known; $\boldsymbol{\varphi}_0(\mathbf{r})$ and $S(\mathbf{r})$ yield the first approximation to the eigenfunction.

Substituting (VII.52) and (VII.53) into Equation (VII.51) and combining the terms of the highest order (R_α^2), we obtain the following system of three algebraic equations for the components of $\boldsymbol{\varphi}_0$:

$$\gamma_2\boldsymbol{\varphi}_0 - i\alpha(\nabla S \times \boldsymbol{\varphi}_0) + (\nabla S)^2\boldsymbol{\varphi}_0 = 0 \tag{VII.54}$$

with $(\nabla S)^2 \equiv \Sigma_i(\partial S/\partial x_i)^2$. Non-trivial solution to (VII.54) exists if the determinant of this system vanishes, which yields the Hamilton–Jacobi equation

$$(\nabla S)^2 \mp \alpha[(\nabla S)^2]^{1/2} + \gamma_2 = 0. \tag{VII.55}$$

The corresponding Hamiltonian, $H(p, q) = p^2 \mp \alpha(q)p$, has the stationary point $q = q_0$, $p = \pm\alpha(q_0)/2$, where q_0 is the α-extremum point, i.e. $\nabla\alpha(q_0) = 0$, $\alpha(q_0) = 1$. Equation (VII.55) is satisfied at the stationary point when

$$\gamma_2 = \tfrac{1}{4}.$$

Now, in non-stationary points (VII.55) yields

$$[(\nabla S)^2]^{1/2} = \tfrac{1}{2}\alpha \pm (\tfrac{1}{4}\alpha^2 - \gamma_2)^{1/2} \tag{VII.56}$$

for the upper sign in (VII.55). The boundary condition at infinity is satisfied if $\mathrm{Im}\, S(r) > 0$, i.e. we should choose the upper sign in (VII.56). It is now easily verified that the solution of Equation (VII.55) is

$$S(r) = r[\pm\tfrac{1}{2} + \tfrac{1}{4}i(1-\alpha)^{1/2}] + O(r^3), \tag{VII.57}$$

where we have shifted the origin to the extremum of $\alpha(\mathbf{r})$ and have adopted the natural normalization condition $S(0) = 0$. As can be seen from (57), the magnetic field is concentrated in the vicinity of the extremum.

(i) *Spherical symmetry*, $\alpha = \alpha(r)$. We are now ready to calculate $\boldsymbol{\varphi}_0$ in this vicinity. To simplify the calculations, we shall restrict our attention to the case when the tensor $\partial^2\alpha/\partial x_i \partial x_j$ is isotropic,

$$\alpha = 1 - r^2/2 + \ldots.$$

Thus, the characteristic length L is defined as

$$L = [-\alpha(0)/\alpha''(0)]^{1/2}.$$

For $S(r)$ we obtain the expression

$$S(r) = \pm\tfrac{1}{2}r + \tfrac{1}{4}ir^2 + O(r^3). \tag{VII.58}$$

The determinant of the system (VII.54) for the components of $\boldsymbol{\varphi}_0$ vanishes.

Hence, the solution to (VII.54) can be written as

$$\boldsymbol{\varphi}_0(\mathbf{r}) = \sigma(\mathbf{r})\mathbf{F}(\mathbf{r}), \tag{VII.59}$$

where $\sigma(\mathbf{r})$ is a scalar function, as yet unknown, and the vector function $\mathbf{F}(\mathbf{r})$ has the following Cartesian components:

$$\mathbf{F}_1(\mathbf{r}) = \{xz \mp iyr; yz \pm ixr; -x^2 - y^2\} \tag{VII.60a}$$

in the vicinity of $r = 0$. In spherical coordinates $\{r, \theta, \phi\}$ where θ is the polar angle between the z-axis and the radius-vector, this expression reduces to

$$\mathbf{F}_1(\mathbf{r}) = r^2 \sin\theta\{0; 1; \pm i\}.$$

In contrast to (VII.58), the function (VII.60a) is not spherically symmetric, the z-axis being preferred. Due to the isotropy of the helicity $\alpha(r)$ we may distinguish any other direction by rewriting the functions $\mathbf{F}$ and σ. Choosing one or another pair of equations from (VII.54) we have

$$\mathbf{F}_2(\mathbf{r}) = \{-y^2 - z^2; xy \mp izr; xz \pm iyr\}, \tag{VII.60b}$$

$$\mathbf{F}_3(\mathbf{r}) = \{xy + izr; -x^2 - z^2; yz \mp ixr\} \tag{VII.60c}$$

for the x- and y-directions respectively. It can easily be seen that (60b) can be transformed to (60a) by

$$\mathbf{F}_2 = \mathbf{F}_1 \frac{xz \pm iyr}{(-x^2 - y^2)}; \qquad \sigma_2 = \sigma_1 \frac{(-x^2 - y^2)}{xz \pm iyr}.$$

A similar transformation converts (VII.60c) into (VII.60a).

Before proceeding to evaluate the function $\sigma(\mathbf{r})$ we remind ourselves that the generated field violates the background spherical symmetry. One may expect the configuration of the field to be somewhat similar to a helical wave. It is the specific orientation of the helical wave which violates the spherical symmetry. There are three linearly independent orientations, therefore the solution for a spherically symmetric distribution, $\alpha = \alpha(r)$ is expected to be triply degenerate. Hence, the function $\sigma(\mathbf{r})$ can be constructed in three different ways. These possibilities correspond to the three forms (VII.60a, b, c) of $\mathbf{F}(\mathbf{r})$. In other words, the triple degeneracy is taken into account by a suitable choice of the three different forms of the function $\mathbf{F}(\mathbf{r})$. Consequently, $\sigma(\mathbf{r})$ has one and the same form for any eigenfuntion belonging to the triplet.

To determine the function $\sigma(\mathbf{r})$ and the coefficient γ_1, the equation for the next approximation, $\boldsymbol{\varphi}_1$, should be used:

$$\begin{aligned}&[\gamma_2 - i\alpha\nabla S\times + (\nabla S)^2]\boldsymbol{\varphi}_1 \\ &\quad = -[\gamma_1 - \alpha\nabla\times - \nabla\alpha\times + i\nabla S\times + i\nabla\times(\nabla S\times)]\boldsymbol{\varphi}_0\end{aligned} \tag{VII.61}$$

(Maslov, 1965; Maslov and Fedorjuk, 1981). Here we took advantage of the condition $\boldsymbol{\varphi}_0 . \nabla S = 0$ which is a leading-order consequence of the equation $\nabla \cdot \mathbf{B} = 0$. Note that in order to obtain $\sigma(\mathbf{r})$ and γ_1 it is sufficient to satisfy the latter equation only to leading order.

The operator of the left-hand side of (VII.61) is identical to the operator of Equation (VII.54). Since this operator is degenerate a solution $\boldsymbol{\varphi}_1$ to Equation (VII.61) exists only when the right-hand side of (VII.61) is orthogonal to the eigenvector of the adjoint operator of (VII.54), i.e. if

$$\mathbf{F}^*[\gamma_1 - \alpha\nabla\times - \nabla\alpha\times + i\nabla S\times + i\nabla\times(\nabla S\times)]\sigma\mathbf{F} = 0,$$

where $\mathbf{F}$ is any function from the triplet (VII.60) and the asterisk denotes the complex conjugate. After some algebra, this equation reduces to

$$(\gamma_1 + \tfrac{7}{2})\sigma + (\mathbf{r} \cdot \nabla\sigma) = 0 \qquad \text{(VII.62)}$$

for a spherically symmetric distribution of the helicity, where the terms of two lowest orders in r_i are kept. Solutions to Equation (VII.62) are homogeneous functions of an arbitrary positive integer of order l. Keeping in view that the generated magnetic field should be differentiable at the point of the extremum of $\alpha(r)$ (i.e. at $r = 0$), we find a set of permissible l's. The function $S(r)$ which appears in (VII.52), is not differentiable at the origin [see (VII.58)]. However, one can construct two types of real differentiable asymptotic solutions $\mathbf{B}$ with help of the function $\boldsymbol{\varphi}_0$ for a suitable choice of σ. The first is obtained when $\sigma = f_l(\mathbf{r})$, where f_l is a homogeneous differentiable function of the coordinates of order l. In this case we have from (VII.62):

$$\gamma_1^{(1)} = -(\tfrac{7}{2} + l), \qquad l = 0, 1, 2, \ldots. \qquad \text{(VII.63)}$$

The corresponding asymptotic solution to Equation (VII.51) is merely the real part of (VII.52):

$$\begin{aligned}\mathbf{B} = \{&xz \cos \tfrac{1}{2} R_\alpha r + yr \sin \tfrac{1}{2} R_\alpha r;\ yz \cos \tfrac{1}{2} R_\alpha r - xr \sin \tfrac{1}{2} R_\alpha r;\\ &-(x^2 + y^2) \cos \tfrac{1}{2} R_\alpha r\}\ f_l \exp[(\tfrac{1}{4} R_\alpha^2 + \gamma_1^{(1)} R_\alpha)t - \tfrac{1}{4} R_\alpha r^2],\end{aligned} \qquad \text{(VII.64)}$$

where f_l is a homogeneous polynomial in x, y, z of degree l.

The second type of asymptotic solution arises when $\sigma = r^{-1} f_m$ and, consequently,

$$\gamma_1^{(2)} = -(\tfrac{7}{2} + m - 1), \qquad m = 0, 1, 2, \ldots. \qquad \text{(VII.65)}$$

The corresponding differentiable asymptotic solution is the imaginary part of (VII.52) and has the form

$$\begin{aligned}\mathbf{B} = \{&-yr \cos \tfrac{1}{2} R_\alpha r + xz \sin \tfrac{1}{2} R_\alpha r;\ xr \cos \tfrac{1}{2} R_\alpha r + yz \sin \tfrac{1}{2} R_\alpha r;\\ &-(x^2 + y^2) \sin \tfrac{1}{2} R_\alpha r\}\ r^{-1} f_m \exp[(\tfrac{1}{4} R_\alpha^2 + \gamma_1^{(2)} R_\alpha)t - \tfrac{1}{4} R_\alpha r^2],\end{aligned} \qquad \text{(VII.66)}$$

where f_m is a homogeneous polynomial in x, y, z of degree m. For instance,

$$f_{m(l)} = \text{const}, \qquad \text{for } m(l) = 0,$$

$$f_{m(l)} = ax + by + cz, \qquad \text{for } m(l) = 1, \text{etc.}$$

The asymptotic forms (VII.64) and (VII.66) correspond to the function $\mathbf{F}_1(\mathbf{r})$ from the triplet (VII.60). The remaining functions (VII.60b, c) yield analogous expressions.

The mode with $m = 0$ in (VII.65) grows most rapidly. For this mode $\gamma_1 = -\frac{5}{2}$ and

$$\gamma_{\max} = \tfrac{1}{4} R_\alpha^2 - \tfrac{5}{2} R_\alpha + O(1). \tag{VII.67}$$

The eigenvalues γ are real if we restrict ourselves to the first two orders in R_α. At higher orders the operator of (VII.51) is, however, not self-adjoint and the eigenvalues can acquire imaginary parts. But as can be seen from above, the imaginary part of any eigenvalue is at least R_α^2 times smaller than its real part. Hence, the period of oscillations of the magnetic field generated by the α^2-dynamo is at least R_α^2 times larger than its growth time.* Note that the method used here permits us to calculate eigenvalues with greater relative accuracy than the simultaneously derived eigenvectors.

From $\gamma_{\max} = 0$ we can estimate the threshold (critical) value of R_α:

$$(R_\alpha)_{\text{cr}} = 10 + O(0.1). \tag{VII.68}$$

The asymptotic solution corresponding to the eigenvalue (VII.67) is triply degenerate and has the form

$$\mathbf{B} = \operatorname{Im}[r^{-1}\mathbf{F}(\mathbf{r}) \exp(\pm \tfrac{1}{2} i R_\alpha r)] \exp(\gamma_{\max} t - \tfrac{1}{4} R_\alpha r^2), \tag{VII.69}$$

where $F(r)$ is any function from (VII.60a, b, c). In spherical coordinates $\{r, \theta, \phi\}$, where θ is the polar angle between r and the z-axis, the solution corresponding to (VII.60a) reduces to

$$\mathbf{B} = r \sin\theta \{0; \sin \tfrac{1}{2} R_\alpha r; \cos \tfrac{1}{2} R_\alpha r\} \exp(\gamma_{\max} t - \tfrac{1}{4} R_\alpha r^2).$$

The remaining asymptotic eigenfunctions from the triplet can be likewise simply expressed with each in its particular spherical coordinate system (the polar angle being defined with respect to the x- or y-axis respectively).

The generated field is concentrated within a distance of order $R_\alpha^{-1/2}$ from the point at which α attains its extremal value. However, we should emphasize that the field vanishes at this very point $\mathbf{B}(0) = 0$.

The degree of degeneracy of the eigenvalues (VII.63) and (VII.65) increases with l (and m) due to the increase of the number of linearly independent

* Numerical experiments by Baryshnikova and Shukurov (1987) reveal another type of oscillatory solution where oscillations are probably due to the effects of boundaries of the generation region.

homogeneous polynomials f_l (and f_m) of degree l (and m). The critical value of R_α also increases.

In the first approximation in R_α, the generated field is force-free near the α-extremum point, i.e.

$$\nabla \times \mathbf{B} \approx \tfrac{1}{2} R_\alpha \mathbf{B}.$$

This is true for solutions of any l (or m).

It is clear that the solution **B** is no longer triply degenerate if the tensor $\partial^2 \alpha / \partial x_i \partial x_j$ is anisotropic.

If the distribution of $\alpha(r)$ possesses a number of local extrema, the generation proceeds nearly independently around each (provided the separation of the extremum points is considerably greater than $R_\alpha^{-1/2}$). In particular, when α is, say, an odd function of z, the dipole and quadrupole modes should be characterized by nearly identical critical values of R_α. This remarkable peculiarity has been revealed numerically by Steenbeck and Krause (1966) and Roberts (1972) and discussed by Proctor (1977a, b). Note also that numerical calculations of Steenbeck and Krause (1966) and Roberts (1972) demonstrated that the field generated by the α^2-dynamo is concentrated near the extrema of α and vanishes at these points. The critical values of R_α in their models (with the anisotropic tensor $\partial^2 \alpha / \partial x_i \partial x_j$) are of the order of ten, which is in good agreement with our result (VII.68). (When comparing the results of their numerical calculations with our analysis, one should pay proper attention to the definition of the characteristic scale on which the Reynolds number is based.)

(ii) *Axial symmetry*, $\alpha = \alpha(s)$. Let us now consider the situation when $\alpha(\mathbf{r})$ attains its extremum on some line, e.g. $s = 0$. The discussion which has led to the expression (VII.57) for $S(\mathbf{r})$ is still valid, with the only difference that now

$$\alpha = 1 - \tfrac{1}{2} s^2 + \ldots, \qquad s^2 = x^2 + y^2,$$

$$S = \pm \tfrac{1}{2} s + \tfrac{1}{4} i s^2 + O(s^3).$$

Due to the existence of the preferred direction, z, the triple degeneracy is now removed and there is only one vector function

$$\mathbf{F}(\mathbf{r}) = \{ iy; -ix; \pm s \}$$

which satisfies the Equation (VII.54). Calculations which are quite similar to the ones carried out above bring us to the following equation for σ and γ_1:

$$(\gamma_1 + 2)\sigma + (\mathbf{s} \cdot \nabla \sigma) = 0.$$

The solution with $\sigma = s^{-1}$ grows most rapidly and

$$\gamma_{\max} = \tfrac{1}{4} R_\alpha^2 - R_\alpha,$$

$$(R_\alpha)_{\text{cr}} = 4 + O(1).$$

The corresponding asymptotic solution,

$$\mathbf{B} = \mathrm{Re}[s^{-1}\mathbf{F}(\mathbf{r})\exp(iR_\alpha s + \gamma_{\max} t)]$$

is given by

$$\mathbf{B} = \{\mp y \sin \tfrac{1}{2} R_\alpha s;\ \ \pm x \sin \tfrac{1}{2} R_\alpha s;\ \ \pm s \cos \tfrac{1}{2} R_\alpha s\}\ \ s^{-1} \exp(\gamma_{\max} t - \tfrac{1}{4} R_\alpha s^2).$$

In cylindrical polar coordinates it reduces to

$$\mathbf{B} = \{0;\ \ \pm \sin \tfrac{1}{2} R_\alpha s;\ \ \pm \cos \tfrac{1}{2} R_\alpha s\}\ \ \exp(\gamma_{\max} t - \tfrac{1}{4} R_\alpha s^2).$$

We shall not write the expressions for the eigenvectors with arbitrary l and m, because this can be done in a manner similar to the one used during the derivation of (VII.64) and (VII.66). The magnetic field generated by the two-dimensional distribution of mean helicity is asymptotically force-free in the vicinity of the symmetry axis, i.e. $\nabla \times \mathbf{B} \simeq \tfrac{1}{2} R_\alpha \mathbf{B}$ just as in the three-dimensional situation.

(iii) *Planar symmetry*, $\alpha = \alpha(x)$. A detailed discussion of the case when $\alpha(\mathbf{r})$ attains an extremum on some surface (plane) can be found in Isakov *et al.* (1981) and Shukurov *et al.* (1983, 1985). The calculations are quite similar to those carried out above. Now

$$\alpha = 1 - \tfrac{1}{2} x^2 + \tfrac{1}{6} \alpha_3 x^3 + \tfrac{1}{24} \alpha_4 x^4 + \ldots \qquad \text{(VII.70)}$$

and

$$S = \pm \tfrac{1}{2} x + \tfrac{1}{4} i x^2 + O(x^3).$$

The greatest eigenvalue

$$\gamma_{\max} = \tfrac{1}{4} R_\alpha^2 - \tfrac{1}{2} R_\alpha + O(1)$$

is doubly degenerate. This degeneracy arises due to the invariance with respect to rotation in the plane $x = \text{const}$. The corresponding asymptotic solutions are

$$\mathbf{B} = \{0;\ \ \sin \tfrac{1}{2} R_\alpha x;\ \ \cos \tfrac{1}{2} R_\alpha x\}\ \ \exp(\gamma_{\max} t - \tfrac{1}{4} R_\alpha x^2),$$

$$\mathbf{B} = \{0;\ \ -\cos \tfrac{1}{2} R_\alpha x;\ \ \sin \tfrac{1}{2} R_\alpha x\}\ \ \exp(\gamma_{\max} t - \tfrac{1}{4} R_\alpha x^2).$$

The critical value of R_α is given by

$$(R_\alpha)_{\mathrm{cr}} = 2 + O(1).$$

Few following terms of asymptotic expansion have been calculated in this one-dimensional case for the principal mode (Shukurov *et al.*, 1985). In particular, this result yields the oscillation frequency of the solution, Im γ. With inclusion of higher order terms, the asymptotic expansion of the greatest eigenvalue is given by

$$\gamma_{\max} = \tfrac{1}{4} R_\alpha^2 - \tfrac{1}{2} R_\alpha - \tfrac{1}{16}(1 - \alpha_4 - \tfrac{11}{9} \alpha_3^2) +$$

$$+ i \frac{\alpha_3}{64} (1 - \alpha_4 - \tfrac{11}{9} \alpha_3^2) R_\alpha^{-1} + O(R_\alpha^{-2}).$$

The imaginary part of the eigenvalue appears only in the fourth approximation. The ratio of the oscillation frequency to the growth rate is given by

$$\frac{\operatorname{Im}\gamma}{\operatorname{Re}\gamma} \approx \frac{1}{10\, R_\alpha^3}$$

(for $\alpha_3 = \alpha_4 = 1$).

These asymptotic forms for $R_\alpha \gg 1$ can be complemented by a simple asymptotic solution for $R_\alpha \ll 1$. Eigenvalues for $R_\alpha \ll 1$ follow from the perturbation theory when free decaying modes with $R_\alpha = 0$ are the unperturbed state. Note that in order for the unperturbed problem to have a discrete spectrum, boundary conditions should be introduced, e.g. vanishing of the field at the boundaries of a slab. The perturbation $\delta\gamma$ to the greatest eigenvalue is given by

$$\delta\gamma = -\tfrac{1}{2} iR_\alpha \int_{-1}^{1} \alpha' \sin^2 \pi x \, \mathrm{d}x,$$

which yields for $\alpha(x)$ of the chosen form (VII.70)

$$\delta\gamma = -i \frac{\alpha_3}{18}\left(1 - \frac{3}{2\pi^2}\right) R_\alpha + O(R_\alpha^2) \qquad \text{for } R_\alpha < 1$$

(Shukurov *et al.*, 1985).

We have derived the asymptotic solutions of the α^2-dynamo problem for a non-uniform distribution of the mean helicity. The cases when the helicity attains its extremal values at isolated points, lines and surfaces and possesses the corresponding symmetry are considered. It is shown that the threshold values of R_α are rather close to each other for all possible configurations of $\alpha(\mathbf{r})$, i.e. $(R_\alpha)_{\text{cr}} = 10, 4, 2$ respectively. Therefore, it is clear that the anisotropy of the tensor $\partial^2\alpha/\partial x_i \partial x_j$ at the α-extremum points cannot considerably change the value of $(R_\alpha)_{\text{cr}}$. Moreover, we can now place certain limits on such a variation. For instance, $10 > (R_\alpha)_{\text{cr}} > 2$ when the helicity has an extremum at some isolated point, even when the tensor $\partial^2\alpha/\partial x_i \partial x_j$ is anisotropic there.

The results obtained here throw light on some qualitative peculiarities of the spherical numerical models of Steenbeck and Krause (1966) and Roberts (1972). The critical values of R_α for dipole and quadrupole modes are very close to each other, because the generation of magnetic field proceeds nearly independently around each extremum point of the mean helicity (one in the northern hemisphere and the other in the southern one), the points being separated by a distance much larger than $R_\alpha^{-1/2}$. Thus, the eigenvalue (the growth rate) and the critical value of R_α are only slightly modified by field reversal near either extremum, such a reversal being equivalent to a change from dipole symmetry to quadrupole symmetry (or vice versa).

Our results reveal a further interesting feature, not properly emphasized previously: the generated field vanishes at the α-extremum (isolated) point, but the field maximum is reached in its vicinity. The same is true for the components of **B** which are orthogonal to the axis of symmetry. This result is in agreement with numerical calculations [see, e.g., Figs. 16.5 and 16.6 in Krause and Rädler (1980)].

Strictly speaking, the described asymptotic solutions are valid only for $R_\alpha \gg 1$. However, one-dimensional computations of Baryshnikova and Shukurov (1987) show that these solutions have reasonable accuracy for the values of R_α as low as $(R_\alpha)_{cr}$ and can be applied to the dynamo systems with moderate R_α. On the other hand, boundary effects can introduce qualitatively new, rapidly oscillating solutions.

VII.11. Magnetic Fields Within Spiral Arms

The turbulent dynamo mechanism in differentially rotating spiral galaxies, which was discussed above, produces large-scale magnetic fields with a dominant azimuthal component. A typical pitch angle of magnetic lines to the circumferential direction is 10°—20°. Meanwhile, the large-scale fields observed within the spiral arms are closely aligned with the arms that generally have somewhat smaller pitch angles, typically about 10°. In addition, magnetic fields within the arms are considerably stronger than in interarm regions. In this section we discuss how these differences can be explained in those cases where large-scale spiral shock waves are present at the spiral arms edges (Ruzmaikin, 1976).

The most popular understanding of a spiral arm describes it as a density wave which rotates as a solid body at the angular velocity that is smaller than that of the galactic disc. For the Galaxy, the angular velocity of the spiral pattern is estimated as $\Omega_p \approx 13.5$ km s^{-1} kpc^{-1} while the disc rotates with $\Omega \approx 25$ km s^{-1} kpc^{-1} at $r = 10$ kpc radius. The gas meets the arm at its inner edge. Since the difference between linear rotation velocities of the spiral arms and the disc exceeds the average sound speed (~ 10 km s^{-1}) over a wide range of radii, the spiral shock wave forms along the inner edge of the spiral arm; due to effective cooling, the shock is nearly isothermal (Roberts, 1969; Kaplan and Pikelner, 1974). According to well developed theoretical considerations (Polovin, 1960), the shock wave intensifies the component of magnetic field which is tangential to the spiral arm. When the field upstream from the wave is purely tangential, the intensification factor of the magnetic field is equal to the density compression factor. In the general case, magnetic lines that enter the shock wave are 'refracted' so that the normal field component remains unchanged while the tangential component is intensified. The ratio of tangential field components before the shock and immediately after it is given by

$$B_{t2} = B_{t1}\frac{\rho_2}{\rho_1}(1-\eta)\left(1-\frac{\rho_2}{\rho_1}\eta\right)^{-1} \qquad \text{(VII.71)}$$

with

$$\eta = \frac{B_{\eta 1}^2}{4\pi\rho_1 V_{n1}^2}$$

where ρ_2/ρ_1 is the density compression, B_t and B_n are the tangential and the normal components of the magnetic field, respectively, and V_n is the normal component of the regular velocity. Subscripts 1 and 2 refer to pre-shock and post-shock values, respectively. Relation (VII.71) follows from the continuousness of the momentum flux and tangential electric field at the shock front and takes into account the subsequent one-dimensional expansion.

It may seem that the tangential component of magnetic field can be additionally amplified by the next spiral arm and repetitive passage of spiral shock waves can provide an effective field generation (in the galactic lifetime the spiral pattern makes several tens of rotations). However, the field does not remain constant between consecutive spiral arms but rather recovers, along with the gas density, its pre-shock value. Indeed, the magnetic field can be considered frozen into the gas, and magnetic lines in the interarm gas are parallel to the streamlines of the gas (see Roberts and Yuan, 1970). Meanwhile, the gas velocity field, having experienced the shock wave action (the jump in the normal velocity component accompanied by the change in the sign of the derivative of the tangential velocity), returns to the initial state before the next spiral arm approaches (Roberts, 1969). When the gas orbits are closed and stationary, the above effect of magnetic field generation is completely absent. However, the gas flow is only quasi-stationary and in fact the gas spirals toward the galactic center (Kaplan and Pikelner, 1974). Let us consider the consequent amplification of the large-scale magnetic field. This effect is negligible in the Milky Way galaxy, but it may be of importance in some cases when the spiral pattern is loosely wound.

Let us begin with estimation of the radial displacement undergone by a gas volume element after one rotation in the reference frame connected with the density wave (Roberts and Shu, 1972). Let V_n and V_t be the normal and tangential components of the gas velocity, respectively ($V_{n1} = r(\Omega - \Omega_p) \sin i$ and $V_{t1} = r(\Omega - \Omega_p) \cos i$ with i the pitch angle of the spiral arm), $\sigma \simeq 3\,\mathrm{M}_\odot\,\mathrm{pc}^{-2}$ is the gas surface density, $a \simeq 8\ \mathrm{km\ s}^{-1}$ is the isothermal sound velocity. At the shock front, V_n and σ experience a jump, while the other quantities are continuous. After the passage of the shock wave V_n and σ return to nearly their initial values. However conservation of $V^2/2 + a^2 \log \sigma + (\Psi - r^2\Omega^2 p/2)$, where $\Psi(r)$ is the gravitational potential ($\nabla\Psi_1 = r\Omega^2$), along the streamline that does not intersect a shock wave dictates that the combination $V_t^2/2 + \Psi - r^2\Omega_p^2/2$ should be changed, at the shock front, by the jump $[V_n^2 + a^2 \log \sigma]_{2,1}$. From this we determine, in the approximation of weak departures from the stationary flow, the radial displacement of the volume element after one full rotation

$$\Delta r = -\frac{m\,[V_n^2/2 + a^2 \log \sigma]_{2,1}}{(\Omega - \Omega_p)\,|\,\mathrm{d}r^2\Omega\,|\,\mathrm{d}r\,|}$$

where m (= 2) is the number of spiral arms.

Very approximately,

$$-\Delta r \approx \xi r^3 \sin^2 i; \qquad \xi \approx 10^{-2}\,\mathrm{kpc}^{-2}$$

for our Galaxy (Roberts and Shu, 1972). Dependence of the radial displacement Δr on the galactocentric distance r indicates that volume elements do not only move along spirals toward the center but also approach each other along the radius. This leads to amplification of azimuthal magnetic field. In addition, a pair of volume elements residing at different radii recede from one another along the azimuth due to non-uniformity of rotation, which leads to amplification of the azimuthal component in the expense of the radial one. To evaluate the total variation of magnetic field we use the induction equation with the dissipation neglected,

$$\frac{\partial \mathbf{B}}{\partial t} = \mathrm{rot}(\mathbf{V} \times \mathbf{B}).$$

We assume that the magnetic field and velocity are both two-dimensional (r and φ components), and spiral arms are tightly wound ($\sin i \ll 1$). Averaging the induction equation over φ and letting

$$V_\varphi \approx r\Omega(r),$$

$$V_r \approx \frac{\Delta r}{\Delta t} \approx -\frac{\xi r^3(\Omega - \Omega_p)\sin^2 i}{2\pi},$$

gives the following equation for the azimuthal field component:

$$\frac{\partial B_\varphi}{\partial t} = -\frac{\partial}{\partial r}(V_r B_\varphi) + B_r r \frac{\mathrm{d}\Omega}{\mathrm{d}r}$$

with $B_r \approx$ const.

The solution to this equation grows exponentially as

$$B_\varphi(r, t) = \left[B_{\varphi 0}(r_0) + \tau r \frac{\mathrm{d}\Omega}{\mathrm{d}r} B_r \right] \exp(t/\tau) - \tau r \frac{\mathrm{d}\Omega}{\mathrm{d}r} B_r, \qquad \text{(VII.72)}$$

where the amplification time is

$$\tau = -\left(\frac{\partial V_r}{\partial r}\right)^{-1} \approx \left(\frac{3\Delta r}{r}\right)^{-1} \frac{2\pi}{\Omega - \Omega_p}.$$

At the radius $r = 10$ kpc we have $2\pi/(\Omega - \Omega_p) \simeq 5 \times 10^8$ years which gives $\tau \simeq 10^{10}$ years. Therefore, this amplification effect is negligible in our Galaxy. In the limit $t \ll \tau$ the exponent in (VII.72) can be expanded, $\exp(t/\tau) \approx 1 + t/\tau$ and the picture reduces to the well known wrap-up of B_φ from B_r by differential rotation. We should bear in mind, however, that $\tau \propto (\sin i)^{-2}$ and the great value of τ is mainly due to small pitch angle of the spiral arms, $\sin i \ll 1$. In those galaxies where spiral arms are only weakly wound, especially in barred spirals where i approaches 90°, this effect of magnetic field amplification may be essential. The quantitative estimate requires knowledge of gas trajectories in barred spirals.

Note also that a purely azimuthal magnetic field in barred spirals would experience a rather weak 'refraction' and amplification at spiral shock waves since the field in this case is perpendicular rather than parallel to the spiral arms. In other words, the observed enhancement factor of the magnetic field within the arms may be used for an estimation of the magnetic lines pitch angle when the pitch angle of spiral arms is also known.

Thus, we reach the conclusion that spiral shock waves in the Galaxy (and many other galaxies) affect only the shape of magnetic lines, aligning them with spiral arms within the arms, but hardly influencing the field strength averaged over the whole disc. However, in galaxies with loosely wound spiral patterns, e.g. in barred spirals, the effect of magnetic field amplification due to the unclosed (non-circular) form of gas orbits may be essential.

Let us now turn to the problem which, in a certain sense, is inverse to that just discussed: how do magnetic fields affect the spiral pattern? According to Roberts and Yuan (1970), the role of magnetic fields in the formation of spiral density waves reduces to contribution $H^2/4\pi$ into total pressure in the interstellar medium. This makes evident the result of these authors: the presence of a magnetic field reduces the strength of the shock wave, i.e. the density enhancement factor, and makes weaker non-linear gas dynamic effects. Interstellar magnetic fields contribute approximately one third of the total pressure in the interstellar medium, the remaining part being equally divided between turbulent motions and cosmic rays. Therefore, magnetic fields play a noticeable role: due to them the density compression in the spiral shock wave is lower by approximately a factor of 1.5, as compared with non-magnetic shocks. It is interesting that the results of Roberts and Yuan allow them to obtain an upper estimate on the magnetic field strength admissible in their model. The density compression in spiral shock waves must be sufficiently high in order to ensure rapid star formation in spiral arms. The required compression factor is estimated as 3:1. From this Roberts and Yuan estimate that the shock waves would not be sufficiently strong if $H \gtrsim 6\ \mu$G.* The exact value of this upper limit is, of course, model dependent, but the closeness of the obtained value to the fields actually observed in our Galaxy and other spiral

* Note that this estimate refers rather to the total interstellar magentic field which is estimated as $H \simeq 5\ \mu$G in the Galaxy, rather than to the large-scale magnetic field component $B \simeq 2\ \mu$G.

galaxies (see Table IV.3) emphasizes the necessity of careful examination of magnetic effects in analysis of the spiral structure of galaxies.

To conclude this section we make one more point. In the previous two sections we have shown that in some galaxies nonaxisymmetric fields having two-armed spiral structure could be generated in an *axisymmetric* disc, i.e. even *before* the spiral pattern. Thus, the question arises concerning the relation between bisymmetric magnetic fields and the spiral structure: did they form simultaneously and, if not, which one was the first?

VII.12. Non-linear Effects in the Galactic Dynamo

When the velocity field is fixed, the dynamo equations are linear in magnetic field which implies infinite growth of self-excited magnetic field. Only back-action of magnetic force onto the motion results in a stationary state. The practically important question is: have the large-scale magnetic fields in galaxies had enough time to reach the stationary state? The characteristic time of exponential growth of the field is only several times shorter than the galactic lifetime and the field back-action on motions may prove to be quite weak, provided the initial, seed field is sufficiently weak. Nevertheless, the estimate of the stationary field strength is of paramount importance from the fundamental point of view.

Note first that the stationary state of the large-scale magnetic field can hardly be that corresponding to the equipartition between magnetic and turbulent kinetic energies. Indeed, the field growth is due, in particular, to the mean helicity of interstellar turbulence which can be probably suppressed far before magnetic force produces significant modifications of the total turbulent velocity field. A qualitative analysis of the influence of magnetic field on the mean helicity in the galactic disc was first performed by Vainshtein and Ruzmaikin (1972).

Assuming that the influence of the large-scale magnetic field on the motion reduces, to the lowest approximation, solely to modification of the mean helicity of the flow,

$$\alpha = \alpha_h + \alpha_m(\mathbf{B}),$$

where α_h is the hydrodynamic helicity considered above and α_m is the additional helicity due to the magnetic force. The differential rotation $V(r) = r\Omega(r)$ and the turbulent diffusivity are assumed to be unaffected. For the sake of simplicity, we restrict ourselves to axisymmetric case, $\partial/\partial\varphi = 0$, and take into account the thinness of the galactic gaseous disc, $\partial/\partial z \gg \partial/\partial r$. Consider the dynamo equation (VII.15) for the stationary state $\partial\mathbf{B}/\partial t = 0$:

$$\operatorname{rot}(\mathbf{V} \times \mathbf{B} + \alpha\mathbf{B} - \beta \operatorname{rot} \mathbf{B}) = 0, \tag{VII.73}$$

or, equivalently,

$$\mathbf{V} \times \mathbf{B} + \alpha \mathbf{B} - \beta \operatorname{rot} \mathbf{B} = \nabla \Phi. \tag{VII.74}$$

Under adopted assumptions, the φ-component of Equation (VII.73) and r and φ components of Equation (VII.74) reduce to the following approximate equations:

$$-\Omega B_r + \beta \frac{\partial^2 B_\varphi}{\partial z^2} = 0, \tag{VII.75}$$

$$-\Omega_r B_z + \beta \frac{\partial B_\varphi}{\partial z} = 0, \tag{VII.76}$$

$$\alpha B_\varphi - \beta \frac{\partial B_r}{\partial z} = 0, \tag{VII.77}$$

which yield the order of magnitude estimates of the ratios between field components (note that $\beta \simeq l_0 v_0/3 \simeq h v_0/10$)

$$\frac{B_r}{B_\varphi} \simeq \frac{\beta}{\Omega h^2} \simeq \frac{v_0}{10\,\Omega h} < 1,$$

$$\frac{B_z}{B_\varphi} \simeq \frac{\beta}{\Omega r h} \simeq \frac{h}{r} \frac{B_r}{B_\varphi}.$$

Thus, the azimuthal field component is dominant in a thin disc far from the center.

An equation for B_φ can be obtained by eliminating B_r from (VII.75) and (VII.77),

$$(\alpha_h + \alpha_m) B_\varphi - \frac{\beta^2}{\Omega} \frac{\partial^3 B_\varphi}{\partial z^3} = 0, \tag{VII.78}$$

and a boundary condition at the disc surface should be prescribed.

To find out the function $\alpha_m(\mathbf{B})$ the Navier—Stokes equation should be considered with magnetic force included (see Zeldovich *et al.*, 1983). When the magnetic back-action is still weak, this function is quadratic in B. Upon adopting the estimate $\alpha_m \simeq -\kappa B_r B_\varphi / 4\pi\rho v$, where $\kappa < 1$ is the dimensionless parameter, the following crude estimate follows from (VII.78) for the steady-state field strength:

$$\frac{B_\varphi^2}{4\pi\rho v^2} \simeq \frac{1}{10\,\kappa}.$$

Noting that the value of κ is rather uncertain, we can conclude the magnetic energy density is close to, but less than, the equipartition value.

Solution of the non-linear dynamo problem remains a fascinating theoretical problem. It is already clear, however, that the above picture of magnetic field saturation in the galactic disc is an oversimplification. It includes only the field saturation across the disc. In a real situation, after the field growth has been halted in the localization region of the linear eigenfunction, the turbulent diffusion would transport the field (which is still maintained by the dynamo) toward weak tails of the eigenfunction. A similar transfer occurs both along the radius and across the disc. As a result, a slow spreading occurs of localized magnetic structures formed at the linear stage. Therefore, the large-scale magnetic field concentrations observed in spiral galaxies (e.g. the ring in M31) are a testimony of their magnetic youth. Note that localized field structures may arise due to another kind of process, non-linear self-excited oscillations like the solar activity cycle. In this case magnetic field is rebuilt sufficiently frequently, even in the non-linear regime, as to make non-essential the non-linear spreading of the structure. We do not expect that non-linear oscillatory regimes can be established in galactic discs because linear solutions of the dynamo problem for the disc geometry are not oscillatory when the dynamo numbers are moderate and the generation proceeds not far from the threshold. Meanwhile, the oscillatory character of non-linear solutions in a spherical shell, which describe the solar activity cycle, stems from oscillatory linear solutions (see, e.g., Zeldovich *et al.*, 1983). Linear oscillatory modes appear in the disc geometry only for dynamo numbers as high as a few hundreds (Isakov *et al.*, 1981) which may, in principle, lead to non-linear oscillations of the fields in galaxies with very effective generation and in central parts of normal galaxies.

We wish to draw the reader's attention to the fact that some hints on the described processes of establishing the radial field distribution are present even in the simplest equation (VII.78). This is the third-order equation, while the initial form (VII.73) seems to suggest the appearance of the fourth-order equation after one of unknown functions is eliminated from a pair of second-order equations. Consequently, the properly posed boundary value problem for Equation (VII.78) should not, in the general case, have solutions satisfying all boundary conditions. Consistent analysis of all such difficulties would show the necessity of including the radial dependences of the solutions. The indicated degeneracy of the induction equation with vanishing growth rate is well known in linear theory, where it was considered by Ruzmaikin *et al.* (1979) and Arnold *et al.* (1982).

VII.13. Generation of Fluctuation Fields

Chaotic, small-scale magnetic fields play in the interstellar and intergalactic medium a role no less important than that of the mean, large-scale fields. For instance, interpretation of the synchrotron emission requires knowledge of a higher fractional momentum of the total field, $\langle H_{(\gamma + 1)/2} \rangle$ where $\gamma \simeq 2.6$—2.8 is the energy spectral index of relativistic electrons. According to the Faraday rotation observations, in the Galaxy the fluctuation **b** and the mean **B** fields are related by

$$\frac{\langle b^2 \rangle^{1/2}}{B} = 1\text{—}2,$$

and the total field $\mathbf{H} = \mathbf{B} + \mathbf{b}$ has the root mean square strength $\langle H^2 \rangle^{1/2} = (1.5\text{—}2)B$.

Still more delicate properties of fluctuation fields are required by the theories of interstellar turbulence, of cosmic ray confinement and propagation, and of star formation.

There are two ways in which small-scale fields come into being in a turbulent conductive medium. First, they inevitably accompany the large-scale magnetic field, no matter whether the latter is produced by an external source or sustained by the mean field turbulent dynamo. Second, the self-excitation of small-scale magnetic fields is possible in a turbulent medium independently of the presence of either mean field or mean helicity.

Let us begin with small-scale fields that arise owing to the action of turbulent motions on the embedded large-scale magnetic field. The mean field may be either steady or growing in time due to the dynamo action. In the latter case we consider the quasi-stationary situation since the growth time of the large-scale field ($\sim 10^9$ years in the Galaxy) much exceeds the turbulent correlation time ($\sim 10^7$ years).

The essence of the generation process is the tangling of magnetic lines by turbulent motions which causes the magnetic energy to be transferred from large-scale ones into smaller ones. The mean helicity of the turbulence initiates the inverse spectral energy flux, but this proceeds at a much lower rate and the large-scale field can be considered quasi-stationary.

Consider the initial state in which the energy density of the large-scale magnetic field is small as compared with the kinetic energy density of hydrodynamic turbulence. Two asymptotic spectral distributions of magnetic fields arise in this case. At high wave numbers k, where magnetic dissipation is essential, the energy income to the given scale from the large-scale field is balanced by Ohmic dissipation. Since the revolution time of Kolmogorov vortices, $\tau(k) \simeq (kv(k))^{-1} \propto k^{-2/3}$ decreases with k, the magnetic energy input at a given scale is determined by the direct tangling of the large-scale field by vortices at this scale rather than by magnetic energy cascading along the spectrum. This means that in the wave-number domain where the magnetic dissipation is essential the order-of-magnitude equality holds

$$|\operatorname{rot} \mathbf{v} \times \mathbf{B}| \simeq \nu_m |\Delta \mathbf{b}|,$$

which gives in the wave-number space

$$kv(k)B \simeq \nu_m k^2 b(k),$$

where $\mathbf{v}$ is the turbulent velocity field. Thus we determine the energy density spectrum of magnetic fields, $M(k) \simeq k^{-1}b^2(k)$, as

$$M(k) \simeq \frac{B^2}{\nu_m^2} k^{-2}E(k), \qquad \text{(VII.79)}$$

where $E(k) \simeq k^{-1}v^2(k)$ is the spectral kinetic energy density of the turbulence. For the Kolmogorov hydrodynamic turbulence $E(k) = \varepsilon^{2/3}k^{-5/3}$ and

$$M(k) = \varepsilon^{2/3} \frac{B^2}{\nu_m^2} k^{-11/3},$$

which spectrum was first obtained by Golitsyn (1960).

At larger scales, $k < k_m = k_0 R_m^{3/4}$, where k_0 is the dominant, energy-range turbulent scale and $R_m = v_0/(k_0\nu_m)$ is the magnetic Reynolds number, the spectral energy transfer time $\tau(k)$ is shorter than the dissipation time $\tau_d = \nu_m k^2$. In this case the energy production $kv(k)B$ is balanced by the energy cascade $b^2(k)/\tau(k)$ of magnetic field at a given scale, which gives the following spectrum:

$$M(k) = B^2k^{-1}, \qquad b(k) = B \qquad \text{(VII.80)}$$

(Ruzmaikin and Shukurov, 1982) for the Kolmogorov hydrodynamic turbulence. Of course, this spectrum can be obtained from Golitsyn's result (VII.79) when the Ohmic diffusion ν_m is replaced by the turbulent one, $\beta(k) = v(k)/k$. From the other arguments, the spectrum (VII.80) has been obtained by Tchen (1965) and Nakano (1976). The resulting spectrum is shown in Figure VII.20a.

When the large-scale magnetic field grows in time, the equipartition between magnetic and kinetic energies $M(k_m) = E(k_m)$ is eventually reached first at the scale $k_m = k_0 R_m^{3/4}$ (see Figure VII.20b). Afterwards, the inertial range appears and begins to spread along the interval of wave-numbers, over which the spectrum has the form

$$M(k) \simeq E(k) = (a_0\varepsilon)^{1/2}k^{-3/2},$$

where $a_0 = B/(4\pi\rho)^{1/2}$ is the Alfvén velocity (Iroshnikov, 1963; Kraichnan, 1965). This spectrum is discussed at length in Section VI.3. When the large-scale field comes to a steady state, the stationary spectrum is settled, which is shown in Figure VII.20. For a given mean field B the inertial range begins at the wave number

$$k_e = k_0(v_0/a_0)^3$$

(Ruzmaikin and Shukurov, 1982). The short-wavelength boundary of the inertial range is still determined by the balance between the energy cascade rate along the inertial-range spectrum and the dissipation rate, $\tau_m = \tau_d$, where τ_m is given by

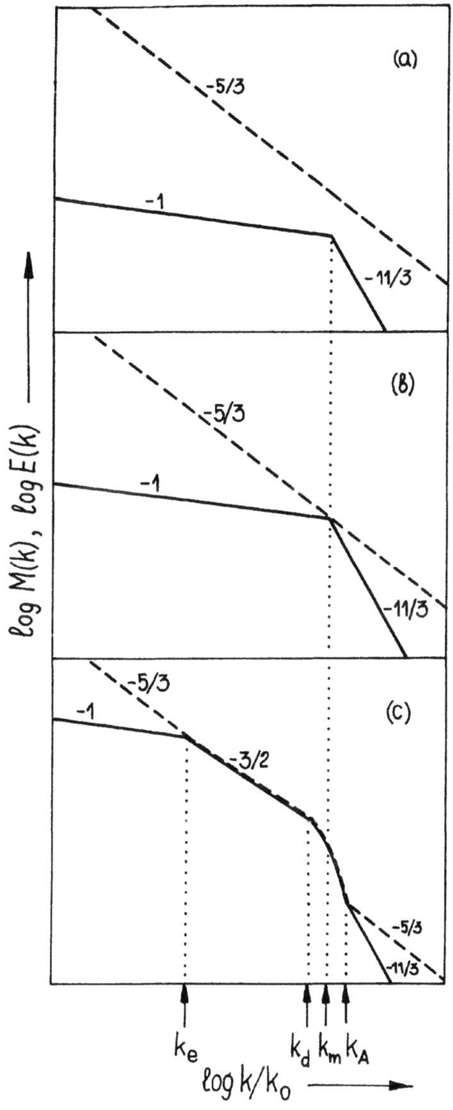

Fig. VII.20. Spectra of magnetic (solid lines) and kinetic (broken lines) energy densities in weak MHD turbulence. Numbers give the spectral indices. (a): The linear regime, $M(k) \ll E(k)$. Spectrum $M(k) \propto k^{-1}$ is due to tangling of the large-scale magnetic field by Kolmogorov turbulence. (b): With growth of the magnetic energy equipartition is first reached at $k = k_m$. (c): Developed inertial range of MHD turbulence. All indicated indices are asymptotical values and are valid only far from all intersections with other spectral ranges.

(VI.4). This boundary, the dissipation wave number, is thus determined as

$$k_d = k_0 R_m^{2/3} (v_0/a_0)^{1/3}.$$

For $B = \langle b^2 \rangle^{1/2}$, i.e. when the equipartition between kinetic and magnetic energies spreads down over the whole spectrum up to the outer, energy-range scale we have $v_0 = a_0$ and $k_d = k_0 R_m^{2/3}$.

At still smaller scales, $k > k_d$, the energy of MHD turbulence rapidly dissipates and both $M(k)$ and $E(k)$ decay quasi-exponentially with wave number but remain equal until the Alfvén frequency $\omega(k) = ka_0$ exceeds the inverse dissipation time (Kraichnan and Nagarajan, 1967), that is for $k < k_A = a_0/\nu_m$. For $k > k_A$ the linear spectra $E(k) \propto k^{-5/3}$ and $M(k) \propto k^{-11/3}$ are re-established until those wave numbers at which kinematic viscosity ultimately damps the hydrodynamic turbulence.

Let us now estimate the scales typical of the isotropic MHD turbulence in the interstellar medium. In dense molecular and hydrogen clouds and in the warm intercloud medium the dissipation of magnetic fields is due to ambipolar diffusion (Mestel and Spitzer, 1956; Pikelner, 1966, p. 55; Kulsrud and Pearce, 1969; McIvor, 1977; Ruzmaikin and Shukurov, 1982). The dissipation is associated with collisions of ions moving in the magnetic field with neutral atoms. The role of neutral atoms in current dissipation in a partially ionized plasma dominates collisions with charged particles when

$$F^2 \omega_i \tau_i > 1,$$

where F is the mass fraction of neutral atoms, $\omega_i = eB/m_i c$ is the ion Larmor frequency, τ_i is the product of ion free path time and $(m_i + m_a)/m_a$, and m_i and m_a are the masses of ions and atoms, respectively. For pure hydrogen plasma $(m_i = m_a = m)$ this inequality takes the form

$$x < 1\text{--}10^{-3} \left(\frac{n}{1\ \mathrm{cm}^{-3}} \right)^{1/2} \left(\frac{T}{10^4\ \mathrm{K}} \right)^{1/4} \left(\frac{H}{1\ \mu\mathrm{G}} \right)^{-1/2} \left(\frac{\sigma_{ia}}{10^{-14}\ \mathrm{cm}^2} \right)^{1/2},$$

where $x = n_i/n$ is the ionization degree, n_i and n are the ion number density and the total number density, respectively, T is the gas temperature, and σ_{ia} is the collision cross section of ions and atoms. Thus, ambipolar diffusion is essential both in interstellar clouds ($n_e = n_i \simeq 0.05\ \mathrm{cm}^{-3}$, $n \simeq 20\ \mathrm{cm}^{-3}$, $T \simeq 100$ K, $B \simeq 15\ \mu$G, $\sigma_{ia} \simeq 10^{-14}\ \mathrm{cm}^2$) and in the warm intercloud medium ($n_e \simeq 0.03\ \mathrm{cm}^{-3}$, $n \simeq 0.2\ \mathrm{cm}^{-3}$, $T \simeq 10^4$ K, $B \simeq 2\ \mu$G, $\sigma_{ia} \simeq 10^{-14}\ \mathrm{cm}^{-3}$). To estimate the ambipolar diffusion coefficient, determine the velocity of ion motion under the action of magnetic force,

$$n_i m_i \frac{V_m}{\tau_{ia}} \simeq \frac{H^2}{8\pi l},$$

where l is the scale of the magnetic field H, V_m is the diffusion velocity of ions, and τ_{ia} is the free path time of ions between collisions with atoms. Upon writing $\tau_{ia} = l_{ia}/\bar{v}_i$ where $l_{ia} = 1/(n_a \sigma_{ia})$ is the ion free path, $n_a \approx n$, and $\bar{v}_i = (3kT/m_i)^{1/2}$ is the ion thermal velocity (here k is the Boltzmann constant), we obtain the following estimate for the ambipolar diffusion velocity:

$$V_m \simeq \frac{H^2}{8\pi l n_i \sigma_{ia} n (3kTm_i)^{1/2}}.$$

The corresponding diffusion coefficient is given by

$$\nu_m \simeq \tfrac{1}{3} V_m l \simeq \frac{H^2}{24\pi n_i n \sigma_{ia} (3kTm_i)^{1/2}}.$$

In terms of convenient units, this expression takes the form

$$\nu_m \simeq 7 \times 10^{19} \frac{\text{cm}^2}{s} \left(\frac{0.03\ \text{cm}^{-3}}{n_e}\right) \left(\frac{1\ \text{cm}^{-3}}{n}\right) \left(\frac{10^4\ \text{K}}{T}\right)^{1/2} \times$$

$$\times \left(\frac{H}{1\ \mu G}\right)^2 \left(\frac{10^{-14}\ \text{cm}^2}{\sigma_{ia}}\right).$$

According to Dalgarno (1960) the cross section of collisions of a proton with hydrogen atom is 10^{-14} cm^2 at the temperature $T = 10^4$ K and weakly depends on temperature up to $T = 10^6$ K when $\sigma_{H+H} \approx 0.5 \times 10^{-14}$ cm^2. At lower temperatures $T < 500$ K the cross section σ_{H+H} has been determined by Osterbrock (1961) to be 1.5×10^{-14} cm^2.

On average over galactic disc we may put $H \simeq 2\ \mu$G, $n \simeq 1$ cm^{-3}, $T \simeq 10^4$ K and the ambipolar diffusion coefficient in pure hydrogen plasma is estimated as $\nu_m \simeq 3 \times 10^{20}$ cm^2 s^{-1}.

In dense clouds the role of carbon ions may be considerable which enhances ambipolar diffusion (McIvor, 1977).

Dissipation of hydromagnetic waves in a rarefied gas of the hot component of interstellar medium involves much more complicated physical processes (Cesarsky, 1980). According to Foote and Kulsrud (1979), the waves propagating obliquely to the field suffer collisionless damping

$$\nu_m \simeq \frac{\overline{v_i^2}}{\omega_i} \simeq 2 \times 10^{16}\, \frac{\text{cm}^2}{s} \left(\frac{T}{10^6\,\text{K}} \right) \left(\frac{1\,\mu\text{G}}{H} \right).$$

In the hot tunnels $T \simeq 10^6$ K, $n_e \approx n \simeq 2 \times 10^{-3}$ cm^{-3}, and the field strength from the $n^{2/3}$ scaling gives the estimate $H \simeq 3 \times 10^{-2}\ \mu$G. For these values of parameters, $\nu_m \simeq 5 \times 10^{17}$ cm^2 s^{-1}.

The parameters of isotropic spectra of turbulent magnetic fields in various phases of interstellar medium are given in Table VII.1 (see also McIvor, 1977; Ruzmaikin and Shukurov, 1982).

TABLE VII.1
Statistical properties of small-scale interstellar magnetic fields

	l_0 pc	v_0 km s^{-1}	b_0 μG	a_0 km s^{-1}	ν_m cm^2 s^{-1}	R_m	l_d pc	l_e pc
Neutral clouds	10	5.6	18.0	5.6	5×10^{21}	5×10^4	0.05	10
Intercloud medium	100	10	0.8	4.0	3×10^{20}	10^6	0.015	6
Hot tunnels	100	10	0.03	1.5	5×10^{17}	5×10^8	0.0003	0.4
Galactic disc in average	100	10	2.5	5.5	3×10^{20}	10^6	0.01	20

Remarks to Table VII.1

l_0 is the energy-range scale of turbulence; v_0 is the turbulent velocity and b_0 the magnetic field at this scale; a_0 is the Alfvén velocity; ν_m is the magnetic diffusivity; R_m is the magnetic Reynolds number; l_d is the dissipation turbulent scale; l_e is the scale at which the inertial range $l_d < l < l_e$ begins. The value $b_0 \simeq 0.03\ \mu$G for hot tunnels is probably an underestimate (see Section III.7).

Comparing the estimates of Table VII.1 with observational data of Armstrong *et al.* (1981) (see Figure VI.13) it can be easily seen that observations indicate a much greater extension of the spectrum toward small scales. Why does the turbulence persist at those scales where the dissipation seems to be fatally strong? On the one hand, turbulent motions and fields can have different energy sources at the scale ranges from tens to tenths of parsecs and from 10^{13} to 10^{10} cm. Note also that for $k > k_A \simeq 1/3 \times 10^{12}$ cm in the hot tunnels and $k > k_A \simeq 1/10^{15}$ cm in the intercloud medium the inertial Kolmogorov hydrodynamical spectrum and the magnetic dissipation spectrum $M(k) \propto k^{-11/3}$ are re-established. However, the level of turbulence in this range is much lower as compared with larger wavelengths $k < k_d$. Armstrong *et al.* (1981) presumes that the spectrum is continuous over the tremendous scale range from 100 pc down to 10^{10} cm but this may be an artefact of their data reduction (see Section VI.3).

A possible explanation of survival of MHD turbulence at the scales where isotropic turbulence would die away may be associated with pronounced anisotropy of MHD turbulence at small scales (Higdon, 1984).

As we have mentioned in the beginning of this section, there is another source of small-scale fields in turbulent media: direct generation of fluctuation fields by the turbulence without participation of the mean helicity. The necessary condition for generation small-scale, zero-mean magnetic fields in this case is a sufficiently high value of the magnetic Reynolds number, probably in excess of 100. Magnetic Reynolds numbers are typically much higher in astrophysical plasmas. Some results from this field will be discussed in Chapters VIII and IX in applications to galaxy clusters, radiogalaxies and jets.

VII.14. Seed Fields

Any dynamo is the mechanism of magnetic field *amplification* rather than *creation.* It can be seen from the obvious fact that the induction equation with zero initial and boundary conditions has only identical zero as the solution. Therefore, a consistent picture must include the source of the galactic seed magnetic fields. We propose to associate this source with magnetic fields ejected by supernovae and young stars with strong mass outflow. The ejections form together a small-scale random magnetic field. However, there is a probability that the sum of many small-scale fields gives a weak large-scale field.

Consider first the other possible sources of the seed fields of galactic scale. A well known battery mechanism (Biermann, 1950; see also Kemp, 1983) is based on inclination of level surfaces of the gas density and pressure (or temperature) which arises in differentially rotating objects, in particular, in protogalaxies. This mechanism can provide the large-scale magnetic field of order 10^{-15} G in a protogalaxy in 10^{10} years (Harrison, 1969). The field of order 10^{-21} G may be created in a rotating protogalaxy due to stronger interaction of protogalactic plasma electrons, as compared with protons, with the relic microwave background (Mishustin and Ruzmaikin, 1971). Similar processes may be operative even before recombination, provided the cosmological plasma was in a state of turbulent motion (Harrison, 1970, 1973; Bairlein, 1978).

In order to amplify exponentially, $B = B_0 \exp \Gamma t$, initial field $B_0 = 10^{-21}$ G up to the modern galactic value $B = 10^{-6}$ G in $T = 10^{10}$ years, the growth rate $\Gamma = \Gamma_* \beta / h^2$ where Γ_* is the dimensionless growth rate of order unity, should be at least $1.7\, \beta/h^2 \simeq 1.7/(5 \times 10^8$ years). The characteristic growth time, 3×10^8 years, should not exceed the rotation period of the Galaxy! Meanwhile, even in the most favourable model the growth rate of the Galactic large-scale magnetic field does not exceed $1.5/(5 \times 10^8$ years) (Ruzmaikin *et al.*, 1985) while a more realistic estimate gives $0.3/(5 \times 10^8$ years) (see §7).

In addition, the mentioned mechanisms create mainly the fields of dipole configuration because they are intrinsically associated with rotation of the object (Vainshtein *et al.*, 1980, §7.6). Battery mechanisms provide only a linear growth of the created magnetic field in time. For this reason, the field cannot grow sufficiently strong in a finite, however large, time of the order of 10^{10} years.

The hypothetical quasi-uniform relic magnetic field (see Section V.3) can be also invoked as the seed. But such a field remains as yet only hypothetical; it was never detected. Modern observations only place a severe upper limit on its strength,

$$B_0 < 3 \times 10^{-11}\ \mathrm{G}$$

(Vallée, 1983c). In order to amplify such a field to the observed value, the growth rate $1/(5 \times 10^8$ years) is required. Although this latter value does not directly contradict the dynamo theory, it seems to be excessive. As long as observational evidence in favour of existence of even such weak relic fields is lacking, we believe that the galactic seed fields have another origin.

A few attempts have been made to explain the observed Galactic magnetic field by magnetic ejections from supernovae (Syrovatskii, 1970), young stars (Michel and Yahil, 1973), and star clusters (Layzer *et al.*, 1979). As we have discussed in Section V.4, the resulting field has too large, of order 10^3, ratio of the random field component to the large-scale one. Meanwhile, the field observed now has nearly equal random and regular components.

However, the mechanism of stellar ejections provides a remarkable possibility of creating the seed field for the galactic turbulent dynamo which can intensify the initially weak large-scale component up to the required level (Ruzmaikin *et al.*, 1986). It is clear that the required amplification factor is as low as 10^3 in the case of the supernovae-ejected seed field!

When considering the evolutionary problem of the large-scale magnetic field, one should expand the tangled seed field $\mathbf{B}_0$ over eigenfunctions $\mathbf{B}_n$ of the linear problem for field excitation.* In other words, we should evaluate the following integrals:

$$C_n = \frac{1}{\int B_n^2 \mathrm{d}^3\mathbf{r}} \int \mathbf{B}_0(\mathbf{r})\mathbf{B}_n(\mathbf{r})\, \mathrm{d}^3\mathbf{r} \qquad \text{(VII.81)}$$

over the galactic disc volume $V\ (n = 1, 2, 3, \ldots)$.

The estimate $C_1 = 10^{-3}$ follows under a crude approximation of the lowest eigenfunction by $\mathbf{B}_1 = B\mathbf{e}_\varphi$, where $\mathbf{e}_\varphi$ is the unit azimuthal vector, when the integral

$$C_1 = \frac{1}{BV} \int \mathbf{e}_\varphi \mathbf{B}_0(\mathbf{r})\, \mathrm{d}^3\mathbf{r} = \frac{1}{BV} \sum_{i=1}^{N} \mathbf{e}_\varphi \mathbf{B}_{i0} V_i \qquad \text{(VII.82)}$$

* Strictly speaking, such expansion can be accomplished if the linear eigenvalue problem is self-adjoint or at least has an orthogonal set of eigenfunctions. For an incomplete and non-orthogonal set of eigenfunctions (which is expected in the case of the galactic dynamo problem) the expansion procedure is more complicated but the idea remains the same. Note also that in applications to the spiral galaxies the regimes of weak generation are of interest, when the eigenfunctions are close to the complete and orthogonal set of (free decaying) eigenfunctions of the diffusion equation. Considering crude, order-of-magnitude estimates we do not take these complications into account.

reduces to the sum of a large number N of contributions from the cells of volume V_i filled by uniform random fields B_{i0} (see Figure VII.21) whose root mean square value is

$$b = \langle B_{i0}^2 \rangle^{1/2} = 1\ \mu\text{G}$$

(for the estimate of b see Section V.4). The sum which appears in C_1 is also a random variable whose amplitude (the root mean square value) is $N^{-1/2}$ times the amplitude of summands:

$$\langle C_1^2 \rangle^{1/2} = \frac{b}{BN^{1/2}}\,.$$

This expression gives a rough estimate of the seed amplitude of the $n = 1$ axisymmetric mode

$$B_\varphi|_{t=t_0} = \frac{b}{N^{1/2}} \simeq 3 \times 10^{-4}\ \mu\text{G},$$

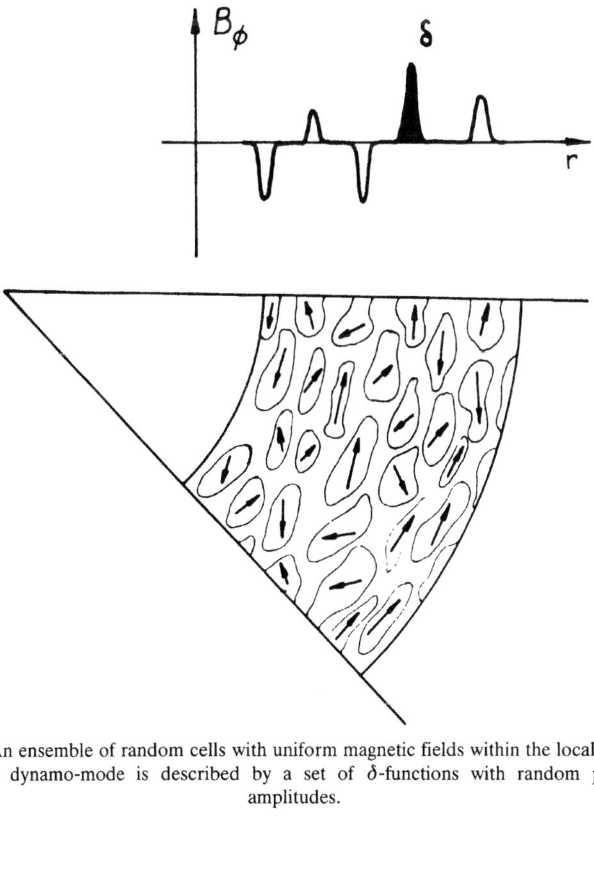

Fig. VII.21. An ensemble of random cells with uniform magnetic fields within the localization region of the lowest dynamo-mode is described by a set of δ-functions with random positions and amplitudes.

since $b \simeq 1\,\mu$G and $N = 12\pi h R^2/(\pi l^3) \simeq 10^6$ is the number of correlation cells of dimension $l \simeq 100$ pc in the galactic disc of half-thickness $h \simeq 400$ pc and radius $R \simeq 15$ kpc.

For a more detailed estimate, one should use the number of correlation cells within the region of localization of the corresponding eigenfunction.The lowest mode in the Galaxy and M31 is concentrated within a ring of $\Delta r \simeq 2$—3 kpc width around $r = 10$ kpc. This gives $N \simeq 300$, which corresponds to

$$B_\varphi|_{t=t_0} \simeq 0.02\,\mu\text{G}.$$

These estimates are suspiciously optimistic. In fact, they are overestimates because we considered the field uniform within each cell (cf. Segalovitz *et al.*, 1976) — see Figure VII.21. This model obviously violates the solenoidality of the magnetic field, div $\mathbf{B}_0 = 0$, at the cell boundaries. It is this violation which results in such high average fields.

A more realistic model considers the field $\mathbf{B}_0$ represented by small-scale field with magnetic lines closed within every cell. By the way, this representation is in better agreement with views on magnetic fields ejected by supernovae and young stars (see Section V.4). With such a configuration, the function $\mathbf{B}_0(\mathbf{r})$ in (VII.81) is rapidly oscillating with the period l and has zero mean. However, the random character of ejections prevents approximation of these oscillations by a periodic function of coordinates with a fixed small period. A relevant approximation is a set of derivatives of delta-functions placed in random positions and having random amplitudes (see Figure VII.22). Replace the eigenfunction B_1 by the uniform azimuthal field inside the localization region and put it zero outside, that is make more narrow the integration region. This gives

$$C_1 \simeq \frac{1}{\int B_1^2\,\mathrm{d}^3\mathbf{r}} \sum_{i=1}^{N} \int \mathbf{B}_1(\mathbf{r})\mathbf{b}_i\, l\delta'(\mathbf{r}-\mathbf{r}_i)\,\mathrm{d}^3\mathbf{r},$$

where b_i and $\mathbf{r}_i$ are random field values and positions respectively, and l is the correlation scale. Adopting $\mathrm{d}B_1/\mathrm{d}r \simeq B_1/\Delta r$, where $\Delta r \simeq 3$ kpc is the radial width of the lowest mode localization region (see §7), we obtain by integration by parts

$$C_1 \simeq \frac{b}{B_1 N^{1/2}} \frac{l}{\Delta r} \simeq 2 \times 10^{-3}\,\mu\text{G},$$

i.e. the initial field is $\Delta r/l \simeq 30$ times smaller than in the model of uniform cells. The origin of the factor $N^{-1/2}$ is quite evident. The additional small factor $l/\Delta r \simeq \frac{1}{30}$ arises due to solenoidality of magnetic field.

The required growth rate for the obtained seed fields is of order $0.35/(5 \times 10^8$ years). This growth rate is in good agreement with estimates of the dynamo theory.

We have assumed above that the ejected magnetic loops have normal vectors

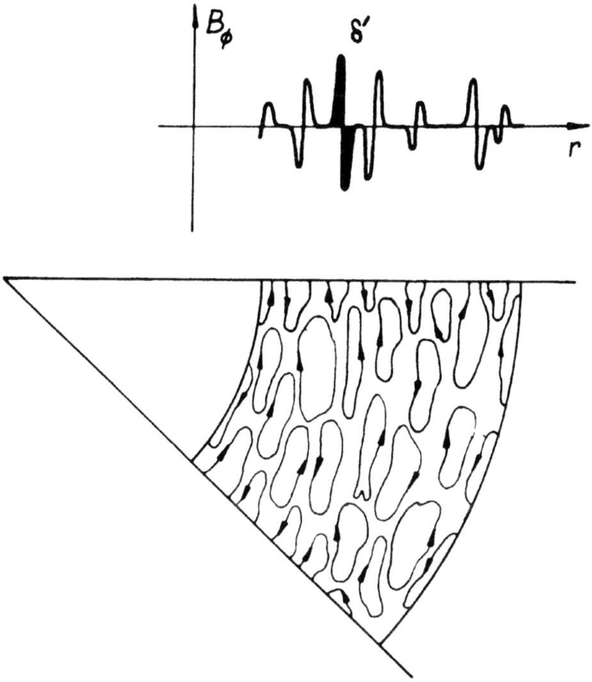

Fig. VII.22. An ensemble of random magnetic loops within the localization region of the lowest dynamo-mode is described by a set of derivatives of the δ-function.

along the z direction. However, accounting for random orientations of the loops does not change the final estimate. Indeed, the factor $\frac{1}{3}$ would appear due to random orientations. However, an independent contribution of the loops into the radial field component B_{1r} should be also taken into account. Since for the eigenfunction $B_{1r} \simeq |D|^{-1/2} B_{1\varphi} \simeq 0.3\, B_{1\varphi}$ where D is the dynamo number, this contribution gives approximately $0.3 \times \frac{1}{3} \simeq 0.1$ of the above estimate. For those loops whose normal vector is directed along the radius the characteristic scale is rather the disc thickness, over which only eight cells have space to reside, than the radial eigenfunction width with 30 cells on it. But supernovae explode preferentially near the central disc plane where the vertical gradient of the quadrupole eigenfunction is small. Therefore, the contribution from such loops is $8^2/30 \simeq 2$ times smaller than from those with the vertical normal vector. Now addition of all these small contributions returns us to the estimate of C_1 given above.

Note that superposition of the seed field from random fields gives a pronounced preference to higher modes of the dynamo: the higher the mode number is and, hence, the more frequently it changes the sign along the radius or the azimuth, the higher is the probability for it to be in phase with the random field $\mathbf{B}_0(\mathbf{r})$. As long as the radial scale of the nth mode can be estimated as $\Delta r/n$, its initial amplitude is estimated as

$$B_\varphi^{(n)}\big|_{t=t_0} \simeq \frac{b}{N^{1/2}}\, n\, \frac{l}{\Delta r}$$

and grows with n. The growth rates of axisymmetric modes decrease with n approximately as

$$\Gamma_n \simeq \Gamma_0 - Cn,$$

where Γ_0 and C are certain constants (see §6). Therefore, at any moment there is a dominant mode whose number decreases with time. In the limit $t \gg C^{-1}$ the lowest mode of fixed sign dominates, if non-linear effects do not fix earlier the combination of several modes.

Let us estimate the number of the mode currently dominant in the Galaxy, presuming that non-linear effects are negligible. If non-linear effects are essential, we obtain an upper limit of n. For the Galaxy

$$\Gamma_n \simeq (0.3 - 0.016)/(5 \times 10^8 \text{ years}).$$

The current amplitude of the nth mode is proportional to $n \exp(\Gamma_n T)$, $T \simeq 10^{10}$ years. This quantity is maximal for

$$n \approx \frac{5 \times 10^8 \text{ years}}{0.016 \times 10^{10} \text{ years}} \approx 3.$$

Thus, field reversals along the radius are quite probable in the Galaxy. Observations indicate two such reversals at $r \simeq 9.5$ kpc (Simard-Normandin and Kronberg, 1979) and at $r \simeq 12$ kpc (Agafonov *et al.*, 1987).

Thus, we may conclude that supernova explosions can provide the seed magnetic field for the turbulent dynamo in spiral galaxies. The magnetic field in the first-generation stars is generated by the stellar turbulent dynamo but with the independent stellar seed field provided, e.g., by the Biermann battery mechanism which is quite effective in stars.

CHAPTER VIII

MAGNETIC FIELDS AROUND GALACTIC DISCS

The existence of magnetic fields in the discs of spiral galaxies is firmly established by both theoretical and observational studies. However, the galactic gaseous discs are not the only objects of galactic size pervaded by magnetic fields. The fields are also detected in gaseous coronae (radiohalos) of galaxies and even in the intergalactic gas of galaxy clusters. In this chapter we briefly discuss the information, scarce though it is, about these 'over-galactic' magnetic fields and views on their origin.

VIII.1. Magnetic Fields in Gaseous Coronae

Interstellar gas in spiral galaxies can be traced far beyond thin discs where massive stars, supernovae, HII regions, and other objects constituting young galactic population are localized. Spiral galaxies are surrounded by rarefied hot gas that forms quasi-spherical coronae extending to ~ 10 kpc heights above the discs. Existence of the corona of the Milky Way was predicted by Pikelner (1953a, b) from analysis of cosmic ray confinement in the Galaxy and by Spitzer (1956) and Pikelner and Shklovsky (1957) from studies of the equilibrium of a hot gas in the disc. Gaseous coronae play an essential role in dynamics of cosmic rays which are probably confined in larger volumes than the volume of galactic discs (Ginzburg, 1984). The sources of relativistic particles and interstellar gas reside largely in the disc and the ionized gas penetrating into the corona inevitably carries upward magnetic fields.

Thus, the presence of magnetic fields in gaseous coronae seems to be quite natural and observations of the coronae are carried out not only through emission and absorption by neutral hydrogen and heavier ions (York, 1982), but through synchrotron emission as well. A weak but still observable synchrotron emission can be traced up to heights of several kiloparsecs above the disc of the Galaxy (Webster, 1978; Phillipps *et al.*, 1981; Beuermann *et al.*, 1985). Synchrotron emissivity of the Galactic corona is some 10 per cent of the disc's value. This implies that magnetic fields in the radiohalo are 3—4 times weaker than in the disc, provided the densities of cosmic ray electrons are nearly the same (see also Badwhar and Stephens, 1977).

Non-thermal continuous radioemission from large heights above the discs has been observed in a number of external spiral galaxies seen edge-on. Among them are NGC 891 (Allen *et al.*, 1978; Beck *et al.*, 1979; Klein *et al.*, 1984), NGC 4631

(Eckers and Sancisi, 1977; Wielebinski and von Kapp-Herr, 1977; Klein *et al.*, 1984; Sukumar and Velusamy, 1985; Werner, 1985) and some others (see, e.g., Hummel *et al.*, 1984a; Sofue *et al.*, 1986 and references therein). Equipartition estimates predict magnetic fields of a few microgauss in radiohalos.

Detection of non-thermal radioemission from gaseous coronae only indicates the presence of magnetic fields there, telling us nothing about their scale and configuration. It seems plausible that coronal magnetic fields have a relatively strong fluctuation component: buyoant magnetic fields rising up from the disc have a characteristic scale not exceeding 1 kpc (Parker, 1979). However, there are also indications of large-scale magnetic fields in gaseous coronae. Note in this connection a weak polarization of radioemission from the radiohalo of a spiral galaxy NGC 4631 (Werner, 1984, cited by Beck, 1986). Kronberg and Perry (1982) and Welter *et al.* (1984) have investigated statistical properties of residual Faraday rotation measures of quasars, obtained after subtraction of the Galactic contribution, and have found correlation between the values of residual RM and the occurrence of absorption lines in quasar optical spectra. Moreover, the dispersion of the residual RM distribution grows with average distance to the sample of quasars. An idea of such a statistical method was proposed earlier by Komberg *et al.* (1979). The results of this study imply that the Faraday rotation occurs somewhere between quasars and the observer — in gaseous coronae of intervening galaxies and/or in the intergalactic gas of galaxy clusters. Welter *et al.* (1984) estimate the size of gas clouds where the polarization plane rotates as $l = 45$ kpc, the density of thermal electrons as around 10^{-3} cm^{-3} and the magnetic field as around 2 μG. Estimated sizes and magnetic fields seem to be excessive for galactic coronae, but the possibility of the presence in the coronae of magnetic fields, whose scale is measured in kpc, deserves special attention. The fields of such scale can hardly penetrate into coronae from the discs and should be generated *in situ* directly within coronae.

Consider the possibility of large-scale magnetic field generation in galactic coronae by the turbulent dynamo (Ruzmaikin and Shukurov, 1983). The most popular model of the gas motions above the disc is now presented by the 'galactic fountain': the gas heated to 10^6 K rises from the disc to the corona where it cools at heights of several kiloparsecs and thermal instability then leads to the formation of cool clouds that fall back onto the disc (Shapiro and Field, 1976; Bregman, 1980a, b; Habe and Ikeuchi, 1980). Rotation of the corona (see Section VI.1) makes these motions helical and, combined with differential rotation, they can generate a regular magnetic field. The mean helicity of the ascending flow can be estimated, through (VI.12), as $\alpha \simeq l^2\Omega/h$. The possibility of large-scale magnetic field generation depends on the value of the dynamo number $|D| = \alpha\Omega h^3/\beta^2$, where we remind that Ω is the angular velocity of rotation, h is the scale height, and $\beta \simeq \frac{1}{3}lv$ is the turbulent magnetic diffusivity. A reasonable estimate of the turbulent velocity seems to be the sound speed, ~ 100 km s^{-1} in the 10^6 K hot gas.

It is immediately seen that the dynamo number does not depend on the turbulent characteristic scale l: $|D| \simeq 9\Omega^2 h^2/v^2$.

Adopting the estimates $\Omega \simeq 10^{-15}$ s^{-1}, $h \simeq 3$ kpc and $v \simeq 100$ km s^{-1} for the turbulent velocity, we obtain $|D| \simeq 7$. This value is nearly equal to the critical one for the dynamo action (see Chapter VII). Hence, generation of large-scale magnetic fields is quite possible in gaseous coronae by the turbulent dynamo. Note also that, in contrast to the disc, the prevalent magnetic mode in coronae may be the dipole one due to quasi-spherical geometry.

To estimate the quasi-stationary strength of the large-scale field we invoke the balance of magnetic force at the basic turbulent scale (where the field still can be considered large-scale) and the Coriolis force which produces helicity and thus leads to the field generation:

$$\frac{B_r B_\varphi}{4\pi l} \sim \rho v \Omega .$$

With estimates of the size of supernova remnants in the corona (which we identify with the turbulence correlation scale) $l \simeq 300$ pc (McKee and Ostriker, 1977) and the gas density $\rho \simeq 1.7 \times 10^{-27}$ g cm^{-3}, and putting according to the dynamo theory, $B_r/B_\varphi \sim (\alpha/h\Omega)^{1/2} \sim l/h$, we now obtain the estimate of the stationary value of the large-scale azimuthal field:

$$B_\varphi \simeq 1.4\,\mu\text{G}.$$

The radial field B_r is $h/l \simeq$ ten times weaker as compared with the azimuthal one.

Radiohalos are observed not only around spiral galaxies. Magnetic field has been studied in a radio and X-ray emitting region around a giant elliptical galaxy M87 which lies in the center of the Virgo galaxy cluster (see Figure II.2). The radius of the radio emitting region is about 45 kpc. Radioemission is polarized with intrinsic degree of polarization 25% (Andernach *et al.*; 1979; Dennison, 1980a). This indicates the presence of a large-scale field with the ratio of chaotic field b to the large-scale component B estimated as $b/B \simeq 1.5$ (see formula (III.6)). Polarization observations at several wavelengths have shown that the Faraday rotation measure is about 50 rad m^{-2} for the outer parts of the halo (Andernach *et al.*, 1979) reaching $\simeq 400$ rad m^{-2} for central regions (Schmidt *et al.*, 1978). These rotation measures are produced predominantly within the halo itself.The contribution of our Galaxy is small: 11 radiosources situated near M87 on the sky have rotation measures ranging from 0.4 to 8.3 rad m^{-2}.

Thermal X-ray emission, observed in the same region where the synchrotron emission originates, allows estimation of the thermal electrons density, which gives $n_e \simeq 5 \times 10^{-3}$ cm^{-3} for an isothermal model. With the observed Faraday rotations, this gives the estimate of the longitudinal large-scale magnetic field $B_{\parallel} \simeq 2\,\mu$G although after account for radial variations of the electron density the estimate may become as low as $B_{\parallel} \simeq 0.5\,\mu$G (Dennison, 1980a).

The large-scale nature of magnetic field in the M87 radiohalo is also suggested by the smooth variation of radioemission polarization over the halo image (Andernach *et al.*, 1979).

Existing data allow an estimation of energies stored in magnetic field and relativistic particles (Dennison, 1980a). For the isothermal and uniform spherical model of the radio/X-ray halo these energies differ only by a factor of three. In a more realistic model with electron density decreasing with radius, the total particle energy, 10^{59} erg, is 10^3 times larger than the magnetic energy. This example emphasizes once more unreliability of the equipartition estimates.

The regular character of the magnetic field around M87 indicates its intrinsic association with the galaxy itself. Note in this connection that M87 may possess a massive unseen halo extending to a few hundred kiloparsecs radius (Sarazin, 1986).

VIII.2. Magnetic Fields in Clusters of Galaxies

Ionized gas is detected in the intergalactic space of rich clusters of galaxies through its X-ray emission which is interpreted as thermal bremsstrahlung of a hot ($\sim 10^8$ K) rarefied ($\sim 10^{-3}$ cm^{-3}) gas (Sarazin, 1986). The total mass of the gas is approximately the same as the mass of all galaxies in a cluster. In addition, some clusters emit non-thermal radioemission which is interpreted as having a synchrotron origin. The spectrum of the radioemission for the Coma cluster of galaxies is shown in Figure VIII.1 (Jaffe, 1977). These observations are well fitted by the power-law spectrum

$$S_\nu = 25\,(\nu/100\text{ MHz})^{-1.2}\text{ Jy}$$

between frequencies 10 and 1400 MHz. The size of radioemitting region is modest in comparison with the overall dimension of the cluster. Its diameter is estimated as 1 Mpc/$\mathcal{H}_{50}$, where $\mathcal{H}_{50}$ is the Hubble constant measured in units 50 km s^{-1} Mpc^{-1}, at all frequencies (see Jaffe, 1977; Sarazin, 1986), i.e. about twice the diameter of the cluster core. Meanwhile, the isothermal distribution of galaxies in this cluster can be traced up to the radius 4 $\mathcal{H}_{50}^{-1}$ Mpc. Radio and X-ray emitting regions have nearly equal sizes.

The region which is bright in synchrotron emission is often called "the cluster radiohalo". Radiohalos are not an inevitable property of galaxy clusters, they have been detected only in a few cases. The best studied is the halo of the Coma cluster. Radiohalos have been detected in clusters A401, A754, A1367, A2142, A2255, A2256, A2319 and, perhaps, in the Perseus cluster (see Dennison, 1979; Sarazin, 1986 and references therein). Recall for comparison that Abel's catalogue contains about 2000 clusters. The number of clusters with detected radiohalos is miniscule. However, the observed radioemission is weak and it was suggested that radiohalos are actually typical for X-ray emitting clusters (Lea and Holman, 1978).

However, though the radiohalos are small in comparison with the clusters as a

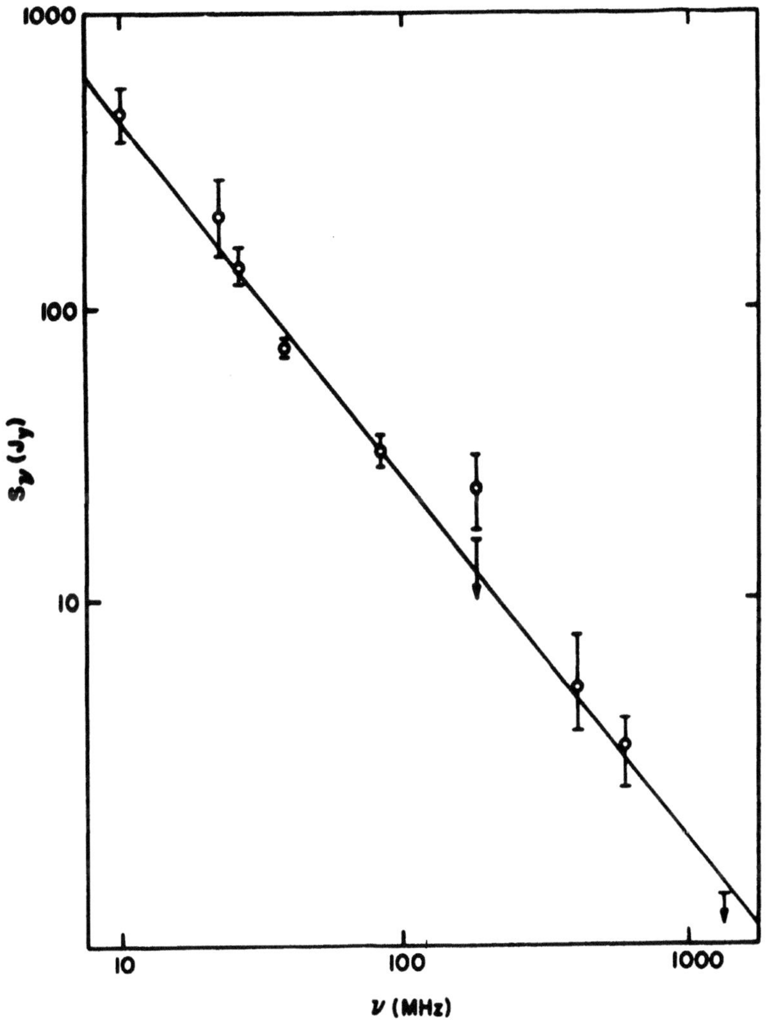

Fig. VIII.1. Spectrum of radioemission from the cluster of galaxies in Coma (Jaffe, 1977).

whole, they appear gigantic when compared with galactic dimensions — 100 times larger than the diameter of a typical galaxy. However, we shall see below that the scale of magnetic fields in radiohalos is not so large and is comparable with the galactic size.

The existence of extended synchrotron-emitting regions poses two principal problems. First, what is the origin of relativistic electrons that fill so vast a volume? Second, what is the origin of the magnetic fields?

A number of processes have been suggested to account for relativistic electron population. Among them are diffusion and convective transport of electrons from a cluster's radiogalaxies at velocities exceeding the Alfvén speed (Jaffe, 1977; Holman *et al.*, 1979). Electrons may be of secondary origin, being produced in interaction with relativistic protons which diffuse from radiogalaxies, with the thermal gas (Dennison, 1980b).

The strength of magnetic fields in radiohalos required by the models of diffusive transport of energetic electrons, is at least 2 μG. In the framework of the secondary-electron models, energy equipartition arguments also give 2 μG fields. The lower limit of 0.3 μG on the magnetic field strength in the Coma cluster follows from the upper limit on the X-ray emission produced via inverse Compton scattering of the microwave background on relativistic electrons (Perola and Reinhardt, 1972; Gursky and Schwartz, 1977; see Section III.1 above). Such fields are sufficiently strong to produce a noticeable Faraday rotation for the density of thermal electrons $n_e \simeq 10^{-3}$ cm^{-3} and path length $L \simeq 1$ Mpc. The magnetic field of 2 μG, uniform throughout a radiohalo, would produce the rotation measure $RM \simeq 2 \times 10^3$ rad m^{-2}. However, the observed rotation measures do not exceed 100 rad m^{-2} (see Jaffe, 1980; Vallée *et al.*, 1986). This last value implies that the magnetic field averaged over the line of sight should not exceed $\simeq 0.1$ μG. It follows then that magnetic field is chaotic and a rough estimate of its scale l can be obtained under the assumption that the line of sight passes through L/l cells with a uniform field $b \simeq 1$ μG in every cell. This configuration gives the field variance $\langle b^2 \rangle^{1/2} \simeq b(L/l)^{-1/2}$ and, therefore, $l \simeq L(\langle b^2 \rangle^{1/2}/b)^2 \simeq 10$ kpc. The scale of the radiohalo's magnetic field turns out to be comparable with the size of a galaxy.

Such fluctuation fields can be revealed with help of statistical analysis of rotation measures for the sources which are seen through galaxy clusters. As far as the magnetic field is chaotic, the observed rotation measures themselves do not differ, on average, for the sources seen through clusters and the 'field' sources (Dennison, 1979). Differences are expected in variances of statistical distributions, around the common mean value, of rotation measures for these two groups of sources (Komberg *et al.*, 1979). Such a statistical study has been performed by Kronberg and Perry (1982, for 39 sources) and by Welter *et al.* (1984, for 116 sources) who considered quasars with absorption lines in their spectra (see preceeding Section), and by Lawler and Dennison (1982) who observed 24 sources projected on galaxy clusters and 46 sources situated far from clusters.

The distribution of rotation measures proved to be slightly wider for the sources seen through clusters. However, the widening is weak and has a low confidence level of 80%. The observed widening yields an estimate

$$\frac{b}{(L/l)^{1/2}} \simeq 0.07\ \mu\text{G},$$

where b is the field inside the cell of the size l. The upper limit at 94% confidence level is obtained as $b/(L/l)^{1/2} \lesssim 0.2\ \mu$G. These estimates are somewhat smaller than was expected, but they are still compatible with the existence of magnetic fields of 1 μG, or slightly weaker, at the scale $\gtrsim$ 20 kpc (Lawler and Dennison, 1982).

The results of Welter *et al.* (1984) indicate fields of 2 μG at 50 kpc scale, but it is not clear to which objects these fields should be attributed: to galaxy clusters, galactic gaseous coronae, or some other absorbing gas clouds laying on the path to quasars. Vallée *et al.* (1986) have obtained indications of chaotic magnetic fields in the cluster A2319; the scale of the field is only few times smaller than the size of the halo.

What seems to be established is the existence in galaxy clusters of fluctuation magnetic fields whose scale seems to be of the order of galactic dimensions.

The fields of strength 10^{-6} G, observed in galaxy clusters, evidently cannot be explained by contraction of any primordial magnetic field whose strength is below 10^{-10} G (see Chapter V). However, two possible origins of these fields seem to be acceptable. On the one hand, the observed field may turn out to be a superposition of the fields provided by expanded remnants of strong radio sources that formerly belonged to the cluster (Perola and Reinhardt, 1972). On the other hand, as long as observations indicate the existence, in galaxy clusters, of magnetic fields whose scale is far below the size of a cluster, an attempt seems reasonable to connect these fields with turbulent motions in intergalactic gas. Which sources can feed such turbulence? Convective motions associated with super-adiabatic temperature gradient seem to be improbable (see Sarazin, 1986). Galaxies in a cluster move at velocities $\simeq$ 1000 km s^{-1} close to the sound speed in intergalatic gas. Thus, strong turbulent wakes arise behind individual galaxies, that may create magnetic fields and motions at wavelength 20—40 kpc in the central part $r \lesssim$ 150 kpc of a cluster (Jaffe, 1980). Another investigated possibility of the turbulence generation is the Kelvin—Helmholtz instability at the boundaries between galaxies and intergalactic gas (Nepveu, 1985).

The origin of chaotic fields in clusters of galaxies may be due to generation by turbulent motions of intergalactic gas. The absence of a pronounced overall rotation of galaxy clusters implies that the mean helicity of the motions is negligibly small. Thus, the regular magnetic field, of scale comparable with the cluster size, cannot be generated. However, fluctuation fields with zero mean can be readily generated.

Self-excitation of the magnetic field remains a characteristic property of random fluid motions even when they are not helical on average. This is warranted by the general theorem which states that, for sufficiently high magnetic Reynolds number, a magnetic field in a random flow grows exponentially at a rate that does not depend on magnetic diffusivity in the limit $R_m \to \infty$ (Molchanov *et al.*, 1985). The theorem generalizes earlier views on the possibility of magnetic field intensification by turbulence (Batchelor, 1950; Schlüter and Biermann, 1950; Kraichnan and Nagarajan, 1967). Existing estimates indicate that chaotic magnetic fields grow when the magnetic Reynolds number exceeds $R_{m,\,\mathrm{cr}} \simeq 10^2$ (Artamonova and Sokoloff, 1986; Kleeorin *et al.*, 1986).

Spatial distribution of the generated field is intermittent: rare but intense magnetic ropes stand out against a more or less uniform background of fluctuation fields. In the limit of large R_m these ropes store the major portion of magnetic energy. Generation of magnetic energy $\langle b^2 \rangle$ and correlation properties of generated fields have been studied by Kazantsev (1967), Kraichnan and Nagarajan (1967), Pouquet *et al.* (1976), Léorat *et al.* (1981), and Novikov *et al.* (1983). Here we note only that the radius of the ropes is of the order of the skin-layer thickness, $\sim lR_m^{-1/2}$, where l is the turbulent correlation scale and R_m is the magnetic Reynolds number. Meanwhile, the length of the ropes may reach a value as high as $\sim l$. Some ropes are closed while the others are tangled and decay at great distances. The characteristic growth time of the field is of the order of v/l, where v is the turbulent velocity. Note also that the larger R_m, the more diverse are geometrical forms of the ropes. The simplest configuration (type $N = 1$) is a closed loop. The next simplest is a closed coil of magnetic lines (type $N = 2$), etc. The critical Reynolds number grows with complexity of the magnetic configuration: the condition for excitation of the type-N rope reads $R_m > (R_{m,\,\mathrm{cr}})^N$ (Kleeorin *et al.*, 1986). The ropes are convected at velocities of the order of v, the turbulent velocity. In the linear approximation, their lifetime much exceeds the turbulent correlation time. However, a realistic lifetime estimate requires non-linear analysis and stability considerations.

Specific models of magnetic field generation in clusters of galaxies are needed. The application of above-mentioned ideas to radiogalaxies is briefly discussed in the next chapter.

CHAPTER IX

PROBLEMS OF MAGNETIC FIELD GENERATION IN GALACTIC NUCLEI, QUASARS AND RADIOGALAXIES

The recurring theme of the preceding chapters are magnetic fields that permeate the discs of spiral galaxies and their immediate surroundings. In many ways, still more fascinating are the magnetic phenomena in other astronomical species of galactic order: nuclei of galaxies, quasars and radiogalaxies. In these objects violent activity is associated with internal magnetic fields. Existence of these fields has been firmly established for a long time but a solution to the problem of their origin is still at its embryonic state. At this stage of development, only very general ideas will be discussed here. We concentrate on the basic question: whether or not the observed or probable motions in these objects can form and maintain magnetic fields?

IX.1. Center of the Galaxy

The center of our Galaxy is the site of violent activity (see Kardashev, 1985; Hyland, 1986). Recent observations indicate that magnetic fields play an essential role in a region of 100 pc radius around the galactic center (Yusef-Zadeh *et al.*, 1984; Morris and Yusef-Zadeh, 1985; Seiradakis *et al.*, 1985; Sofue and Handa, 1984; Inoue *et al.*, 1984; Tsuboi *et al.*, 1985, 1986; Quinn and Sussman, 1985). High-resolution observations with the Very Large Array have shown that the radiostructure with linear dimensions 180×210 pc^2 consists of separate thin, long filaments whose width is of order 1 pc and length exceeds 30 pc. This peculiarity indicats a dominant role of magnetic field in the formation of this structure. The configuration of the magnetic field has been revealed by polarization observations. The measured degree of polarization is high and reaches a maximum value of 60%. Figure IX.1 shows the observed intensity of polarized radioemission with superimposed vectors of transverse magnetic field. Figure IX.2 depicts the distribution of Faraday rotation measures that characterize the longitudinal (line-of-sight) field component. Positive *RM*s, which correspond to the field directed toward us, are shown unshaded and negative *RM*s, shown shaded, indicate the field directed away from us. The mean rotation measure in the southern part of this structure is 1500 rad m^{-2}. Adopting the electron density as $n_e = 30$ cm^{-3} and the path length over which *RM* is produced as $l = 5$ pc yields the estimate of the longitudinal field component as 10 μG (Tsuboi *et al.*, 1985).

These results suggest the magnetic configuration comprises a combination of

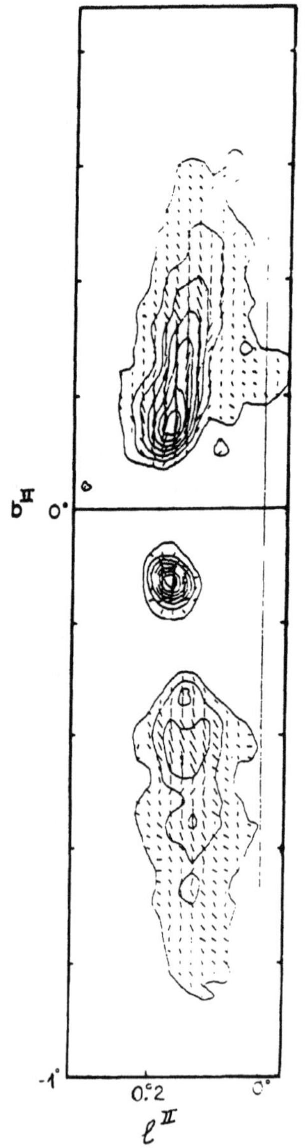

Fig. IX.1. Distribution of intensity of polarized radioemission from the central region of the Galaxy with superposed orientations of the transverse magnetic field. Contour interval is 60 mJy/beam (Tsuboi *et al.*, 1986).

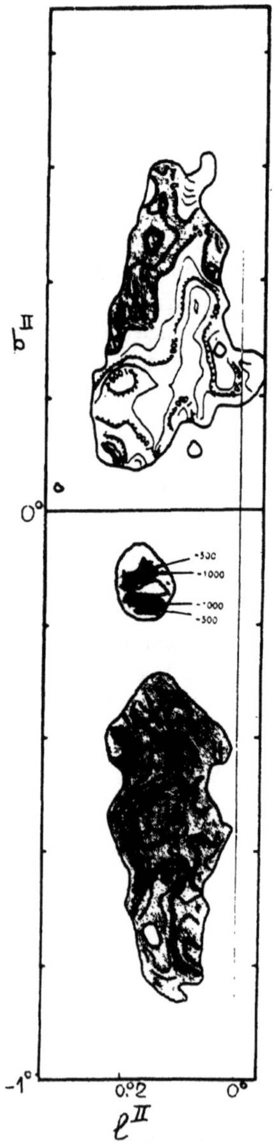

Fig. IX.2. Distribution of the Faraday rotation measure over the central region of the Galaxy. Shaded areas are for negative *RM* (line-of-sight component of magnetic field directed away from us) and non-shaded areas are for positive *RM*. Contour interval is 250 rad/m^2 (Tsuboi *et al.*, 1986).

poloidal and azimuthal fields with the azimuthal component being predominantly odd with respect to the Galactic equator (Sofue *et al.*, 1987). Noticeable asymmetry with respect to the equator, $b^{II} = 0°$, may be due to the presence of a weaker symmetric component of azimuthal field. The overall picture of the poloidal field looks like a dominant dipole field perturbed by the quadrupole one. Magnetic fields of such geometry can be readily generated by helical turbulence supplemented by differential rotation. Conditions in the center of the Galaxy are very favourable for magnetic field generation: effective dynamo numbers are very large there, exceeding several hundreds (Ruzmaikin and Shukurov, 1981). In such conditions dipole and quadrupole magnetic modes have nearly the same growth rates. Dipole and quadrupole modes are excited by the $\alpha\omega$-dynamo at nearly the same rate when

$$|D| \gtrsim 200,$$

where D is the dynamo number (Isakov *et al.*, 1981; Zeldovich *et al.*, 1983). The α^2-dynamo mechanism is notable for a small difference between generation rates of dipole and quadrupole modes for all values of relevant parameters (see Sokoloff *et al.*, 1983; Section VII.10).

These preliminary considerations deserve to be developed into detailed models of magnetic field generation in the center of the Galaxy.

Even stronger fields are revealed in the Galactic center at smaller scales. Large polarization (7%) of infrared emission from a region within a parsec of the dynamic center is explained by Aitken *et al.* (1986) by thermal emission from dust grains aligned by a magnetic field. It suggests that discrete sources in the Galactic center contain ionized gas clumps with hydrogen density $\simeq 10^5$ cm^{-3} at temperature $\simeq 10^4$ K and dimension $\simeq 0.1$ pc. The magnetic field necessary to provide the required alignment through the Davis—Greenstein mechanism is $B \gtrsim 10^{-2}$ G.

IX.2. Quasars and Active Galactic Nuclei

Quasars and active galactic nuclei emit enormous energy up to 10^{48} ergs per second over a wide wavelength range from very compact regions, as indicated by the rapid radiation flux variations on timescales as short as minutes.

An energy source which combines a powerful output with extreme compactness is a massive black hole surrounded by an accretion disc (Lynden-Bell, 1969; Rees, 1984). The energy output by magnetized gas accreting onto a non-rotating black hole is not sufficiently effective for the explanation of observed luminosities (Shvartsman, 1971), although a rotating black hole can provide a more powerful energy source. Blandford and Znajek (1977) have proposed a model of an electromagnetic mechanism of energy extraction from a rotating black hole. An excellent representation of this model has been given by MacDonald and Thorne (1981).

In the framework of the Newtonian theory, the matter flowing around a gravitating object is not sensitive to the rotation of the central body. However, the relativistic theory of gravitation reveals the effect of entrainment of a test particle by rotation of the central body. Far from the gravitating center, the resulting angular velocity decreases with distance as

$$\omega \simeq a r_g / r^3,$$

where $r_g = 2GM/c^2 = 3(M/M_\odot)$ km is the gravitational radius and a is the angular momentum per unit mass of the black hole. Therefore, the effect is pronounced only near the inner edge of the accretion disc, i.e. at a distance no closer than 1.5 r_g from the black hole.

Consider a black hole that rotates inside a gaseous disc whose angular velocity of rotation $\mathbf{\Omega}$ is parallel to $\boldsymbol{\omega}$ and which is magnetized. Note that according to Ginzburg's theorem (Ginzburg, 1964; Ginzburg and Ozernoi, 1964) black holes are precluded from having intrinsic magnetic fields. The field should be maintained by an external source. Such a source can be provided by the hydromagnetic dynamo in an accretion disc (Takahara, 1979).

Suppose for simplicity that the field is static and axisymmetric near the black hole. The field frozen into the disc gas rotates at angular velocity Ω. Therefore, the system of the disc and the black hole can act as a homopolar inductor. In rest reference frame the electric field is $\mathbf{E} = -\mathbf{v} \times \mathbf{H}/c$, where $\mathbf{v}$ is the rotation velocity at radius r determined by the angular velocity difference $\omega - \Omega$.

Consider the closed contour which is shown by the broken line in Figure IX.3. The section 1—2—2′ passes through the near vicinity of the black hole, while the section 3—4 lies in a remote region where $\omega \simeq 0$. The remaining two segments pass along magnetic lines and, thus, do not contribute to the electromotive force, which is given by

$$\text{e.m.f.} = \int \mathbf{E}\, d\mathbf{l} = \int_1^2 + \int_2^{2'} + \int_3^4 .$$

The sections 2′—3 and 4—1 make no contribution to e.m.f. As was shown by Bičák and Janiš (1985), the contribution from the section 2—2′, which passes in the intermediate vicinity of the horizon of the black hole, is small and tends to zero when the hole's angular momentum tends to the maximal possible value. The segment 1—2 makes a contribution of order $(\omega - \Omega)\Phi/2\pi c$, where Φ is the magnetic flux through the area swept out by the segment. The potential difference over the section 3—4 is $\Omega\Phi/2\pi c$. The resulting e.m.f. is of the order $\omega\Phi/2\pi c$. After calculation of impedance Z at these parts of the contour according to the Kirchhoff's law, one can estimate the power output thus extracted from the black hole as

$$L \simeq B^2 r_g^2 c (a/a_{\max})^2$$

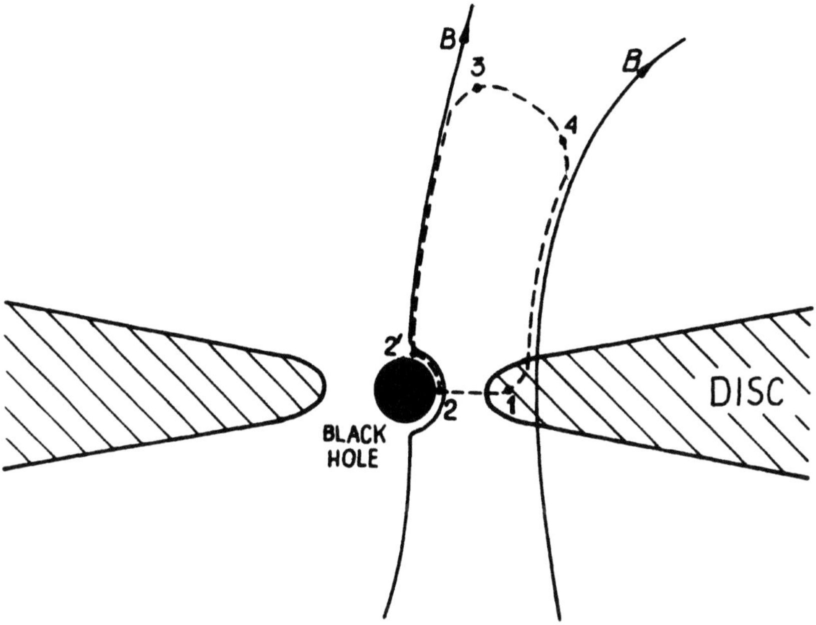

Fig. IX.3. The diagram illustrates the homopolar induction effect near the rotating black hole (after MacDonald and Thorne, 1981).

(Rees *et al.*, 1982), where $a_{\max}$ is the maximal specific angular momentum of a black hole.

Substituting $M = 10^8 M_\odot$ we see that a field of order 10^4 G is necessary for the power output 10^{48} erg s^{-1}.

Let us now estimate the magnetic field component parallel to the rotation axis that can be produced by the hydromagnetic dynamo in the accretion disc. Suppose, as is commonly accepted (Shakura and Sunyaev, 1973), that the angular momentum of accreting matter is transported outward by magnetic stresses:

$$-\dot{M}r^2\Omega \simeq 2\pi r^2 W_{r\phi},$$

where $2\pi W_{r\phi} = hB_rB_\phi$ (Takahara, 1979), h is the half-thickness of the disc, and Ω is the Keplerian angular velocity of rotation.

We can make the following order of magnitude estimates of the ratios of the field components for the field generated by differential rotation and the helical turbulence:

$$\frac{B_r}{B_\phi} \sim \left(\frac{\alpha}{\Omega h}\right)^{1/2}, \quad \frac{B_z}{B_r} \sim \frac{h}{r}.$$

(Zeldovich *et al.*, 1983; see also Chapter VII above), where α is the mean helicity of the disc turbulence. Therefore,

$$B_z \sim \frac{h}{r}\left(\frac{\dot{M}\,\Omega}{h}\right)^{1/2} \varepsilon^{1/4},$$

where $\varepsilon = \alpha/\Omega h \simeq 10^{-1}$—$10^{-2}$. Substituting $M = 10^8\, M_\odot$, $\dot{M} = 1\, M_\odot$/year, and $h = 10^{13}$ cm (Takahara, 1979) we see that B_z indeed can be as large as 10^3—10^4 G. This promising estimate needs to be further substantiated by detailed models of field generation.

We should note in conclusion that recent observations have given further evidence for the case of a massive black hole at the center of our Galaxy (Lo *et al.*, 1985; Crawford *et al.*, 1985; Serabyn and Lacy, 1985).

IX.3. Radiogalaxies

Nearly ten per cent of elliptical galaxies have powerful extended radiosources associated with them. Dimensions of these radiosources range from several tens of kiloparsecs to a few megaparsecs, i.e. from dimensions of normal galaxies up to the scales of rich galaxy clusters (Miley, 1980; De Young, 1984). These are the largest single objects in the Universe.

The most striking property of extended radiosources is their double structure of emission distribution (Figure IX.4). This suggests a natural explanation of their appearance by ejection of energetic matter from a central compact object (galactic nucleus or quasar). The surface brightness of many sources is concentrated in two bright isolated lobes without apparent connection with the central source. In other sources, the bridges are visible that extend from the core toward the lobes. Narrow bridges which have the highest brightness near the core are called jets, with a clear hint on the secondary dynamical origin of these objects. A greater number of radiosources are rather featureless, diffuse, and without prominent lobes.

Emission of extended extragalactic radiosources is interpreted as synchrotron emission of relativistic electrons (Section III.3). Interferometric observations have yielded not only the surface brightness but also distributions of spectral index, polarization degree, position angle and the Faraday rotation measure for many sources. These data have allowed to obtain magnetic field maps for many sources (see Miley, 1980).

For some radiosources, lower estimates of magnetic field strength can be obtained from an analysis of the interaction of the microwave relic background with relativistic electrons (see Section III.1). The inverse Compton scattering of microwave photons on energetic electrons can provide photons with energies in the X-ray range. For a known energy distribution of electrons, comparison of

(a)

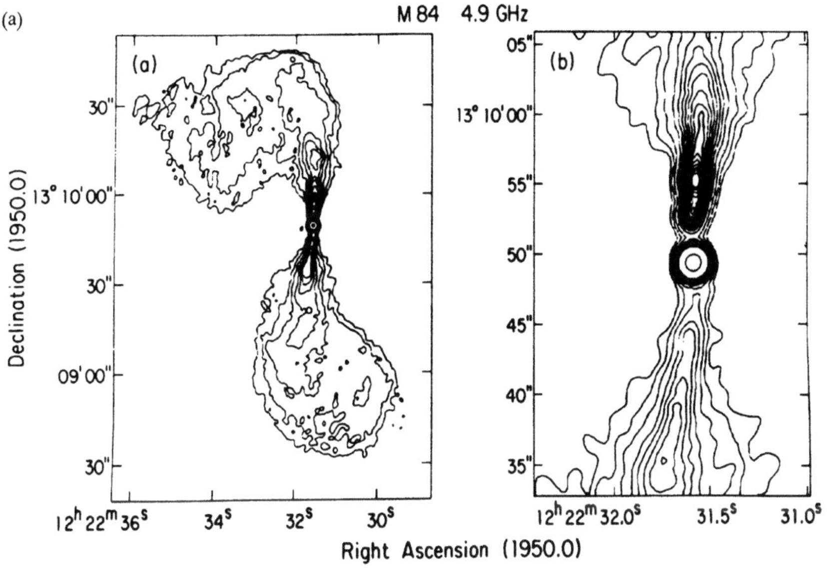

(b)

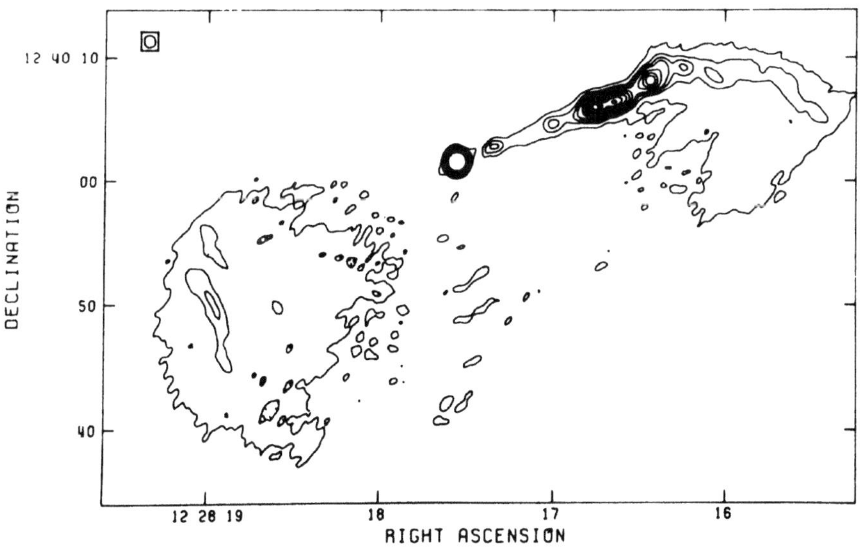

Fig. IX.4. Double structure is a typical feature of radiogalaxies. (a): A weak radiogalaxy M84. The right panel shows detail of the central region with a jet (from Bridle and Perley, 1984). (b): The jet and radio core of M87, a giant elliptical galaxy in the center of the Virgo galaxy cluster (from Miley, 1980).

X-ray and radio fluxes gives the following field estimates:

$$B \gtrsim \begin{cases} 1.6 \times 10^{-7}\,\mathrm{G} & \text{for Cen A,} \\ 1.6 \times 10^{-6}\,\mathrm{G} & \text{for Cyg A,} \\ 2 \times 10^{-6}\,\mathrm{G} & \text{for Vir A} \end{cases}$$

(see the review by De Young, 1984). These estimates agree with equipartition between energies stored in magnetic field and relativistic electrons.

Energy losses by the electrons through synchrotron radiation and inverse Compton scattering dictate very frustrating requirements: population of relativistic electrons must be reproduced more or less continuously in a considerable fraction of the source volume.

The minimal energy required for generation of the observed synchrotron emission with lumnosity J can be estimated as

$$E_{\min} \propto R^{9/7} J^{4/7}, \tag{IX.1}$$

where R is the size of the region occupied by particles and fields (see Section III.1). The numerical factor in this relation depends on the frequency and spectral index and is about 10^4 under typical conditions. Numerically, $E_{\min} \simeq 10^{56}$–10^{61} erg. The relation (IX.1) implies the equipartition between energy of relativistic particles and magnetic energy. The corresponding relation for magnetic field reads

$$H_{\mathrm{eq}} \propto R^{-6/7} J^{2/7}.$$

Typical values of M thus estimated range from 10^{-3} to 10^{-6} G.

Magnetic fields necessary to produce the observed synchrotron emission of radiosources are 1—10 μG on average over the source volume. Such fields cannot be explained by ejection or continuous outflow from the central compact object (De Young, 1980). Consider, for example, the field M_j carried out by the jet of radius $r_j \simeq 1$ kpc at the velocity $v_j \simeq 10^8$ cm s^{-1} during the lifetime of a radiosource, $t_s \simeq 10^8$ years. For the field M_s in the radiosource whose volume is $V_s \simeq 10^6$ kpc^3 we have

$$M_s^2 V_s \simeq M_j^2 \pi r_j^2 v_j t_s.$$

This gives $M_s^2/M_j^2 \simeq 3 \times 10^{-4}$. Similar estimates follow from other models. Meanwhile, observations give for this ratio estimate as high as 10^{-2}—1. Therefore, magnetic field must be generated within the radiosource itself.

Turbulent plasma motions can be an effective generator of magnetic fields. The MHD turbulence can be maintained e.g. by instabilities at the boundary between the jet and the ambient medium (Benford *et al.*, 1980; Begelman *et al.*, 1984).

While a considerable progress has been achieved in the problem of magnetic field generation by a turbulent conductive medium (see review of Molchanov *et al.*, 1985), development of detailed models for specific objects meets serious mathematical and computational difficulties. Some simplification is provided by the separate analysis of the mean and fluctuation fields. In order to generate the mean magnetic field, the turbulence must be helical on average (see Chapter VII). Therefore, the presumption of zero mean helicity allows us to investigate the behaviour of fluctuation fields with zero average value. After the additional assumption of incompressible flow, the problem is reduced to a solution of the system of induction equation and Navier–Stokes equation with driving force **f** and pressure P:

$$\left(\frac{\partial}{\partial t} + \mathbf{v}\nabla\right)\mathbf{H} = (\mathbf{H}\nabla)\mathbf{v} + \nu_m \Delta \mathbf{H},$$

$$\left(\frac{\partial}{\partial t} + \mathbf{v}\nabla\right)\mathbf{v} = \frac{1}{4\pi\rho}(\mathbf{H}\nabla)\mathbf{H} + \nu\Delta\mathbf{v} - \frac{\nabla P}{\rho} + \mathbf{f}. \tag{IX.2}$$

The simplest way is to proceed to hierarchy of moment equations. Usually only the second moments are considered (the first moments — the averages — vanish for zero mean helicity). Derivation of the equation for the second moment, or autocorrelation function, takes the simplest form for the flow with short time-correlation, as was first suggested by Kazantsev (1967; Molchanov *et al.*, 1985). Numerical analysis of the differential equation resulting in the case of the prescribed velocity field was performed by Novikov *et al.* (1983).

Another way exploits the closure of moment equations hierarchy in the quasi-normal approximation (when the fourth moments can be expressed through the second ones) with account for a finite correlation time for non-linear energy transfer between vortices of different scales (Pouquet *et al.*, 1976). This procedure yields the following system of integral equations in the Fourier space:

$$\left(\frac{\partial}{\partial t} + 2\nu_m k^2\right) M(k) = \int K(k, p, q)\,\mathrm{d}p\,\mathrm{d}q,$$

$$\left(\frac{\partial}{\partial t} + 2\nu k^2\right) E(k) = F(k) + \int T(k, p, q)\,\mathrm{d}p\,\mathrm{d}q,$$

where E and M are the spectra of kinetic and magnetic energy, respectively, $F(k)$ is the spectrum of the driving force in Equations (IX.2), and the kernels T and K describe non-linear energy transfer between hydrodynamic and magnetic modes.

This system of equations has been integrated by De Young (1980) in application to extended radiosources. The force driving the turbulence was chosen as having flat spectrum $F(k)$ = const (= 0.1) in the inertial range of wave numbers ranging from $k_0 = 1$ (corresponding to 1 kpc wavelength) to $k_d = k_0 Re^{3/4}$, where $Re = v_0/\nu k_0$ is the Reynolds number. A convenient unit to measure v_0 is the sound speed, 10^3 km s^{-1}. The turnover time of the energy-range vortex, $t_0 = 1$ kpc/10^3 km s^{-1} $\simeq 10^6$ years, is chosen as unit time. The lifetime of the radiosource, 10^8 years, in dimensionless units is expressed as $t = 100$. The initial value of magnetic energy is chosen to be 10^{-6} in order to have $M(k) \sim 1$ at the final stage, which is close to equipartition with the kinetic energy. In physical units, this means that initial seed field 10^{-9} G is amplified up to the value 10^{-6} G. However, the energy requirements to the source of turbulence remain uncomfortably high.

Figure IX.5 shows temporal variations of total kinetic and magnetic energies for vanishing mean helicity of the turbulent flow. We can see that in approximately 20 characteristic times (10^6—10^7 years) a stationary state is achieved, in which the magnetic energy is slightly less than the half the kinetic energy. The spectral energy distribution is presented in Figure IX.6 according to the paper of De Young.

Inclusion of the mean helicity of turbulence makes greater the generated total magnetic energy, which becomes close to the kinetic energy. In addition, spectral distribution of the magnetic field continuously shifts toward larger scales. At the final moment $t = 100$, the maximum of $M(k)$ is being achieved at $k \sim 0.1$ with noticeable amplitude at even smaller wave numbers.

IX.4. Jets

Jets are observed in radiosources as thin objects that extend from a central galaxy toward extended bright regions (Figure IX.7). For the first time a jet was discovered in a giant elliptic galaxy M87 in optical range by Curtis as early as in 1918 (see Figures II.2 and IX.4b). Nowadays, jets are extensively studied in radio as well. They probably exist in the majority of radiosources, but are clearly seen on both sides of the central galaxy only in weak sources. The jets appear less often in the sources of intermediate luminosity where they are one-sided in the majority of cases and almost never are they observed in powerful radiosources. The hypothesis of wide occurrence of jets is based on observations that in radiosources the nucleus and outer emitting regions often lie on an almost straight line even when a jet is not seen. Jets are often surrounded by cocoons of lower brightness which are believed to be formed by a plasma outflow from the jet's head. Typical lengths of jets range from $\simeq 2$ kpc in Virgo A to $\simeq 260$ kpc in NGC315. Smaller jets are observed in some Galactic objects, e.g. SS433 (Margon, 1982) and Sco X-1 (Achterberg *et al.*, 1983) and, possibly, in the center of the Galaxy (Uchida *et al.*, 1985).

It seems that a jet represents a conductive channel along which the mass, momentum, energy, and magnetic flux are carried from the nucleus toward the outer components, lobes.

How are the jets formed and why are they so well collimated over such enormous scales? In many cases, the gaseous pressure in an ambient medium is

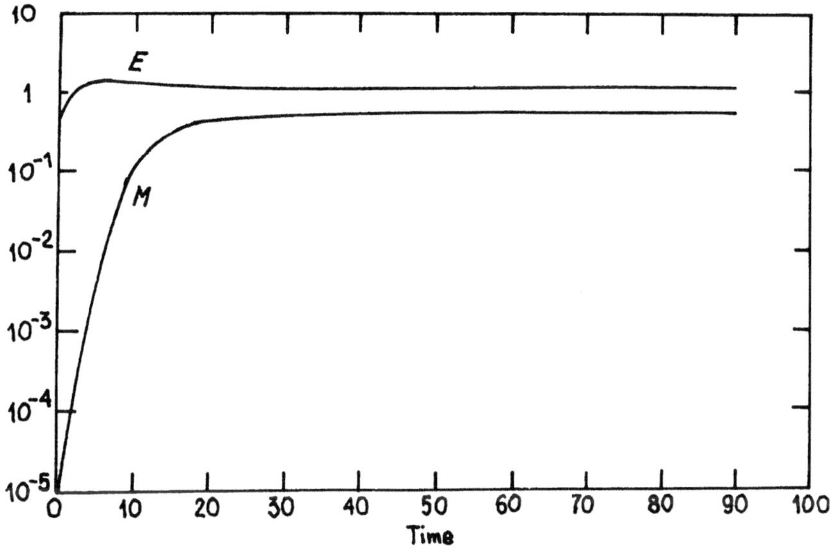

Fig. IX.5. The time dependence of the total magnetic and kinetic energies for zero mean helicity (after DeYoung, 1980).

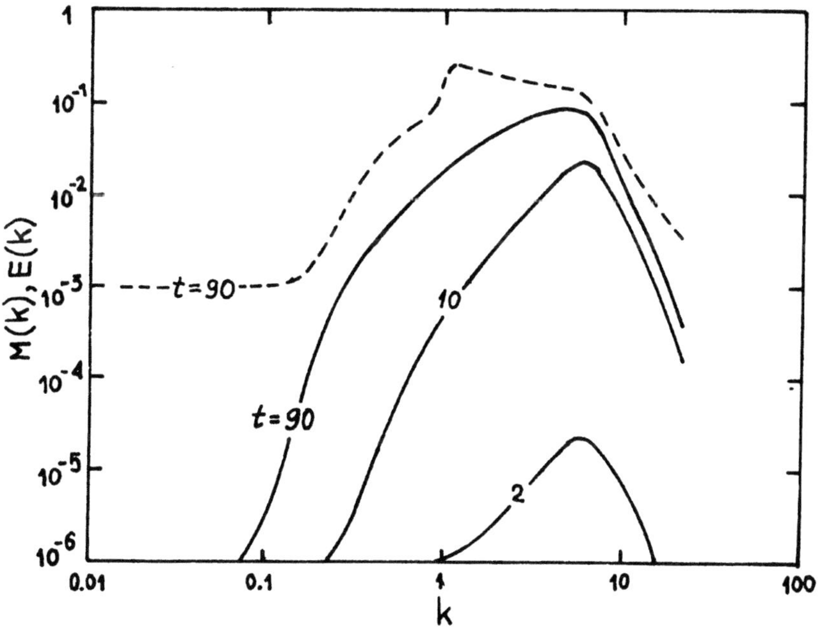

Fig. IX.6. The time development of magnetic (solid lines) and kinetic energy spectra for zero mean helicity (after De Young, 1980). The broken line shows kinetic energy spectrum at $t = 90$.

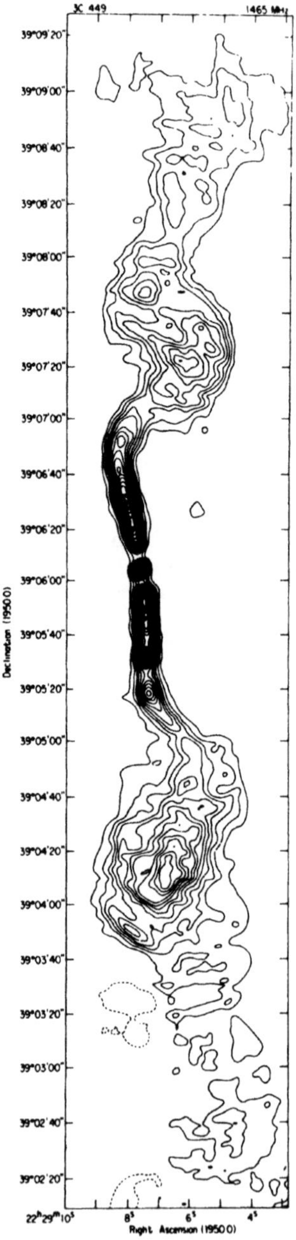

Fig. IX.7. The jet in 3C449 (from Miley, 1980).

sufficient to explain the observed shape of a jet. But why is this state stable? In some jets the inner pressure undoubtedly exceeds the ambient value. A detailed analysis of the problems of jets, and extensive references, can be found in the review of Begelman *et al.* (1984).

In connection with synchrotron emission of jets is the important problem of the origin of ultra-relativistic electrons. Modern views attribute electron acceleration, for example, to Cherenkov interaction of electrons with plasma waves reflected by the front of a strong shock wave (Galeev, 1984). The shock waves appear to be due to ejection from the central source of plasma condensations. The velocity of condensations is estimated as 0.6 c. The shock wave, formed by the incursion of the first condensation into an unperturbed medium with density $n \sim 10^{-2}$ cm^{-3}, is insufficient for electron acceleration. However, the following condensations move in a gas of lower density and provide necessary acceleration. Electrons acquire a universal energy spectrum that is a power law with spectral index $\gamma_e = 2$. Acceleration proceeds along magnetic lines. Galeev (1984) also calculates the spectrum of synchrotron emission of electrons accelerated by a shock wave. This spectrum consists of power-law sections $I_\nu \propto \nu^{-\alpha}$ with certain indices α.

Below we consider briefly only the magnetic aspects of the jet problem.

Magnetic field explains the synchrotron emission of jets and probably plays an essential role in their collimation and stability. The collisional free path of charged particles is large in the jet plasma, but their Larmor radius is less than the jet's width. Therefore, the magnetohydrodynamic approximation is applicable with account for anomalous resistivity.

Magnetic field morphology in jets can be crudely described as follows (see Figure IX.8). Near the nucleus of weak sources the field is predominantly parallel to the jet axis, while further along the jet it becomes nearly perpendicular. In powerful sources the field configuration is different and the magnetic field is everywhere aligned with the jet axis (Bridle, 1982).

The simplest qualitative estimate can be made under the assumption of uniform plasma flow and in neglect of resistivity, i.e. for a frozen-in magnetic field when the magnetic flux $\int \mathbf{B} \cdot d\mathbf{S}$ is conserved. Choosing the area element in this integral to lie transversely to the jet and in the plane of longitudinal cross-section one obtains, respectively.

$$B_{\parallel} \propto R^{-2}, \qquad B_{\perp} \propto R^{-1},$$

where R is the jet's radius. This simple result shows that the magnetic field inside a jet becomes predominantly transverse with distance from the central source and the corresponding growth of R. Polarization observations actually show that this picture is relevant for weak jets, e.g. NGC315, NGC6251, 3C31, 3C449. However, the longitudinal field dominates everywhere in strong jets (e.g., 4C32.69).

The longitudinal field component would be dominant when there exists a

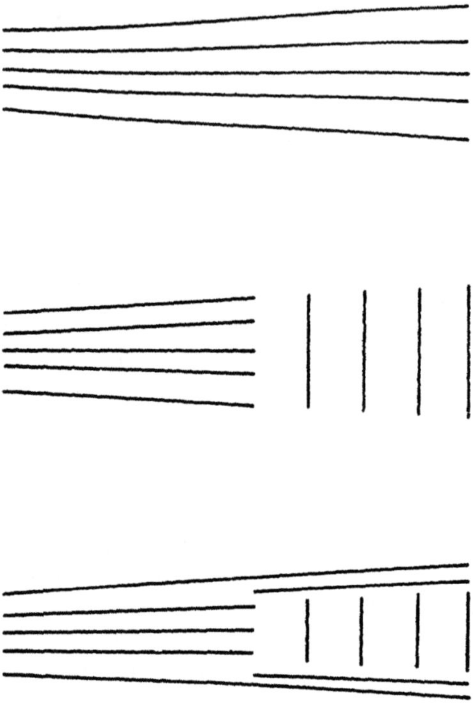

Fig. IX.8. A schematic representation of magnetic field configurations in astrophysical jets.

sufficiently strong transverse gradient of longitudinal plasma velocity, similarly to the water flow in a river where velocities are the highest at the middle and decline toward the banks. The induction equation gives

$$\left(\frac{\partial}{\partial t} + \mathbf{v}\nabla\right) B_{\parallel} = (\mathbf{B}_{\perp}\nabla_{\perp})v_{\parallel},$$

where we have neglected magnetic diffusion and omitted the term $(\mathbf{B}_{\parallel}\nabla_{\parallel})\mathbf{v}_{\parallel}$. This equation shows that $B_{\parallel}$ is continuously generated from $\mathbf{B}_{\perp}$. It is clear that the direction of longitudinal field thus generated changes to the opposite in every cross-section at the central symmetry axis of the jet.

Another effect of the regular velocity field on the magnetic field may be amplification due to the overall differential rotation of the plasma inside the jet (Ruzmaikin *et al.*, 1987). However, we have no observational indications of such a flow with helical streamlines in observed jets.

The role of turbulent plasma flows, which are expected in jets due to high Reynolds numbers, has been considered by De Young (1980) and discussed in the previous section. Here we emphasize only that turbulent generation of magnetic field proceeds both in a flow with non-zero mean helicity $\langle \mathbf{v} \cdot \mathrm{rot}\, \mathbf{v} \rangle$ and in the absence of the mean helicity. The fields with zero mean value are generated in the latter case. A model of turbulent generation of regular magnetic fields in expanding jets is presented by Gvaramadze *et al.* (1988). Magnetic fields that can arise due to interaction of the jet plasma with radiation field are considered by Rose (1987).

Solution of the consistently formulated problem of magnetic field (mean plus fluctuations) generation in jets with account for proper boundary conditions and flow dynamics remains an outstanding problem in the physics of astrophysical jets.

CHAPTER X

CONCLUDING REMARKS

The origin of galactic magnetism now seems to be firmly established. Only motions of an ionized, conductive medium are able to maintain and amplify magnetic fields at galactic scales. This process of maintenance and generation of the magnetic field is called the hydromagnetic dynamo; the same term is usually applied to the motion itself, which provides the dynamo action. In spiral galaxies, the basic elements of the dynamo are the differential rotation of the gaseous disc and chaotic motions of the interstellar gas that become helical under influence of rotation and in presence of the density gradient. The hydromagnetic dynamo is a universal process that also generates magnetic fields in the liquid cores of planets and the convective zones of stars. An advantageous peculiarity of the galactic dynamo is the fact that in galaxies the velocity field can be observed directly. Moreover, in our Galaxy we even reside *within* a dynamo machine.

The dynamo requires presence of some weak, initial, seed magnetic field. The authors conceive that the seed fields in galaxies are produced by ejections of magnetic fields from supernovae and young stars inside galaxies. We cannot, however, exclude the hypothetical weak intergalactic magnetic field as a seed for the galactic dynamo.

Up to now, the kinematic dynamo of the large-scale magnetic field has been developed and understood in sufficient detail that gives us information about conditions for the field generation along with configuration of the growing field. The kinematic dynamo does not take into account action of the generated field upon the motions. Thus, the nonlinear dynamo theory should be developed in order to obtain the stationary field strength. This problem is somewhat mitigated by the fact that the nonlinearity is probably weak in real galaxies – galaxies are relatively young from the magnetic point of view since the characteristic time of exponential growth of a weak seed magnetic field is generally as large as few tenths of the galactic lifetime.

Another problem that awaits a final solution considers properties of the fluctuative, chaotic component of interstellar magnetic fields, their correlation properties and spatial distribution.

The dynamo theory can be further applied to astrophysical jets and radio-galaxies. Also poorly understood are the processes of magnetic field generation in centers of galaxies, galactic coronae and in the intergalactic gas of galaxy clusters.

The observational aspect of the current progress in investigation of galactic magnetic fields is due primarily to the achievements of radioastronomy. With great radiotelescopes, the large-scale magnetic structures were recently detected in several spiral galaxies. Further accumulation of observational data should be accompanied by proper advance in methods of data processing and interpretation.

Comparison of the observational and theoretical results definitely requires observational determination of such non-magnetic characteristics of spiral galaxies as sizes and shapes of *ionized* gaseous discs and the mean helicity of interstellar turbulence.

Optical investigation of the large-scale galactic magnetic fields, that align interstellar dust particles, has recently entered a new stage of development due to application of CCD devices. Qualitatively new results of Zeeman observations that now allow us to obtain maps of magnetic field distribution may play a prominent role in the study of chaotic interstellar magnetic fields.

In the studies of the galactic magnetism, we are now witnessing the end of an initial stage of revelation of the origin of galactic magnetic fields and derivation of first observational examples. This book is devoted primarily to the results obtained at this stage and perspectives of further developments. Now we expect further, deeper investigation of galactic magnetic fields and their role in the world of galaxies.

REFERENCES

Aannestad, P. A. and Greenberg, J. M.: 1983, 'Interstellar Polarization, Grain Growth, and Alignment', *Astrophys. J.* **272**, 551–562.
Aannestad, P. A. and Purcell, E. M.: 1973, 'Interstellar Grains', *Ann. Rev. Astron. Astrophys.* **11**, 309–362.
Abbott, D. C.: 1982, 'The Return of Mass and Energy to the Interstellar Medium by Winds from Early-type Stars', *Astrophys. J.* **263**, 723–735.
Abrahamian, M. G.: 1975, 'The Ellipsoidal Figures of Equilibrium of Interstellar Medium inside the Spheroidal Galaxies', *Astrofizika* **11**, 487–498.
Abrahamian, M. G. and Kaplan, S. A.: 1975, 'The Equilibrium Figures of Interstellar Medium in Spheroidal Galaxies', *Astrofizika* **11**, 121–133.
Abrahamian, M. G. and Oganesian, R. S.: 1977, 'Interstellar Medium in Spheroidal Galaxies', *Astrofizika* **13**, 253–262.
Achterberg, A., Blandford, R. D., and Goldreich, P.: 1983, 'Magnetic Focusing in the Sco X-1 Radio Source', *Nature* **304**, 607–609.
Agafonov, G. I., Sokoloff, D. D., and Ruzmaikin, A. A.: 1987, 'The Magnetic Field in the Perseus Arm', *Astron. J.* (SSSR) (in press).
Aitken, D. K., Roche, P. F., Bailey, J. A., Briggs, G. P., Hough, J. M., and Thomas, J. A.: 1986, 'Infrared Spectropolarimetry of the Galactic Centre: Magnetic Alignment in the Discrete Sources', *Monthly Notices Roy. Astron. Soc.* **218**, 363–384.
Albert, C. E.: 1983, 'Neutral Interstellar Gas in the Lower Galactic Halo', *Astrophys. J.* **272**, 509–539.
Alfvén, H.: 1937, 'Versuch zu einer Theorie über die Entstehung der kosmischen Strahlung. I, II', *Zs. f. Phys.* **105**, 579–588; **107**, 319–333.
Alfvén, H.: 1949, 'On the Solar Origin of Cosmic Radiation', *Phys. Rev.* **75**, 1732–1735.
Alfvén, H.: 1950, *Cosmical Electrodynamics*, Oxford.
Alfvén, H. and Herlofson, N.: 1950, 'Cosmic Radiation and Radio Stars', *Phys. Rev.* **78**, 616.
Allen, R. J., Baldwin, J. E., and Sancisi, R.: 1978, 'Radio Continuum Observations of the Edge-on Disc Galaxy NGC 891', *Astron. Astrophys.* **62**, 397–409.
Andernach, H., Baker, J. R., von Kap-Herr, A., and Wielebinski, R.: 1979, 'The Radio Continuum Halo of M87', *Astron. Astrophys.* **74**, 93–99.
Andreassian, R. R.: 1980, 'On the Structure of the Galactic Magnetic Field', *Astrofizika* **16**, 707–713.
Andreassian, R. R.: 1982, 'Investigation of the Magnetic Field of the Galaxy', *Astrofizika* **18**, 255–262.
Armstrong, J. W., Cordes, J. M., and Rickett, B. J.: 1981, 'Density Power Spectrum in the Local Interstellar Medium', *Nature* **291**, 561–563.
Arnold, V. I., Zeldovich, Ya. B., Ruzmaikin, A. A., and Sokoloff, D. D.: 1982, 'Stationary Magnetic Field in a Periodic Flow', *Dokl. Akad. Nauk SSSR* **266**, 1357–1361.
Artamonova, O. V. and Sokoloff, D. D.: 1986, 'Asymptotic Analysis of the Fast Growth of the Magnetic Field Second Moment in a Mirror-symmetric Flow', *Vestn. Moskov. Univ., Ser. 3 Fizika i Astron.* **27**, 8–13.
Asseo, E. and Sol, H.: 1987, 'Extragalactic Magnetic Fields', *Phys. Rep.* **48**, 307–436.
Avedisova, V. S.: 1985, 'Galactic Rotation Curve from Diffuse Nebulae', *Pisma Astron. J. (SSSR)* **11**, 898–905.
Baade, W.: 1963, *Evolution of Stars and Galaxies*, Harvard Univ. Press, Cambridge, Mass.
Babcock, H. W.: 1947, 'Zeeman Effect in Stellar Spectra', *Astrophys. J.* **105**, 105–191.

Badhwar, G. D. and Stephens, S. A.: 1977, 'Hydrostatic Equilibrium of Gas, Extent of Cosmic Ray Confinement, and Radio Emission in the Galaxy', *Astrophys. J.* **212**, 494—506.

Baierlein, R.: 1978, 'The Amplification of Magnetic Fields During the Radiation Era', *Monthly Notices Roy. Astron. Soc.* **184**, 843—870.

Baiesi-Pillastrini, G. C., Palumbo, G. G. C., and Vettolani, G.: 1983, 'Galaxies Rotation Curves: A Catalogue', *Astron. Astrophys. Suppl.* **53**, 373—381.

Baker, P. L.: 1973, 'A Statistical Investigation of Neutral Hydrogen Line Profiles', *Astron. Astrophys.* **23**, 81—92.

Baker, P. L.: 1979, 'On the Magnetic Field in Interstellar Clouds', *Astron. Astrophys.* **75**, 54—60.

Baker, P. L. and Burton, W. B.: 1975, 'Investigation of Low-latitude Hydrogen Emission in Terms of a Two-component Interstellar Gas Model', *Astrophys. J.* **198**, 281—297.

Bania, T. M. and Lockman, F. J.: 1985, 'The Vertical Distribution of Galactic HI: The Arecibo-Green Bank Survey'. In: *The Milky Way Galaxy. Proc. IAU Symp. No. 106* (eds. H. van Woerden, R. J. Allen, and W. B. Burton), p. 173, D. Reidel, Dordrecht.

Barabanov, A. V. and Zasov, A. V.: 1979, 'On the Central Minimum of Stellar Disc's Density of Some Galaxies', *Astron. J. (SSSR)* **56**, 252—257.

Barlow, M. J.: 1978a, 'The Destruction and Growth of Dust Grains in Interstellar Space — I. Destruction by Sputtering', *Monthly Notices Roy. Astron. Soc.* **183**, 367—395.

Barlow, M. J.: 1978b, 'The Destruction and Growth of Dust Grains in Interstellar Space — II. Destruction by Grain Surface Reactions, Grain-grain Collisions and Photodesorption', *Monthly Notices Roy. Astron. Soc.* **183**, 397—415.

Barlow, M. J.: 1978c, 'The Destruction and Growth of Dust Grains in Interstellar Space — III. Surface Recombination, Heavy Element Depletion and Mantle Growth', *Monthly Notices Roy. Astron. Soc.* **183**, 417—434.

Barrow, J. D., Juszkiewicz, R., and Sonoda, D. H.: 1985, 'Universal Rotation: How Large Can It Be?', *Monthly Notices Roy. Astron. Soc.* **213**, 917—943.

Barvainis, R. and Predmore, C. R.: 1984, 'Time Variability of the 87 GHz Linear Polarization and Flux Density of Active Extragalactic Objects', *Astrophys. J.* **282**, 402—411.

Barvainis, R. and Wootten, A.: 1986, 'Upper Limits on the Polarization of Ammonia Emission in OMC1', *Preprint*, NRAO-86/164.

Baryshnikova, Yu. S., Ruzmaikin, A. A., Sokoloff, D. D., and Shukurov, A. M.: 1987a, 'The Turbulent Dynamo in Galaxies'. In: *Proc. Int. Conf. Plasma Phys.*, Vol. 4, Kiev, 1987, pp. 231—234.

Baryshnikova, Yu. S., Ruzmaikin, A. A., Sokoloff, D. D., and Shukurov, A. M.: 1987b, 'Generation of Large-scale Magnetic Fields in Spiral Galaxies', *Astron. Astrophys.* **177**, 27—41.

Baryshnikova, Yu. S. and Shukurov, A. M.: 1987, 'Oscillatory α^2-dynamo: Numerical Investigation', *Astron. Nachr.* **308**, 89—100.

Basko, M. M. and Polnarev, A. G.: 1980, 'Polarization and Anisotropy of the Relict Radiation in an Anisotropic Universe', *Monthly Notices Roy. Astron. Soc.* **191**, 207—216.

Batchelor, G. K.: 1950, 'On the Spontaneous Magnetic Field in a Conducting Fluid in Turbulent Motion', *Proc. Roy. Soc. London* **A201**, 405—416.

Batchelor, G. K.: 1953, *The Theory of Homogeneous Turbulence*, Cambridge Univ. Press, Cambridge.

Beck, R.: 1982, 'The Magnetic Field in M31', *Astron. Astrophys.* **106**, 121—132.

Beck, R.: 1983, 'Magnetic Fields and Spiral Structure'. In: *Internal Kinematics and Dynamics of Galaxies. Proc. IAU Symp. No. 100* (ed. E. Athanassoula), pp. 159—160. D. Reidel, Dordrecht.

Beck, R.: 1986, 'Interstellar Magnetic Fields', *IEEE Trans. Plasma Sci.* PS-**14**, 740—747.

Beck, R. and Gräve, R. (eds.): 1987, *Interstellar Magnetic Fields. Proc. Workshop Held at Schloss Ringberg, Tegernsee, Sept. 8—12, 1986*, Springer Verlag, Berlin.

Beck, R., Biermann, P., Emerson, D. T., and Wielebinski, R.: 1979, 'Radio Continuum Observations of NGC 891 and NGC 253 at 8.7 GHz', *Astron. Astrophys.* **77**, 25—30.

Beck, R., Klein, U., and Krause, M.: 1985, 'M81 at High Radio Frequencies', *Astron. Astrophys.* **152**, 237—249.

Beck, R., Wielebinski, R., and Berkhuijsen, E. M.: 1980, 'Distribution of Polarized Radio Emission in M31', *Nature* **283**, 272—275.

Beech, M.: 1985a, 'Primordial Magnetic Fields and Symmetry Restoration', *Astrophys. Space Sci.* **114**, 345–350.

Beech, M.: 1985b, 'Is the Intergalactic Magnetic Field Primordial?', *Astrophys. Space Sci.* **116**, 207–212.

Begelman, M. C., Blandford, R. D., and Rees, M. J.: 1984, 'Theory of Extragalactic Radio Sources', *Rev. Mod. Phys.* **56**, 255–351.

Benford, G., Ferrari, A., and Trussoni, E.: 1980, 'Turbulence-related Morphology in Extragalactic Radio Sources', *Astrophys. J.* **241**, 98–110.

Berge, G. L. and Seielstad, G. A.: 1967, 'New Determinations of the Faraday Rotation for Extragalactic Radio Sources', *Astrophys. J.* **148**, 367–376.

Bergeron, J. and Souffrin, S.: 1971, 'Heating of HI Regions by Hard UV Radiation: Physical State of the Interstellar Matter', *Astron. Astrophys.* **11**, 40–52.

Berkhuijsen, E. M.: 1971, 'A Survey of the Continuum Radiation at 820 MHz Between Declinations − 7° and +85°. II. A Study of the Galactic Radiation and the Degree of Polarization with Special Reference to the Loops and Spurs', *Astron. Astrophys.* **14**, 359–386.

Beuermann, K., Kanbach, G., and Berkhuijsen, E. M.: 1985, 'Radio Structure of the Galaxy: Thick Disk and Thin Disk at 408 MHz', *Astron. Astrophys.* **153**, 17–34.

Bičák, J. and Janiš, V.: 1985, 'Magnetic Fluxes Across Black Holes', *Monthly Notices Roy. Astron. Soc.* **212**, 899–915.

Biermann, L.: 1950, 'Über den Ursprung der Magnetfelder auf Sternen und im interstellaren Raum', *Z. Naturforsch.* **5a**, 65–71.

Bietenholz, M. F. and Kronberg, P. P.: 1984, 'Is There Really Evidence for Universal Rotation?', *Astrophys. J. Lett.* **287**, L1–L2.

Binney, J. J.: 1976, 'Is the Flattening of Elliptical Galaxies Necessarily Due to Rotation?', *Monthly Notices Roy. Astron. Soc.* **177**, 19–30.

Binney, J.: 1980, 'The Rotation of Elliptical Galaxies', *Monthly Notices Roy. Astron. Soc.* **190**, 421–439.

Binney, J.: 1982, 'Dynamics of Elliptical Galaxies and Other Spheroidal Components', *Ann. Rev. Astron. Astrophys.* **20**, 399–430.

Birch, P.: 1982, 'Is the Universe Rotating?', *Nature* **298**, 451–454.

Bisnovatyi-Kogan, G. S., Ruzmaikin, A. A., and Sunyaev, R. A.: 1973, 'Condensation of Stars and Formation of a Magnetic Field in Protogalaxies', *Sov. Astron.* **17**, 137–139, 1973.

Blandford, R. D. and Znajek, R. L.: 1977, 'Electromagnetic Extraction of Energy from Kerr Black Holes', *Monthly Notices Roy. Astron. Soc.* **179**, 433–456.

Blitz, L.: 1979, 'The Rotation Curve of the Galaxy to $R = 16$ kiloparsecs', *Astrophys. J. Lett.* **213**, L115–L119.

Blitz, L., Fich, M., and Stark, A. A.: 1980, 'The Galactic Rotation Curve to $R = 18$ kpc'. In: *Interstellar Molecules. Proc. IAU Symp. No. 87* (ed. B. H. Andrew), pp. 213–220. D. Reidel, Dordrecht.

Blumenthal, G. R., Faber, S. M., Primack, J. R., and Rees, M. J.: 1984, 'Formation of Galaxies and Large-scale Structure with Cold Dark Matter', *Nature* **311**, 517–525.

Boer, K. S. de and Savage, B. D.: 1983, 'Absorption by Halo Gas in the Direction of M13', *Astrophys. J.* **265**, 210–215.

Bohlin, R. C., Savage, B. D., and Drake, J. F.: 1978, 'A Survey of Interstellar HI from Ly_α Absorption Measurements. II', *Astrophys. J.* **224**, 132–142.

Bolshev, L. N. and Smirnov, Yu. V.: 1965, *Tables of Mathematical Statistics*, Nauka, Moscow (in Russian).

Bolton, J. G. and Wild, J. P.: 1957, 'On the Possibility of Measuring Interstellar Magnetic Fields by 21-cm Zeeman Splitting', *Astrophys. J.* **125**, 296–297.

Bosma, A.: 1981, '21-cm Line Studies of Spiral Galaxies. II. The Distribution and Kinematics of Neutral Hydrogen in Spiral Galaxies of Various Morphological Types', *Astron. J.* **86**, 1825–1846.

Braginsky, S. I.: 1964a, 'Self-excitation of a Magnetic Field During the Motion of a Highly Conducting Fluid', *JETP* **47**, 1084–1098 (*Sov. Phys. JETP* **20**, 726–735, 1964).

Braginsky, S. I.: 1964b, 'Theory of Hydromagnetic Dynamo', *JETP* **47**, 2178–2193 (*Sov. Phys. JETP* **20**, 1462–1471, 1964).
Brecher, K. and Blumenthal, G.: 1970, 'On the Origin of Cosmic Magnetic Fields', *Astrophys. Lett.* **6**, 169–173.
Bregman, J. N.: 1980a, 'The Galactic Fountain of High-velocity Clouds', *Astrophys. J.* **236**, 577–591.
Bregman, J. N.: 1980b, 'X-ray Coronae Around Galaxies', *Astrophys. J.* **237**, 681–685.
Bregman, J. D., Troland, T. H., Forster, J. R., Schwarz, U. J., Goss, W. M., and Heiles, C.: 1983, 'Aperture Synthesis Observations of the 21-cm Zeeman Effect', *Astron. Astrophys.* **118**, 157–162.
Bridle, A. H.: 1969, '10 MHz Absorption in the Local Spiral Arm', *Nature* **221**, 648–649.
Bridle, A. H.: 1982, 'Systematics of Large-scale Radio Jets'. In: *Extragalactic Radio Sources. Proc. IAU Symp. No. 97* (eds. D. S. Heeschen, C. M. Wade), pp. 121–128. D. Reidel, Dordrecht.
Bridle, A. H. and Perley, R. A.: 1984, 'Extragalactic Radio Jets', *Ann. Rev. Astron. Astrophys.* **22**, 319–358.
Bridle, A. H. and Venugopal, V. R.: 1969, 'Distribution and Temperature of Interstellar Electron Gas', *Nature* **224**, 545–547.
Brinks, E. and Burton, W. B.: 1984, 'A High Resolution Hydrogen Survey of Messier 31. II. The Warped Flaring Hydrogen Layer', *Astron. Astrophys.* **141**, 195–214.
Broten, N. W., MacLeod, J. M., and Vallée, J. P.: 1985, 'An Observational Trend for Large (100–300 pc) Interstellar Magnetic (1–10 μ Gauss) Bubbles', *Astrophys. Lett.* **24**, 165–171.
Brouw, W. N. and Spoelstra, T. A. Th.: 1976, 'Linear Polarization of the Galactic Background at Frequencies Between 408 and 1411 MHz', *Astron. Astrophys. Suppl.* **26**, 129–146.
Buczilowski, U. R.: 1985, 'Radiokontinuumsstrahlung der Spiralgalaxie M33', Ph.D. Thesis, Rheinischen Friedrich-Wilhelms-Universität, Bonn.
Buczilowski, U. R.: 1987, 'Magnetic Fields in M33 and NGC 6946'. In: *Proc. of the Workshop 'Interstellar Magnetic Fields'*, 8–12 Sept. 1986, Schloss Ringberg/Tegernsee, pp. 54–56. Springer Verlag, Berlin.
Burbidge, G. R.: 1956, 'On Synchrotron Radiation from Messier 87', *Astrophys. J.* **124**, 416–429.
Burbidge, G. R.: 1959, 'Estimates of the Total Energy in Particles and Magnetic Field in the Non-thermal Radio Sources', *Astrophys. J.* **129**, 849–852.
Buren, D. van: 1985, 'The Initial Mass Function and Global Rates of Mass, Momentum, and Energy Input to the Interstellar Medium Via Stellar Winds', *Astrophys. J.* **294**, 567–577.
Burke, J. R. and Silk, J.: 1974, 'Dust Grains in a Hot Gas. I. Basic Physics', *Astrophys. J.* **190**, 1–10.
Burn, B. J.: 1966, 'On the Depolarization of Discrete Radio Sources by Faraday Dispersion', *Monthly Notices Roy. Astron. Soc.* **133**, 67–84.
Burns, J. O., Owen, F. N., and Rudnick, L.: 1979, 'The Wide-angle Tailed Radio Galaxy 1159+583: Observations and Models', *Astron. J.* **84**, 1683–1693.
Burton, W. B.: 1976, 'The Morphology of Hydrogen and of Other Tracers in the Galaxy', *Ann. Rev. Astron. Astrophys.* **14**, 275–306.
Burton, W. B. and Gordon, M. A.: 1978, 'Carbon Monoxide in the Galaxy. III. The Overall Nature of its Distribution in the Equatorial Plane', *Astron. Astrophys.* **63**, 7–27.
Burton, W. B., Gordon, M. A., Bania, T. M., and Lockman, F. J.: 1975, 'The Overall Distribution of Carbon Monoxide in the Plane of the Galaxy', *Astrophys. J.* **202**, 30–49.
Bykov, A. M.: 1982, 'Interstellar Turbulence and Shock Waves', *Pisma Astron. J.* (*SSSR*) **8**, 596–599.
Bykov, A. M. and Toptygin, I. N.: 1981, 'Cosmic Rays Acceleration Within Sources by the Supersonic Turbulence', *Izv. Vyssh. Uchebn. Zaved. Fizika* **45**, 474–485 (*Sov. Phys. J.*).
Bykov, A. M. and Toptygin, I. N.: 1985, 'Generation of Turbulence by Shock Waves and Cosmic Ray Diffusion in the Interstellar Medium', *Pisma Astron. J.* (*SSSR*) **11**, 184–189.
Bykov, A. M., and Ushakov, A. Yu.: 1986, 'Statistical Properties of Turbulence Generated by an Ensemble of Shocks', *Preprint No. 1059*, Ioffe Phys. Techn. Inst., Academy of Sciences, Leningrad, USSR.
Caldwell, J. A. R. and Ostriker, J. P.: 1981, 'The Mass Distribution Within Our Galaxy: A Three-component Model', *Astrophys. J.* **251**, 61–87.
Celnick, W., Rohlfs, K., and Braunsfurth, E.: 1979, 'A New Determination of the Thickness of the

Galactic Disk from HII-observations and Discussion of Some Consequences for Galactic Mass Models', *Astron. Astrophys.* **76**, 24—34.

Cesarsky, C. J.: 1980, 'Cosmic-ray Confinement in the Galaxy', *Ann. Rev. Astron. Astrophys.* **18**, 289—319.

Chaisson, E. J. and Vrba, F. J.: 1978, 'Magnetic Field Structures and Strengths in Dark Clouds'. In: *Protostars and Planets* (ed. T. Gehrels), pp. 180—208. Univ. Arizona Press, Tucson.

Chashei, I. V. and Shishov, V. I.: 1980, 'On the Nature of Inhomogeneities of the Interstellar Medium', *Pisma Astron. J. (SSSR)* **6**, 574—578.

Cheng, A. Y. S., Pacholszyk, A. G., and Cook, H.: 1985, 'Circular Polarization of Compact, Extragalactic Radio Sources. I. Synchrotron Emission and Circular Repolarization', *Astrophys. J.* **297**, 639—643.

Chernyshev, V. I. and Stankevich, K. S.: 1975, 'A Measurement of the Spectral Index of the Continuous Cosmic Radioemission at Decimetre Wavelengths', *Izv. Vyssh. Uchebn. Zaved. Radiofizika* **18**, 1706—1708 (*Sov. Radiophys.*).

Chevalier, R. A.: 1977, 'The Interaction of Supernovae with the Interstellar Medium', *Ann. Rev. Astron. Astrophys.* **15**, 175—196.

Chibisov, G. V. and Ptuskin, V. S.: 1981, 'Angular Variations of Nonthermal Radio Emission from the Galaxy Relevant to the Structure of Interstellar Magnetic Field', *Proc. 17th Intern. Cosmic Ray Conf., Paris, 1981*, vol. 2, pp. 233—235. Gif-sur-Ivette.

Chini, R. and Wink, J. E.: 1984, 'The Galactic Rotation Outside the Solar Circle', *Astron. Astrophys.* **139**, L5—L8.

Choudhuri, A. R.: 1984, 'The Effect of Closed Boundary Conditions on a Stationary Dynamo', *Astrophys. J.* **281**, 846—853.

Churilov, S. M. and Shukhman, I. G.: 1982, 'Turbulence in a Self-gravitating Gaseous Disk. II. Turbulent Spectra', *Astron. J. (SSSR)* **59**, 1093—1102 (*Sov. Astron* **26**, 660, 1982).

Clemens, D. P.: 1985, 'Massachusets—Stony Brook Galactic Plane CO Survey: the Galactic Disk Rotation Curve', *Astrophys. J.* **295**, 422—436.

Codina-Landaberry, S. and Magalhães, A. M.: 1976, 'On the Polarizing Interstellar Dust', *Astron. Astrophys.* **49**, 407—413.

Cooke, B. A., Lawrence, A., and Perola, G. C.: 1978, 'Detection of X-ray Emission from the Giant Radio Lobes of Cen A', *Monthly Notices Roy. Astron. Soc.* **182**, 661—671.

Cordes, J. M., Weisberg, J. M., and Boriakoff, V.: 1985, 'Small-scale Electron Density Turbulence in the Interstellar Medium', *Astrophys. J.* **288**, 221—247.

Courant, R. and Hilbert, D.: 1962, *Methods of Mathematical Physics*, Springer Verlag, New York, London.

Cowling, T. G.: 1934, 'Magnetic Fields of Sunspots', *Monthly Notices Roy. Astron. Soc.* **94**, 39—48.

Cox, D. P. and Smith, B. W.: 1974, 'Large-scale Effects of Supernova Remnants on the Galaxy: Generation and Maintenance of a Hot Network of Tunnels', *Astrophys. J. Lett.* **189**, L105—L108.

Crawford, M. K., Genzel, R., Harris, A. I., Jaffe, D. T., Lacy, J. H., Lugten, J. B., Serabyn, E., and Townes, C. H.: 1985, 'Mass Distribution in the Galactic Centre', *Nature* **315**, 467—470.

Crovisier, J. and Dickey, J. M.: 1983, 'The Spatial Power Spectrum of Galactic Neutral Hydrogen from Observations of the 21-cm Emission Line', *Astron. Astrophys.* **122**, 282—296.

Crutcher, R. M., Evans, N. J. II, Troland, T., and Heiles, C.: 1975, 'OH Zeeman Observations of Interstellar Dust Clouds', *Astrophys. J.* **198**, 91—93.

Crutcher, R. M. and Kazès, I.: 1983, 'The Magnetic Field of the NGC 2024 Molecular Cloud: Detection of OH Line Zeeman Splitting', *Astron. Astrophys.* **125**, L23—L26.

Cugnon, P.: 1983, 'Magnetic Alignment of Interstellar Dust Grains for Dominating Magnetic Effects', *Astron. Astrophys.* **120**, 156—163.

Cugnon, P.: 1985, 'Some Considerations on the Theory of the Magnetic Alignment of Dust Grains', *Astron. Astrophys.* **152**, 1—5.

Dagkesamansky, R. D. and Shutenkov, V. R., 1987, 'Intensity Fluctuations of the Background Radioemission and Structure of the Galatic Magnetic Field', *Pisma Astron. J. (SSSR)* **13**, 182—190.

Dalgarno, A.: 1960, 'Low Energy Stopping Power of Atomic Hydrogen', *Proc. Phys. Soc. London* **75**, 374–377.
Davis, J., Jr. and Greenstein, J. L.; 1951, 'The Polarization of Starlight by Alignment of Dust Grains', *Astrophys. J.* **114**, 206–240.
Davydov, A. S.: 1966, *Quantum Mechanics*, NEO Press, Michigan.
Deguchi, S. and Watson, W. D.: 1984, 'Linear Polarization of Molecular Lines at Radio Frequencies', *Astrophys. J.* **285**, 126–133.
Deharveng, J. M. and Pellet, A.: 1975, 'Etude cinématique et dynamique de M31 à partir de l'observation des régions d'émission', *Astron. Astrophys.* **38**, 15–28.
Dennison, B.: 1979, 'On the Intracluster Faraday Rotation. I. Observations', *Astron. J.* **84**, 725–729.
Dennison, B.: 1980a, 'Faraday Rotation in the M87 Radio/X-ray Halo', *Astrophys. J.* **236**, 761–768.
Dennison, B.: 1980b, 'Formation of Radio Halos in Clusters of Galaxies from Cosmic-ray Protons', *Astrophys. J. Lett.* **239**, L93–L96.
DiPrima, R. C. and Swinney, H. L.: 1981, 'Instabilities and Transition in the Flow between Concentric Rotating Cylinders'. In: *Hydrodynamic Instabilities and the Transition to Turbulence* (eds. H. L. Swinney and J. P. Gollub). Ch. 6. Springer, Berlin, Heidelberg, New York.
Dittrich, P., Molchanov, S. A., Sokoloff, D. D., and Ruzmaikin, A. A.: 1984, 'Mean Magnetic Field in Renovating Flow', *Astron. Nachr.* **305**, 119–125.
Dolginov, A. Z., Gnedin, Yu. P., and Silant'ev, N. A.: 1979, *Propagation and Polarization of Radiation in Cosmic Medium*, Nauka, Moscow (in Russian).
Doroshkevich, A. G.: 1965, 'Model of a Universe with a Uniform Magnetic Field', *Astrofizika* **1**, 255–266. (*Astrophysics* **1**, 138–142).
Draine, B. T.: 1983, 'Magnetic Bubbles and High-velocity Outflows in Molecular Clouds', *Astrophys. J.* **270**, 519–536.
Draine, B. T. and Salpeter, E. E.: 1979, 'Destruction Mechanisms for Interstellar Dust', *Astrophys. J.* **231**, 438–455.
Dreiper, N. R. and Smith, H.: 1966, *Applied Regression Analysis*, New York, London, Sydney.
Eckers, R. D. and Sancisi, R.: 1977, 'The Radio Continuum Halo in NGC 4631', *Astron. Astrophys.* **54**, 973–974.
Eichendorf, W. and Reinhardt, M.: 1980, 'Polarization Properties of Extragalactic Radio Sources', *Zesz. Nauk UJ.* No. 570, 7–48; *Acta Cosmologica* **9**, 7–48.
Einasto, J., Haud, U., and Joeveer, M.: 1979, 'The Galactic Circular Velocity near the Sun'. In: *The Large-Scale Characteristics of the Galaxy. Proc. IAU Symp. No. 84* (ed. W. B. Burton), pp. 231–232. D. Reidel, Dordrecht.
Einasto, J., Kaasik, A., and Saar, E.: 1974, 'Dynamic Evidence on Massive Coronas of Galaxies', *Nature* **250**, 309–310.
Einasto, J., Tenjes, P., Barabanov, A. V., and Zasov, A. V.: 1980, 'Central Holes in Disks of Spiral Galaxies', *Astrophys. Space Sci.* **67**, 31–43.
Elmegreen, B. G.: 1982, 'The Parker Instability in a Self-Gravitating Gas Layer', *Astrophys. J.* **253**, 634–654.
Elmegreen, B. G. and Elmegreen, D. M.: 1986, 'Do Density Waves Trigger Star Formation?', *Astrophys. J.* **311**, 554–562.
Elsasser, W. M.: 1946, 'Induction Effects in Terrestrial Magnetism', *Phys. Rev.* **69**, 106–116.
Elvius, A.: 1978a, 'Magnetic Fields in Galaxies as Revealed by the Polarization of Light', *Astrophys. Space Sci.* **55**, 49–57.
Elvius, A.: 1978b, 'Optical Polarization in Galaxies'. In: *Structure and Properties of Nearby Galaxies. Proc. IAU Symp. No. 77* (eds. E. Berkhnuijsen, R. Wielebinski), p. 78. D. Reidel, Dordrecht.
Emerson, D. T.: 1974, 'High Resolution Observations of Neutral Hydrogen in M31. I. The Overall Distribution', *Monthly Notices Roy. Astron. Soc.* **169**, 607–629.
Emerson, D. T.: 1976, 'High Resolution Observations of Neutral Hydrogen in M31-II', *Monthly Notices Roy. Astron. Soc.* **176**, 321–344.
Emerson, D. T.: 1978, 'On the Distribution of CO and HI in M31', *Astron. Astrophys.* **63**, L29–L30.

Emerson, D. T., Kronberg, P. P., Wielebinski, R.: 1978, 'A 2.7. GHz Continium Survey of the M81–M82 System', *Astron. Astrophys.* **63**, 49–52.

Falgarone, E. and Lequeux, J.: 1973, 'A Discussion of the Distribution of Interstellar Matter Close to the Sun', *Astron. Astrophys.* **25**, 253–260.

Feigelson, E. D., Schreier, E. J., Delvaille, J. P., Giacconi, R., Grindlay, J. E., and Lightman, A. P.: 1981, 'The X-ray Structure of Centaurus A', *Astrophys. J.* **251**, 31–51.

Feitzinger, J. V. and Spicker, J.: 1987, 'Stochastic Star Formation, Magnetic Fields and the Fine Structure of Spiral Arms'. In: *Proc. of the Workshop 'Interstellar Magnetic Fields'*, 8–12 Sept. 1986, Schloss Ringberg/Tegernsee, pp. 171–178. Springer Verlag, Berlin.

Fermi, E.: 1949, 'On the Origin of the Cosmic Radiation', *Phys. Rev.* **75**, 1169–1174.

Fich, M. and Blitz, L.: 1984, 'Optical HII Regions in the Outer Galaxy', *Astrophys. J.* **279**, 125–135.

Fleck, R. C., Jr.: 1983, 'Turbulence and the Stability of Molecular Clouds', *Astrophys. J.* **242**, 1019–1022.

Fleck, R. C., Jr.: 1983, 'A Note on Compressibility and Energy Cascade in Turbulent Molecular Clouds', *Astrophys. J. Lett.* **272**, L45–L48.

Foote, E. A. and Kulsrud, R. N.: 1979, 'Hydromagnetic Waves in High β Plasmas', *Astrophys. J.* **233**, 302–316.

Freeman, K. C.: 1970, 'On the Disks of Spiral and SO Galaxies', *Astrophys. J.* **160**, 811–830.

Freeman, K. C.: 1986, 'Dynamics of Disk Galaxies', *Astrophys. Space Sci.* **118**, 337–342.

Frisch, U.: 1985, 'Fully Developed Turbulence and Intermittency'. In *Turbulence and Predictability in Geophysical Fluid Dynamics and Climate Dynamics, LXXXVIII Corso. Soc. Italiana di Fisica*, pp. 71–88.

Fuchs, B. and Thielheim, K. O.: 1979, 'On the Hydrostatic Equilibrium of the Interstellar Gas', *Astrophys. J.* **227**, 801–807.

Fujimoto, M. and Sawa, T.: 1987, 'Bisymmetric Spiral Configuration of Magnetic Fields in Spiral Galaxies. II. A Quasi-Global Theory', *Publ. Astron. Soc. Japan* **39**, 375–392.

Fujimoto, M. and Tosa, M.: 1980, 'Spiral Condensation of Gas in Disk Galaxies by Bisymmetric Twisted Magnetic Fields: Two-dimensional Case', *Publ. Astron. Soc. Japan* **32**, 567–580.

Galeev, A. A.: 1984, 'Generation of Ultra-relativistic Electrons by Shock Waves and their Synchrotron Emission', *JETP* **86**, 1655–1666 (*Sov. Phys. JETP*).

Gardner, F., Morris, D., and Whiteoak, J. B.: 1969, 'The Linear Polarization of Radio Sources Between 11 and 20 cm Wavelength. III. Influence of the Galaxy on Source Depolarization and Faraday Rotation', *Aust. J. Phys.* **22**, 813–838.

Ginzburg, V. L.: 1953, 'The Origin of Cosmic Rays and Radioastronomy', *Usp. Fiz. Nauk (SSSR)* **51**, 343–392.

Ginzburg, V. L.: 1964, 'Magnetic Fields of Collapsing Masses and the Nature of Superstars', *Dokl. Akad. Nauk SSSR* **156**, 43–46 (*Sov. Phys. Dokl.* **9**, 329–332, 1964).

Ginzburg, V. L. (ed.): 1984, Berezinsky, V. S. Bulanov, S. V., Ginzburg, V. L., Dogiel, V. A., and Ptuskin, V. S. *Astrophysics of Cosmic Rays.* Nauka, Moscow (in Russian, English translation in preparation).

Ginzburg, V. L. and Ozernoi, L. M.: 1964, 'On Gravitational Collapse of Magnetic Stars', *JETP* **47**, 1030–1040 (*Sov. Phys. JETP* **20**, 689–698, 1965).

Ginzburg, V. L. and Ptuskin, V. S.: 1985, 'The Origin of Cosmic Rays', *Sov. Sci. Rev. E.: Astrophys. Space Phys.* **4**, 161–265.

Ginzburg, V. L. and Syrovatskii, S. I.: 1964, *The Origin of Cosmic Rays*, Pergamon Press, Oxford.

Goad, J. W.: 1976, 'Kinematics of the Gas in the Nuclear Disk of M81', *Astrophys. J. Suppl.* **32**, 89–113.

Goad, J. W., De Veny, J. B., and Goad, L. E.: 1979, 'Velocities of the Gas in M51', *Astrophys. J. Suppl.* **39**, 439–460.

Golden, R. L., Mauger, B. G., Badwhar, G. D., Daniel, R. R., Lacy, J. L., Stephens, S. A., and Zipse, J. E.: 1984, 'A Measurement of the Absolute Flux of Cosmic-ray Electrons', *Astrophys. J.* **287**, 622–632.

Goldreich, P., Keeley, D. A., and Kwan, J. Y.: 1973, 'Astrophysical Masers. II. Polarization Properties', *Astrophys. J.* **179**, 111–134.

Goldreich, P. and Kylafis, N. D.: 1981, 'On Mapping the Magnetic Field Direction in Molecular Clouds by Polarization Measurements', *Astrophys. J. Lett.* **243**, L75–L78.

Goldreich, P. and Kylafis, N. D.: 1982, 'Linear Polarization of Radio Frequency Lines in Molecular Clouds and Circumstellar Envelopes', *Astrophys. J.* **253**, 606–621.

Golitsyn, G. S.: 1960, 'Fluctuations of the Magnetic Field and Current Density in a Turbulent Flow of a Weakly Conducting Fluid', *Dokl. Akad. Nauk SSSR* **132**, 315–318 (*Sov. Phys. Dokl.* **5**, 536–539, 1960).

Graham, J. A.: 1979, 'The Galactic Distance Scale'. In: *The Large-Scale Characteristics of the Galaxy. Proc. IAU Symp. No. 84* (ed. W. B. Burton), pp. 195–200. D. Reidel, Dordrecht.

Gräve, R. and Beck, R.: 1987, 'The Magnetic Field in IC 342'. In: *Proc. of the Workshop 'Interstellar Magnetic Fields'*, 8–12 Sept. 1986, Schloss Ringberg/Tegernsee, pp. 47–51. Springer Verlag, Berlin.

Greenberg, J. M.: 1968, *Interstellar Grains*. In: *Stars and Stellar Systems*, Vol. 7, Ch. 6. Univ. Chicago Press, Chicago.

Grewing, M. and Walmsley, C. M.: 1974, 'Thermal and Ionization Equilibrium of an X-ray Heated Intercloud Medium', *Astron. Astrophys.* **30**, 281–287.

Guelin, M.: 1974, 'Free Electrons Outside HII Regions'. In: *Galactic Radio Astronomy. Proc. IAU Symp. No. 60* (eds. F. J. Kerr and S. C. Simonson, III), pp. 51–63. D. Reidel, Dordrecht.

Gursky, H. and Schwartz, D. A.: 1977, 'Extragalactic X-ray Sources', *Ann. Rev. Astron. Astrophys.* **15**, 541–568.

Guseinov, O. H., Kasumov, F. K., and Kalinin, E. V.: 1980, 'Radial Distribution of Types I and II Supernovae in Spiral and Elliptical Galaxies', *Astrophys. Space Sci.* **68**, 385–392.

Gvaramadze, V. V. and Lominadze, J. G.: 1986, 'Gas Rotation Above the Galactic Disk'. In: *Plasma Astrophysics. Proc. Joint Varenna-Abastumani School and Workshop, Sukhumi — USSR 18–28 May 1986*, pp. 551–556, ESA Publ. SP-251.

Gvaramadze, V. V., Lominadze, J. G., Ruzmaikin, A. A., Sokoloff, D. D., and Shukurov, A. M.: 1988, 'Turbulent Generation of Magnetic Fields in Astrophysical Jets'. Submitted to *Astrophys. Space Sci.*

Habe, A. and Ikeuchi, S.: 1980, 'Dynamical Behavior of Gaseous Halo in a Disk Galaxy', *Progr. Theor. Phys.* **64**, 1995–2008.

Hale, G. E.: 1908, 'On the Probable Existence of a Magnetic Field in Sun-spots', *Astrophys. J.* **28**, 315–343.

Hall, A. N.: 1980, 'A Reanalysis of Pulsar Dispersion Data', *Monthly Notices Roy. Astron. Soc.* **191**, 751–759.

Hall, J. S. and Mikesell, A. M.: 1949, 'Observations of Polarized Light from Stars', *Astrophys. J.* **54**, 187–188.

Harding, D. S. and Harding, A. K.: 1982, 'The Distribution of Free Electrons in the Inner Galaxy from Pulsar Dispersion Measures', *Astrophys. J.* **257**, 603–611.

Harrison, E. R.: 1969, 'Generation of Galactic Magnetic Fields', *Astrophys. Lett.* **3**, 133–135.

Harrison, E. R.: 1970, 'Generation of Magnetic Fields in the Radiation Era', *Monthly Notices Roy. Astron. Soc.* **147**, 279–286.

Harrison, E. R.: 1973, 'Standard Model of the Early Universe', *Ann. Rev. Astron. Astrophys.* **11**, 155–186.

Haud, U. A.: 1979, 'Rotation Curve of the Galaxy with Account for Expansion of the Gaseous Component', *Pisma Astron. J.* (*SSSR*) **5**, 124–127 (*Sov. Astron. Lett.*).

Haud, U.: 1981, 'Gas Kinematics in M31', *Astrophys. Space Sci.* **76**, 477–490.

Haud, U.: 1984, 'Rotation Curve of Our Galaxy for $R > R_{\odot}$', *Astrophys. Space Sci.* **104**, 337–345.

Haud, U. and Einasto, J.: 1983, 'Mass Distribution and Dark Halos'. In: *Internal Kinematics and Dynamics of Galaxies. Proc. IAU Symp. No. 100* (ed. E. Athanassoula), pp. 81–88. D. Reidel, Dordrecht.

Haves, P.: 1975, 'Polarization Parameters of 183 Extragalactic Radio Sources', *Monthly Notices Roy. Astron. Soc.* **173**, 553–568.

Haves, P., Conway, R. G., and Stannard, D.: 1974, 'The Polarization of Radio Sources at $\lambda = 31$ cm', *Monthly Notices Roy. Astron. Soc.* **169**, 117–131.

Heiles, C.: 1974, 'A Modern Look at "Interstellar Clouds".' In: *Galactic Radio Astronomy. Proc. IAU Symp. No. 60.* (eds. F. J. Kerr and S. C. Simonson, III), pp. 13–44. D. Reidel, Dordrecht.

Heiles, C.: 1976, 'The Interstellar Magnetic Field', *Ann. Rev. Astron. Astrophys.* **14**, 1–22.

Heiles, C. and Stevens, M.: 1986, 'Zeeman Splitting of 18 Centimeter OH Lines toward Cassiopeia A and other Sources', *Astrophys. J.* **301**, 331–338.

Helou, G. and Salpeter, E. E.: 1982, 'Galaxy Spins in the Virgo Cluster', *Astrophys. J.* **252**, 75—80.
Henriksen, R. N. and Turner, B. E.: 1984, 'Star Cloud Turbulence', *Astrophys. J.* **287**, 200—207.
Higdon, J. C.: 1984, 'Density Fluctuations in the Interstellar Medium: Evidence for Anisotropic Magnetogasdynamic Turbulence. I. Model and Astrophysical Sites', *Astrophys. J.* **285**, 109—123.
Hillebrandt, W., Höflich, P., Kafka, P., Müller, E., Schmidt, H. U., and Truran, J. W.: 1987a, 'Evidence for a Finite Electron Neutrino Rest Mass from SN 1987a', *Astron. Astrophys.* **177**, L41—L42.
Hillebrandt, W., Höflich, P., Kafka, P., Müller, E., Schmidt, H. U., and Truran, J. W.: 1987b, 'Indications for Black Hole Formation from Neutrino Observations in SN 1987a', *Astron. Astrophys.* **180**, L20—L22.
Hiltner, W. A.: 1949a, 'Polarization of Radiation from Distant Stars by the Interstellar Medium', *Nature* **169**, 283.
Hiltner, W. A.: 1949b, 'On the Presence of Polarization in the Continuous Radiation of Stars. II', *Astrophys. J.* **109**, 471—478.
Hiltner, W. A.: 1958, 'Interstellar Polarization in M31', *Astrophys. J.* **128**, 9—13.
Ho, C. M. and Huerre, P.: 1984, 'Perturbed Free Shear Layers', *Ann. Rev. Fluid Mech.* **16**, 365—424.
Hodge, P. W.: 1974, 'HII Regions in Galaxies', *Publ. Astron. Soc. Pacific* **86**, 845—860.
Hodge, P. W. and Kennicutt, R. C. Jr.: 1983, 'The Radial Distribution of HII Regions in Spiral Galaxies', *Astrophys. J.* **267**, 563—570.
Hoerner, S. von: 1951, 'Eine Methode zur Untersuchung der Turbulenz der interstellären Materie', *Z. f. Astrophys.* **30**, 17—64.
Hogan, C. J.: 1983, 'Magnetohydrodynamic Effects of a First-order Cosmological Phase Transition', *Phys. Rev. Lett.* **51**, 1488--1491.
Holman, G. D., Ionson, J. A., and Scott, J. S.: 1979, 'Particle Streaming: Is the Alfvén Velocity the Ultimate Speed Limit?', *Astrophys. J.* **228**, 576—581.
Hong, S. S. and Greenberg, J. M.: 1980, 'A Unified Model of Interstellar Grains: A Connection between Alignment Efficiency, Grain Model Size, and Cosmic Abundance', *Astron. Astrophys.* **88**, 194—202.
Hoyle, F.: 1958, 'The Steady State Theory'. In: *La Structure et L'evolution de L'univers: XI Conseil de Physique Solvay, Bruxelles, 9—13 Juin 1958* (ed. R. Stoops), pp. 53—73.
Hoyle, F.: 1969, 'Magnetic Field and Highly Condensed Objects', *Nature* **223**, 936.
Hron, J.: 1987, 'Kinematics of Young Open Clusters and the Rotation Curve of our Galaxy', *Astron. Astrophys.* **176**, 34—52.
Hummel, E., Sancisi, R., and Eckers, R. D.: 1984a, 'Radio Continuum Observations of the Edge-on Spiral Galaxies NGC 4244, 4565, and 5907', *Astron. Astrophys.* **133**, 1—10.
Hummel, E., Smith, P., and Hulst, J. M. van der: 1984b, 'The Large-scale Radio Continuum Structure of the Edge-on Spiral Galaxy NG 253', *Astron. Astrophys.* **137**, 138—144.
Hummel, E., Krause, M., and Beck, R.: 1987, 'The Magnetic Field in the Anomalous Arms in NGC 4258'. In: *Proc. of the Workshop 'Interstellar Magnetic Fields'*, 8—12 Sept. 1986, Schloss Ringberg/Tegernsee, pp. 61—64. Springer Verlag, Berlin.
Hyland, A. R.: 1986, 'The Galactic Centre', *Astrophys. Space Sci.* **118**, 343—364.
Il'in, A. E.: 1987, 'Absorption and Polarization of Light by Dust in the Interstellar Medium: The Interstellar Circular Polarization', *Astrofizika* **27**, No. 3.
Illingworth, G.: 1977, 'Rotation in 13 elliptical galaxies?', *Astrophys. J. Lett.* **218**, L43—L47.
Illingworth, G.: 1983, 'Dynamics of Early-type Galaxies'. In: *Internal Kinematics and Dynamics of Galaxies. Proc. IAU Symp. No. 100* (ed. E. Athanassoula), pp. 257—270. D. Reidel, Dordrecht.
Ilovaisky, S. A. and Lequeux, J.: 1972, 'A Study of Galactic Supernova Remnants: I. Distances, Radio Luminosity Function and Galactic Distribution', *Astron. Astrophys.* **18**, 169—185.
Inoue, M. and Tabara, H.: 1981, 'Structure of the Galacatic Magnetic Field in the Solar Neighborhood', *Publ. Astron. Soc. Japan* **33**, 603—616.
Inoue, M., Takahashi, T., Tabara, H., Kato, T., and Tsuboi, M.: 1984, 'Extremely Large Faraday Rotation at the Radio Arc in the Galactic Center Region', *Publ. Astron. Soc. Japan* **36**, 633—638.

Iroshnikov, R. S.: 1963, 'Turbulence of a Conducting Fluid in a Strong Magnetic Field', *Astron. J. (SSSR)* **40**, 742—750 (*Sov. Astron.* **7**, 566—571, 1964).

Isakov, R. V., Ruzmaikin, A. A., Sokoloff, D. D. and Faminskaya, M. V.: 1981, 'The Asymptotic Properties of the Disc Dynamo', *Astrophys. Space Sci.* **80**, 145—155.

Issa, M. R., Riley, P. A., Strong, A. W., and Wolfendale, A. W.: 1980, 'Distribution of Cosmic-ray Electrons in the Galaxy', *Nature* **287**, 810—812.

Ivanova, T. S. and Ruzmaikin, A. A.: 1985, 'Three-dimensional Model for Generation of the Mean Solar Magnetic Field', *Astron. Nachr.* **306**, 177—186.

Jackson, P. D., FitzGerald, M. P., and Moffat, A. F. J.: 1979, 'Recent Evidence on the Rotation Curve of our Galaxy for $R > R_{\odot}$'. In: *The Large-Scale Characteristics of the Galaxy. Proc. IAU Symp. No. 84* (ed. W. B. Burton), pp. 221—224. D. Reidel, Dordrecht.

Jackson, P. D. and Kellman, S. A.: 1974, 'A Determination of the Galactic HI Half-thickness and a Discussion of Some Dynamical Consequences', *Astrophys. J.* **190**, 53—58.

Jaffe, W. J.: 1977, 'Origin and Transport of Electrons in the Halo Radio Source in the Coma Cluster', *Astrophys. J.* **212**, 1—7.

Jaffe, W. J.: 1980, 'On the Morphology of the Magnetic Field in Galaxy Clusters', *Astrophys. J.* **241**, 925—927.

Jenkins, E. B.: 1978, 'Coronal Gas in the Galaxy. II. A Statistical Analysis of OVI Absorptions', *Astrophys. J.* **220**, 107—123.

Jenkins, E. B. and Meloy, D.: 1974, 'A Survey with "Copernicus" of Interstellar OVI Absorption', *Astrophys. J. Lett.* **193**, L121—L126.

Jokipii, J. R. and Lerche, I.: 1969, 'Faraday Rotation Dispersion in Pulsar Signals and the Turbulent Structure of the Galaxy', *Astrophys. J.* **157**, 1137—1145.

Jokipii, J. R., Lerche, I., and Schommer, R. A.: 1969, 'Interstellar Polarization of Starlight and the Turbulent Structure of the Galaxy', *Astrophys. J. Lett.* **157**, L119—L124.

Jones, R. V. and Spitzer, L., Jr.: 1967, 'Magnetic Alignment of Interstellar Grains', *Astrophys. J.* **147**, 943–964.

Jones, T. W., Rudnick, L., Aller, H. D., Aller, M. F., Hodge, P. E., and Fiedler, R. L.: 1985, 'Magnetic Field Structures in Active Compact Radio Sources', *Astrophys. J.* **290**, 627—636.

Jura, M.: 1982, 'Large-scale Magnetic Fields in Spiral Galaxies?', *Astrophys. J.* **258**, 59—63.

Juszkiewicz, R. and Shukurov, A.: 1986, 'The Primeval Magnetic Field and Isotropy of the Microwave Background' (unpublished).

Kaburaki, O. and Uchida, Y.: 1971, 'Magnetohydrodynamic Wavemode Coupling. Quantum Field-theoretical Approach to Weakly Non-linear Case with Application to Solar Coronal Heating', *Publ. Astron. Soc. Japan* **23**, 405—424.

Kadomtsev, B. B. and Petviashvili, V. I.: 1973, 'On Acoustic Turbulence', *Dokl. Akad. Nauk SSSR* **208**, 794—796 (*Sov. Phys. Dokl.*).

Kalnajs, A.: 1983, in: *Internal Kinematics and Dynamics of Galaxies. Proc. IAU Symp. No. 100* (ed. E. Athanassoula), pp. 87—88. D. Reidel, Dordrecht.

Kapahi, V. K., Subrahmanyan, R., and Singal, A. K.: 1985, 'The Alignment of Distant Radio Sources', *Nature* **313**, 463—465.

Kaplan, S. A.: 1955, 'The Structure, Correlation and Spectral Functions of the Turbulence of Interstellar Gas', *Astron. J. (SSSR)* **32**, 255—264.

Kaplan, S. A.: 1966, *Interstellar Gas Dynamics*, Pergamon Press, Oxford.

Kaplan, S. A., Oganesian, R. S., and Abrahamian, M. G.: 1977, 'Some Questions of Kinetic Theory of Differentially Rotating Interstellar Medium in Galaxies', *Astrofizika* **13**, 263—270.

Kaplan, S. A. and Pikelner, S. B.: 1970, *Interstellar Medium*, Harvard Univ. Press, Cambridge, Mass.

Kaplan, S. A. and Pikelner, S. B.: 1974, 'Large-scale Dynamics of the Interstellar Medium', *Ann. Rev. Astron. Astrophys.* **12**, 113—134.

Kaplan, S. A. and Pikelner, S. B.: 1979, *Physics of Interstellar Medium*, Nauka, Moscow (in Russian).

Kaplan, S. A. and Pronik, V. I.: 1953, 'On the Problem of Turbulent Nature of Motion of Interstellar Gas Clouds', *Dokl. Akad. Nauk SSSR* **89**, 643—646.

Kardashev, N. S.: 1985, 'Phenomenological Model of the Galactic Nucleus', *Sov. Sci. Rev. E: Astrophys. Space Phys.* **4**, 287—315.

Kawabata, K., Fujimoto, M., Sofue, Y., and Fukui, M.: 1969, 'A Large-scale Metagalactic Magnetic Field and Faraday Rotation for Extragalactic Radio Sources', *Publ. Astron. Soc. Japan* **21**, 293—306.

Kazantsev, A. P.: 1967, 'Enhancement of a Magnetic Field by a Conducting Fluid', *JETP* **53**, 1806—1813 (*Sov. Phys. JETP* **26**, 1031—1039, 1968).

Kazès, I. and Crutcher, R. M.: 1986, 'Measurement of Magnetic-field Strengths in Molecular Clouds: Detection of OH-line Zeeman Splitting', *Astron. Astrophys.* **164**, 328—336.

Kellermann, K. I. and Pauliny-Toth, I. I. K.: 1969, 'The Spectra of Opaque Radio Sources', *Astrophys. J. Lett.* **155**, L71—L78.

Kellman, S. A.: 1972, 'The Equilibrium Configuration of the Gaseous Component of the Galaxy', *Astrophys. J.* **175**, 353—362.

Kemp, J. C.: 1982, 'The Biermann Mechanism and Spontaneous Magnetic Field Generation in Stars', *Publ. Astron. Soc. Pacific* **94**, 627—633.

Kiepenheuer, K. O.: 1950, 'Cosmic Rays as the Source of General Galactic Radio Emission', *Phys. Rev.* **79**, 738—739.

King, D. J.: 1986, 'Optical Polarization in the Spiral Galaxy NGC 7331', *Monthly Notices Roy. Astron. Soc.* **220**, 485—491.

Kirko, G. E.: 1985, *Generation and Self-Excitation of Magnetic Fields in Engineering Devices*, Nauka, Moscow (in Russian).

Klee, P.: 1961, *The Thinking Eye*, Lund Humphries, London.

Kleeorin, N. I., Ruzmaikin, A. A., and Sokoloff, D. D.: 1986, 'Correlation Properties of Self-exciting Fluctuative Magnetic Fields'. In: *Plasma Astrophysics. Proc. Joint Varenna-Abastumani School and Workshop, Sukhumi-USSR 18—28 May 1986*, pp. 557—561, ESA Publ. SP-251.

Klein, U., and Gräve, R.: 1987, 'Nonthermal Radio Emission from Dwarf Galaxies: Do They Host Large-scale Magnetic Fields?' In: *Proc. of the Workshop 'Interstellar Magnetic Fields'*, 8—12 Sept. 1986, Schloss Ringberg/Tegernsee, pp. 76—81. Springer Verlag, Berlin.

Klein, U., Beck, R., Buczilowski, U. R., and Wielebinski, R.: 1982, 'A Survey of the Distribution of λ2.8 cm Radio Continuum in Nearby Galaxies', *Astron. Astrophys.* **108**, 176—187.

Klein, U., Wielebinski, R., and Beck, R.: 1984, 'A Survey of the Distribution of λ2.8 cm Radio Continuum in Nearbly Galaxies. V. A Small Sample of Edge-on Galaxies', *Astron. Astrophys.* **133**, 19—26.

Kleiner, S. C. and Dickman, R. L.: 1985, 'Large-scale Structure of the Taurus Molecular Complex. II.: Analysis of Velocity Fluctuations and Turbulence', *Astrophys. J.* **295**, 466—478.

Klimishin, I. A. and Bazilevich, V. M.: 1963, 'Some Results of Statistical Analysis of Interstellar Starlight Polarization Catalogue', *Tsirk. L'vov. Astron. Obs.* No. 39—40, 3.

Knapp, G. R.: 1979, 'The Circular Velocity at the Sun'. In: *The Large-Scale Characteristics of the Galaxy. Proc. IAU Symp. No. 84* (ed. W. B. Burton), pp. 225—230. D. Reidel, Dordrecht.

Knapp, G. R.: 1983, 'Interstellar Matter in Elliptical Galaxies'. In: *Internal Kinematics and Dynamics of Galaxies. Proc. IAU Symp. No. 100* (ed. E. Athanassoula), pp. 297—304. D. Reidel, Dordrecht.

Knapp, G. P., Tremaine, S. D., and Gunn, J. E.: 1978, 'The Global Properties of the Galaxy. I. The HI Distribution Outside the Solar Circle', *Astron. J.* **83**, 1585—1593.

Kobayashi, Y., Jugaku, J., Okuda, H. Sato, S., and Nagata, T.: 1986, 'Infrared Polarimetry of the Stars in the Inner Galaxy', *Astrophys. Space Sci.* **119**, 135—139.

Kodaira, K.: 1974, 'Distribution of the Galactic Supernova Remnants', *Publ. Astron. Soc. Japan* **26**, 255—261.

Kolesnik, I. G. and Yurevich, P. V.: 1985, 'The Rotation Curve of the Galaxy from the OH Survey', *Astrofizika* **22**, 461—472 (*Astrophysics* **22**, No. 3).

Kolmogorov, A. N.: 1941, 'The Local Structure of Turbulence in Incompressible Viscous Fluid for Very Large Reynolds Numbers', *Dokl. Akad. Nauk SSSR* **30**, 299—303.

Kolobov, V. M., Reinhardt, M., and Sazonov, V. N.: 1976, 'A Test of the Isotropy of the Universe', *Astrophys. Lett.* **17**, 183—185.

Komberg, B. V., Ruzmaikin, A. A., and Sokoloff, D. D.: 1979, 'Faraday Rotation in Galaxies and Absorption Lines in Quasars', *Pisma Astron. J. (SSSR)* **5**, 73—76 (*Sov. Astron. Lett.* **5**, 40—42, 1979).

Korchak, A. A. and Syrovatskii, S. I.: 1961, 'Polarization of Radiation and the Structure of the Magnetic Field of Cosmic Sources of Magnetodecelerating Radiation', *Astron. J. (SSSR)* **38**, 885—897 (*Sov. Astron.* **5**, 678—686, 1962).

Kormendy, J.: 1977, 'Brightness Distribution in Compact and Normal Galaxies. III. Decomposition of Observed Profiles into Spheroid and Disk Components', *Astrophys. J.* **217**, 406—419.

Kraichnan, R. H.: 1965, 'Inertial-range Spectrum of Hydromagnetic Turbulence', *Phys. Fluids* **8**, 1385—1386.

Kraichnan, R. H. and Nagarajan, S.: 1967, 'Growth of Turbulent Magnetic Fields', *Phys. Fluids* **10**, 859—870.

Krause, F.: 1971, 'Zur Dynamotheorie magnetischer Sterne: Der "symmetrische Rotator" als Alternative zum "schiefen Rotator"', *Astron. Nachr.* **293**, 187—193.

Krause, F. and Rädler, K.-H.: 1980, *Mean-Field Magnetohydrodynamics and Dynamo Theory*, Pergamon Press, Oxford.

Krause, M., Beck, R., and Klein, U.: 1984, 'The Atypical Galaxy NGC 4258 at High Radio Frequencies', *Astron. Astrophys.* **138**, 385—395.

Krause, M., Beck, R., and Hummel, E.: 1987, 'The Magnetic Field Structure in M81'. In: *Proc. of the Workshop 'Interstellar Magnetic Fields'*, 8—12 Sept. 1986, Schloss Ringberg/Tegernsee, pp. 57—60. Springer Verlag, Berlin.

Kronberg, P. P. and Perry, J. J.: 1982, 'Absorption Lines, Faraday Rotation, and Magnetic Field Estimates for QSO Absorption-line Clouds', *Astrophys. J.* **263**, 518—432.

Kronberg, P. P., Reinhardt, M., and Simard-Normandin, M.: 1977, 'On the Intergalactic Contribution to the Rotation Measures of QSO's', *Astron. Astrophys.* **61**, 771—776.

Kronberg, P. P. and Simard-Normandin, M.: 1976, 'New Evidence of the Origin of Rotation Measures in Extragalactic Radio Sources', *Nature* **263**, 653—656.

Krugljak, Z. B., Kuznetsov, E. A., Lvov, V. S., Nesterikhin, Yu. E., Predtechensky, A. A., Sobolev, V. S., Utkin, E. N., and Zhuravlev, F. A.: 1980, 'Spectrum Evolution at the Transition to Turbulence in a Couette Flow', *Phys. Lett.* **A78**, 269—272.

Kruit, P. C. van der: 1981, 'The Thickness of the Hydrogen Layer and the Three-dimensional Mass Distribution in NGC 891', *Astron. Astrophys.* **99**, 298—304.

Kruit, P. C. van der and Shostak, G. S.: 1982, 'Studies of Nearly Face-on Spiral Galaxies', *Astron. Astrophys.* **105**, 351—358.

Kruit, P. C. van der and Shostak, G. S.: 1983, 'Vertical Motions and the Thickness of HI Disks: Implications for Galactics Mass Models'. In: *Internal Kinematics and Dynamics of Galaxies. Proc. IAU Symp. No. 100* (ed. E. Athanassoula), pp. 69—76. D. Reidel, Dordrecht.

Kulkarni, Sh. R., Blitz, L., and Heiles, C.: 1982, 'Atomic Hydrogen in the Outer Milky Way', *Astrophys. J. Lett.* **259**, L63—L66.

Kulsrud, R.: 1986, 'The Wrap-up Problem of the Galactic Magnetic Field'. In: *Plasma Astrophysics. Proc. Joint Varenna-Abastumani School and Workshop, Sukhumi-USSR, 18—28 May 1986*, pp. 531—537. ESA Publ. SP-251.

Kulsrud, R. M. and Pearce, W. P.: 1969, 'The Effect of Wave-particle Interaction on the Propagation of Cosmic Rays', *Astrophys. J.* **156**, 445—470.

Kuznetsov, V. D.: 1986, 'The Influence of Cosmic Rays on Stability and Large-scale Dynamics of the Interstellar Medium', *Astron. J. (SSSR)* **63**, 446—457.

Kuznetsov, V. D. and Ptuskin, V. S.: 1983, 'Stability of the Equilibrium Distributions of Interstellar Gas, Cosmic Rays, and Magnetic Field in an External Gravitational Field', *Astrophys. Space Sci.* **94**, 5—21.

Kuznetsova, I. P.: 1976, 'On the Faraday Rotation in the Intergalactic Medium', *Astron. J. (SSSR)* **53**, 475—484 (*Sov. Astron.* **20**, 269—274, 1976).

Kuznetsova, I. P.: 1983, 'On the Problem of Linear Polarization of Synchrotron Emission of Cosmic Radiosources', *Proc. XV All-Union Conf. Galactic and Extragalactic Radioastron., Kharkov, 11—13, Oct. 1983*, pp. 74—75. In-t Radiophys. Electron. AS Ukrainian SSR, Kharkov.

Kyazumov, G. A.: 1984, 'Rotation Curves of Normal Galaxies: A Catalogue', *Astron. J. (SSSR)* **61**, 846—853.

Kylafis, N. D.: 1983, 'Linear Polarization of Interstellar Radio-frequency Absorption Lines and Magnetic Field Direction', *Astrophys. J.* **275**, 135—144.

Kylafis, N. D. and Shapiro, P. R.: 1983, 'Polarization of Interstellar Molecular Radiofrequency Absorption Lines', *Astrophys. J. Lett.* **272**, L35—L39.

Laing, R. A.: 1981, 'Magnetic Fields in Extragalactic Radio Sources', *Astrophys. J.* **248**, 87—104.

Landau, L. D. and Lifshitz, E. M.: 1958, *Quantum Mechanics. Non-Relativistic Theory.* Addison-Wesley, Reading, Mass.

Landau, L. D. and Lifshitz, E. M.: 1962, *The Classical Theory of Fields* (2nd edition). Addison-Wesley, Reading, Mass.

Larmor, J.: 1919, 'How Could a Rotating Body Such as the Sun Become a Magnet?' *Rep. 87th Meeting Brit. Assoc., Adv. Sci., Bournemouth, 1919 Sept. 9—13,* pp. 159—160, John Murray, London.

Larson, R. B.: 1979, 'Stellar Kinematics and Interstellar Turbulence', *Monthly Notices Roy. Astron. Soc.* **186**, 479—490.

Larson, R. B.: 1981, 'Turbulence and Star Formation in Molecular Clouds', *Monthly Notices Roy. Astron. Soc.* **194**, 809—826.

Lawler, J. M. and Dennison, B.: 1982, 'On Intracluster Faraday Rotation. II. Statistical Analysis', *Astrophys. J.* **252**, 81—91.

Layzer, D., Rosner, R., and Doyle, H. T.: 1979, 'On the Origin of Solar Magnetic Fields', *Astrophys. J.* **229**, 1126—1137.

Lea, S. M. and Holman, G. D.: 1978, 'X-ray and Radio Emission from Clusters of Galaxies: The Heating of Intracluster Gas by Relativistic Electrons', *Astrophys. J.* **222**, 29—39.

Leahy, D. A. and Vilenkin, A.: 1981, 'Magnetic Field Generation by Rotating Black Holes', *Astrophys. J.* **248**, 13—17.

Legg, M. P. C. and Westfold, K. C.: 'Elliptic Polarization of Synchrotron Radiation', *Astrophys. J.* **154**, 499–514.

Léorat, J., Pouquet, A., and Frisch, U.: 1981, 'Fully Developed MHD Turbulence Near Critical Reynolds Number', *J. Fluid Mech.* **104**, 419—444.

Leung, C. M., Kutner, M. L., and Mead, K. N.: 1982, 'On the Origin and Structure of Isolated Dark Globules', *Astrophys. J.* **262**, 583—589.

Lewis, B. M.: 1984, 'Face-on Galaxies', *Astrophys. J.* **285**, 453—457.

Liszt, M. S.: 1985, 'Determination of Galactic Spiral Structure at Radiofrequencies'. In: *The Milky Way Galaxy. Proc. IAU Symp. No. 106* (eds. H. van Woerden, R. J. Allen, and W. B. Burton), pp. 283–297. D. Reidel, Dordrecht,

Litvak, M. M.: 1971, 'The Meaning of the OH—H_2O Maser Maps', *Astrophys. J.* **170**, 71—80.

Lo, K. Y., Backer, D. C., Eckers, R. D., Kellermann, K. I., Reid, M., and Moran, J. M.: 1985, 'On the Size of the Galactic Centre Compact Radio Source: Diameter <20 AU', *Nature* **315**, 124—126.

Lo, K. Y., Walker, R. C., Burke, B. F., Moran, J. M., Johnston, K. J., and Ewing, M. S.: 1975, 'Evidence for Zeeman Splitting in 1720-MHz OH Line Emission', *Astrophys. J.* **202**, 650—654.

Lockman, F. J.: 1984, 'The HI Halo in the Inner Galaxy', *Astrophys. J.* **283**, 390—397.

Lockman, F. J., Hobbs, L. M., and Shull, J. M.: 1986, 'The Extent of the Local HI Halo', *Astrophys. J.* **301**, 380—394.

Loiseau, N., Hummel, E., Beck, R., and Wielebinski, R.: 1987, 'High Resolution Polarization Observations of M31'. In: *Proc. of the Workshop 'Interstellar Magnetic Fields'*, 8—12 Sept. 1986, Schloss Ringberg/Tegernsee, pp. 42—46. Springer Verlag, Berlin.

Longair, M. S.: 1981, *High Energy Astrophysics,* Sect. 18.2. Cambridge Univ. Press, Cambridge.

Lozinskaya, T. A.: 1980, 'Kinematics of Old Supernova Remnants', *Astron. Astrophys.* **84**, 26—35.

Lozinskaya, T. A. and Kardashev, N. S.: 1963, 'The Thickness of the Gas Disk of the Galaxy from 21 cm Line Observations', *Astron. J. (SSSR)* **40**, 209—215 (*Sov. Astron.* **7**, 161—166, 1963).

Lubimov, V. A., Novikov, E. G., Nozik, V. Z., Tretyakov, E. F., and Kozik, V. S.: 1980, 'An Estimate of the ν_e Mass from the β-spectrum of Tritium in the Valine Molecules', *Phys. Lett.* **B94**, 266—268.

Lüst, R.: 1970, 'The Solar Wind — an Example of a Cosmical Plasma and a Stellar Wind'. In: *Interstellar Gas Dynamics. Proc. IAU Symp. No. 39* (ed. H. J. Habing), pp. 249—262. D. Reidel, Dordrecht.

Lynden-Bell, D.: 1969, 'Galactic Nuclei as Collapsed Old Quasars', *Nature* **223**, 690—694.

MacDonald, D. and Thorne, K. S.: 1981, 'Black-hole Electrodynamics: An Absolute-space Universal-time Formulation', Preprint OAP-616.
Manchester, R. N.: 1972, 'Structure of the Local Magnetic Field, Part I', *Astrophys. J.* **172**, 43—52.
Manchester, R. N.: 1974, 'Structure of the Local Magnetic Field, Part II', *Astrophys. J.* **188**, 637—644.
Margon, B.: 1982, 'Relativistic Jets in SS 433', *Science* **215**, 247—252.
Martin, P. G.: 1974, 'Interstellar Polarization from a Medium with Changing Grain Alignment', *Astrophys. J.* **187**, 461—472.
Martin, P. G.: 1978, *Cosmic Dust. Its Impact on Astronomy*, Clarendon Press, Oxford.
Martin, P. G. and Campbell, B.: 1976, 'Circular Polarization Observations of the Interstellar Magnetic Field', *Astrophys. J.* **208**, 727—731.
Martin, P. G. and Shawl, S. J.: 1979, 'The Wavelength Dependence of Interstellar Polarization in M31', *Astrophys. J. Lett.* **231**, L57—L60.
Martin, P. G. and Shawl, S. J.: 1982, 'An Optical Study of the Magnetic Field in M31', *Astrophys. J.* **253**, 86—90.
Maslov, V. P.: 1965, *Perturbation Theory and Asymptotic Methods*, Moscow Univ. Press, Moscow (in Russian).
Maslov, V. P. and Fedorjuk, M. V.: 1981, *Semi-Classical Approximation in Quantum Mechanics*, D. Reidel, Dordrecht.
Mathewson, D. C. and Ford, V. L.: 1970, 'Polarization Observations of 1800 Stars', *Mem. Roy. Astron. Soc.* **74**, 139—182.
McIvor, I.: 1977, 'The Inertial Range of Weak Magnetohydrodynamic Turbulence in the Interstellar Medium', *Monthly Notices Roy. Astron. Soc.* **178**, 85—100.
McKean, H. P.: 1969, *Stochastic Integrals*, Academic Press, New York.
McKee, C. F. and Ostriker, J. P.: 1977, 'A Theory of the Interstellar Medium: Three Components Regulated by Supernova Explosions in Inhomogeneous Substrate', *Astrophys. J.* **217**, 148—169.
McKee, C. F., Van Buren, D., and Lazareff, B.: 1984, 'Photoionized Stellar Wind Bubbles in a Cloudy Medium', *Astrophys. J. Lett.* **278**, L115—L118.
Mebold, U., Hachenberg, O., and Laury-Micoulaut, C. A.: 1974, 'A Statistical Model for the Intercloud HI-gas. I. Theory', *Astron. Astrophys.* **30**, 329—338.
Melrose, D. B.: 1971, 'On the Degree of Circular Polarization of Synchrotron Radiation', *Astrophys. Space Sci.* **12**, 172—192.
Mestel, L. and Paris, R. B.: 1984, 'Star Formation and the Galactic Magnetic Field', *Astron. Astrophys.* **136**, 98—120.
Mestel, L. and Spitzer, L., Jr.: 1956, 'Star Formation in Magnetic Dust Clouds', *Monthly Notices Roy. Astron. Soc.* **116**, 503—514.
Michel, F. C. and Yahil, A.: 1973, 'Galactic Magnetic Fields: Cellular or Filamentary Structure?', *Astrophys. J.* **179**, 771—780.
Milaneschi, E. and Fabbri, R.: 1985, 'Polarization of the Microwave Background Radiation in Bianchi Type-I Cosmological Models with a Homogeneous Magnetic Field', *Astron. Astrophys.* **151**, 7—15.
Miley, G.: 1980, 'The Structure of Extended Extragalactic Radio Sources', *Ann. Rev. Astron. Astrophys.* **18**, 165—218.
Mishustin, I. M. and Ruzmaikin, A. A.: 1971, 'Occurrence of "Priming" Magnetic Fields During the Formation of Protogalaxies', *JETP* **61**, 441—444 (*Sov. Phys. JETP* **34**, 233—235, 1972).
Mitton, S.: 1972, 'The Polarization Properties of 65 Extragalactic Sources in the 3C Catalogue', *Monthly Notices Roy. Astron. Soc.* **155**, 373—381.
Moffatt, H. K.: 1978, *Magnetic Field Generation in Electrically Conducting Fluids*, Cambridge Univ. Press, Cambridge.
Molchanov, S. A., Ruzmaikin, A. A., and Sokoloff, D. D.: 1983, 'Equations of MHD Dynamo in Random Short-correlated Velocity Field', *Magn. Gidrodin.*, No. 4, 67—72 (*Magnetohydrodynamics*, No. 4, 1983).
Molchanov, S. A., Ruzmaikin, A. A., and Sokoloff, D. D.: 1985, 'Kinematic Dynamo in Random Flow', *Usp. Fiz. Nauk* (*SSSR*) **145**, 593—628 (*Sov. Phys. Usp.* **28**, 307—327, 1985).

Morris, D. and Berge, G. L.: 1964, 'Direction of the Galactic Magnetic Field in the Vicinity of the Sun', *Astrophys. J.* **139**, 1388—1392.

Morris, D. and Tabara, M.: 1973, 'A Study of the Depolarization and Luminosity of Radio Sources', *Publ. Astron. Soc. Japan* **25**, 295—316.

Morris, D. and Whiteoak, J. B.: 1968, 'Observations of the Linear Polarization of Radio Sources at 6 cm Wavelength', *Aust. J. Phys.* **21**, 493—500.

Morris, M. and Lo, K. Y.: 1978, 'A Study of CO Emission from Two Scd Galaxies: IC 342 and NGC 6946', *Astrophys. J.* **223**, 803—810.

Morris, M. and Rickard, L. J.: 1982, 'Molecular Clouds in Galaxies', *Ann. Rev. Astron. Astrophys.* **20**, 517—546.

Morris, M. and Yusef-Zadeh, F.: 1985, 'Unusual Threads of Radio Emission Near the Galactic Center', *Astron. J.* **90**, 2511—2513.

Morton, D. C. and Blades, J. C.: 1986, 'Interstellar Absorption Lines in the Directions of Extra-galactic Objects — II. Analysis of 25 Sight Lines', *Monthly Notices Roy. Astron. Soc.* **220**, 927—948.

Mouschovias, T. Ch.: 1976a, 'Nonhomologous Contraction and Equilibria of Self-gravitating, Magnetic Interstellar Clouds Embedded in an Intercloud Medium: Star Formation. I. Formulation of the Problem and Method of Solution', *Astrophys. J.* **206**, 753—767.

Mouschovias, T. Ch.: 1976b, 'Nonhomologous Contraction and Equilibria of Self-gravitating, Magnetic Interstellar Clouds Embedded in an Intercloud Medium: Star Formation. II. Results', *Astrophys. J.* **207**, 141—158.

Mouschovias, T. Ch.: 1985, 'Star Formation and the Galactic Magnetic Field. The B—ρ relation, flux freezing, and magnetic braking', *Astron. Astrophys.* **142**, 41—47.

Myers, P. C.: 1983, 'Dense Cores in Dark Clouds. III. Subsonic Turbulence', *Astrophys. J.* **270**, 105—118.

Myers, P. C.: 1986, 'Observations of Molecular Cloud Structure and Internal Motions', *Center for Astrophys. Preprint No. 2347* (*Proc. Wyoming Summer School on Interstellar Proc.*, July 1986, to be published by D. Reidel).

Nakano, T.: 1976, 'Effective Magnetic Viscosity Due to Hydrodynamic Turbulence in a Weak Large-scale Magnetic Field', *Publ. Astron. Soc. Japan* **28**, 451—462.

Nee, S. F.: 1980, 'Fluctuation Theory of Starlight Polarization', *Astrophys. J.* **237**, 471—481.

Nee, S. F. and Jokipii, J. R.: 1979, 'Interstellar Polarization in an Irregularly Fluctuating Medium', *Astrophys. J.* **234**, 140—153.

Negroponte, J. and Silk, J.: 1980, 'Polarization of the Primeval Radiation in an Anisotropic Universe', *Phys. Rev. Lett.* **44**, 1433—1437.

Nelson, A. H.: 1985, 'Cosmic Rays and the Parker Instability', *Monthly Notices Roy. Astron. Soc.* **215**, 161—170.

Nepveu, M.: 1985, 'Turbulence in Clusters of Galaxies', *Astron. Astrophys.* **149**, 459—461.

Neugebauer, M., Wu, C. S., and Huba, J. D.: 1978, 'Plasma Fluctuations in the Solar Wind', *J. Geophys. Res.* **83A**, 1027—1034.

Newton, K.: 1980a, 'Neutral Hydrogen and Spiral Structure in M33', *Monthly Notices Roy. Astron. Soc.* **190**, 689—709.

Newton, K.: 1980b, 'Neutral Hydrogen in IC 342-II. The Detailed Structure', *Monthly Notices Roy. Astron. Soc.* **191**, 615—630.

Newton, K.: 1980c, 'Neutral Hydrogen in IC 342-I. The Large-Scale Structure', *Monthly Notices Roy. Astron. Soc.* **191**, 169—184.

Newton, K. and Emerson, D. T.: 1977, 'Neutral Hydrogen in the Outer Regions of M31', *Monthly Notices Roy. Astron. Soc.* **181**, 573—590.

Nikiforov, A. F. and Uvarov, V. B.: 1978, *Special Functions of Mathematical Physics*, Nauka, Moscow (in Russian).

Nishimura, J., Fujii, M., Taira, T., Aizu, E., Hirawa, H., Kobayashi, T., Niu, K., Ohta, I., Golden, R. L., Koss, T. A., Lord, J. J., and Wilkes, R. J.: 1980, 'Emulsion Chamber Observations of Primary Cosmic-ray Electrons in the Energy Range 30—1000 GeV', *Astrophys. J.* **238**, 394—409.

Novikov, V. G., Ruzmaikin, A. A., and Sokoloff, D. D.: 1983, 'The Fast Dynamo in a Reflectionally Invariant Random Velocity Field', *JETP* **85**, 909–918.

Obukhov, A. M.: 1941, 'On Energy Distribution in the Spectrum of a Turbulent Flow', *Dokl. Akad. Nauk SSSR* **32**, 22–24.

Okuda, H.: 1985, 'Near-infrared Studies of the Milky Way'. In: *The Milky Way Galaxy. Proc. IAU Symp. No. 106* (eds. H. van Woerden, R. J. Allen, and W. B. Burton), pp. 123–126. D. Reidel, Dordrecht.

Onderchen, M. P.: 1985, 'Radio Continuum Observations of the Bar and Disk of M83', *Astron. J.* **90**, 1474–1480.

Osborne, J. L., Roberts, E., and Wolfendale, A. W.: 1973, 'Galactic Magnetic Field Irregularities and Their Effect on Cosmic Ray Propagation at Energies Above 10^{17} eV', *J. Phys. A: Math. Nucl. Gen.* **6**, 421–433.

Osterbrock, D. F.: 1961, 'On Ambipolar Diffusion in HI Regions', *Astrophys. J.* **134**, 270–272.

Ostriker, J. P., Peebles, P. J. E. and Yahil, A.: 1974, 'The Size and Mass of Galaxies, and the Mass of the Universe', *Astrophys. J. Lett.* **193**, L1–L4.

Pacholczyk, A. G.: 1970, *Radio Astrophysics. Nonthermal Processes in Galactic and Extragalactic Sources*, W. H. Freeman Co., San Francisco.

Pacholczyk, A. G.: 1973, 'Circular Repolarization in Compact Radio Sources', *Monthly Notices Roy. Astron. Soc.* **163**, 29p–33p.

Pacholczyk, A. G.: 1977, *Radio Galaxies. Radiation Transfer, Dynamics, Stability and Evolution of a Synchrotron Plasmon*, Pergamon Press, Oxford.

Pacholczyk, A. G. and Swihart, T. L.: 1970, 'Polarization of Radio Sources. II. Faraday Effect in the Case of Quasi-transverse Propagation', *Astrophys. J.* **161**, 415–418.

Pacholczyk, A. G. and Swihart, T. L.: 1974, 'Polarization of Radio Sources. V. Absorption Effects on Circular Repolarization in Compact Sources', *Astrophys. J.* **192**, 591–595.

Parker, E. N.: 1955, 'Hydromagnetic Dynamo Models', *Astrophys. J.* **122**, 293–314.

Parker, E. N.: 1963, 'Kinematical Hydromagnetic Theory and Its Applications to the Low Solar Photosphere', *Astrophys. J.* **138**, 552–575.

Parker, E. N.: 1971, 'The Generation of Magnetic Fields in Astrophysical Bodies. II. The Galactic Field', *Astrophys. J.* **163**, 252–278.

Parker, E. N.: 1973a, 'The Generation and Dissipation of Solar and Galactic Magnetic Fields', *Astrophys. Space Sci.* **22**, 279–292.

Parker, E. N.: 1973b, 'Extragalactic Cosmic Rays and Galactic Magnetic Field', *Astrophys. Space Sci.* **24**, 279–288.

Parker, E. N.: 1979, *Cosmical Magnetic Fields. Their Origin and Their Activity*, Clarendon Press, Oxford.

Parker, R. L.: 1966, 'Reconnection of Lines of Force in Rotating Spheres and Cylinders', *Proc. Roy. Soc.* **A291**, 60–72.

Pasha, I. I.: 1985, 'The Direction of Rotation of the Spirals in 109 Galaxies. NGC 3786 and NGC 5426 – the Galaxies with Leading Spiral Arms in Interacting Systems', *Pisma Astron. J.* (*SSSR*) **11**, 3–9.

Paul, J., Casse, M., and Cesarsky, C. J.: 1975, 'Distribution of Gas, Magnetic Fields, and Cosmic Rays in the Galaxy', *Astrophys. J.* **207**, 62–77.

Perola, G. C. and Reinhardt, M.: 1972, 'On the Origin of the Magnetic Field in the Extended Radio and X-ray Source in the Coma Cluster', *Astron. Astrophys.* **17**, 432–436.

Perry, J. F. W. and Helfer, H. L.: 1972, 'Distribution of Neutral Hydrogen in the Perseus Arm', *Astrophys. J.* **174**, 341–360.

Peterson, C. J.: 1978, 'Rotation of 10 Early Type Galaxies', *Astrophys. J.* **222**, 84–94.

Petrovskaya, I. V. and Teerikorpi, P.: 1986, 'Rotation Curve of the Outer Parts of Our Galaxy from Neutral Hydrogen 21 cm Line Profiles', *Astron. Astrophys.* **163**, 39–42.

Phillipps, S., Kearsey, S., Osborne, J. L., Haslam, H., and Stoffel, H.: 1981, 'Distribution of Galactic Synchrotron Emission II', *Astron. Astrophys.* **103**, 405–414.

Piddington, J. H.: 1964, 'The Magnetic Fields and Radio Emission of Galaxies', *Monthly Notices Roy. Astron. Soc.* **128**, 345–360.

Piddington, J. H.: 1970, 'The Galactic Magnetic Field and Cosmic Rays', *Aust. J. Phys.* **23**, 731–750.
Piddington, J. H.: 1972, 'The Origin and Form of the Galactic Magnetic Field. I.: Parker's Dynamo Model', *Cosmic Electrodyn.* **3**, 60–70.
Piddington, J. H.: 1973, 'Dynamo Theories of the Solar and Galactic Magnetic Fields', *Astrophys. Space Sci.* **24**, 259–268.
Piddington, J. H.: 1974, 'The Mode of Creation of Stars, Stellar Systems and the Hubble System of Galaxies', *Astrophys. Space Sci.* **31**, 225–240.
Piddington, J. H.: 1975a, 'Galactic Magnetic Fields and Their Interaction with the Gas and Cosmic Rays', *Astrophys. Space Sci.* **37**, 183–196.
Piddington, J. H.: 1975b, 'Solar Magnetic Fields and Convection. III: Recent Developments in Dynamo and Related Theories', *Astrophys. Space Sci.* **38**, 157–166.
Piddington, J. H.: 1981a, 'The Role of Magnetic Fields in Extra-galactic Astronomy', *Astrophys. Space Sci.* **80**, 457–471.
Piddington, J. H.: 1981b, 'Turbulent Diffusion of Magnetic Fields in Astrophysical Plasmas', *Astrophys. J.* **247**, 293–299.
Pikelner, S. B.: 1953a, 'Interstellar Gas and Magnetic Fields', *Izv. Krym. Astrofiz. Obs.* **10**, 74–96.
Pikelner, S. B.: 1953b, 'Kinematic Characteristics of Interstellar Gas in Connection with Cosmic Ray Isotropy', *Dokl. Akad. Nauk SSSR* **88**, 229–232.
Pikelner, S. B.: 1954, 'A Method of Investigation of Turbulence from Brightness Fluctuations in Nebulae', *Izv. Krym. Astrofiz. Obs.* **11**, 34–38.
Pikelner, S. B.: 1966, *Principles of Cosmic Electrodynamics*, Nauka, Moscow (in Russian).
Pikelner, S. B. and Shajn, G. A.: 1954, 'Brightness Fluctuations in the Orion Nebula as a Possible Consequence of Turbulence', *Izv. Krym. Astrofiz. Obs.* **11**, 22–34.
Pikelner, S. B. and Shklovsky, I. S.: 1957, 'An Investigation of the Properties and Energy Dissipation of the Galactic Halo', *Astron. J. (SSSR)* **34**, 145–158.
Polovin, R. V.: 1960, 'Shock Waves in Magnetohydrodynamics', *Usp. Fiz. Nauk (SSSR)* **72**, 33–52 (*Sov. Phys. Usp.* **3**, 677–688, 1961).
Pottash, S. R.: 1970, 'Mass Loss from Stars'. In: *Interstellar Gas Dynamics. Proc. IAU Symp. No. 39* (ed. H. J. Habing), pp. 272–280. D. Reidel, Dordrecht.
Pouquet, A., Frisch, U., and Léorat, J.: 1976, 'Strong MHD Turbulence and the Nonlinear Dynamo Effect', *J. Fluid Mech.* **77**, 321–354.
Prentice, A. J. R. and ter Haar, D.: 1969, 'On HII Regions and Pulsar Distances', *Monthly Notices Roy. Astron. Soc.* **146**, 423–444.
Price, R. M.: 1974, 'Continuum Radio Emission and Galactic Structure'. In: *Galactic Radio Astronomy. Proc. IAU Symp. No. 60* (eds. F. J. Kerr and S. C. Simonson III), pp. 637–648. D. Reidel, Dordrecht.
Proctor, M. R. E.: 1977a, 'On the Eigenvalues of Kinematic α-effect Dynamos', *Astron. Nachr.* **298**, 19–25.
Proctor, M. R. E.: 1977b, 'The Role of Mean Circulation in Parity Selection by Planetary Magnetic Fields', *Geophys. Astrophys. Fluid Dyn.* **8**, 311–324.
Pryor, C. and Lecar, M.: 1983, 'Can Secondary Infall Produce Flat Rotation Curves?', *Astrophys. J.* **269**, 513–530.
Ptuskin, V. S.: ;1986, 'Hydrodynamics of Interstellar Medium Including Cosmic Rays'. In: *Plasma Astrophysics. Proc. Joint Varenna–Abastumani School and Workshop, Sukhumi, USSR, 18–28 May 1986*, pp. 279–286. ESA Publ. SP-251.
Punin, N. N.: 1922, *Tatlin*, Petrograd (in Russian).
Purcell, E. M.: 1975, 'Interstellar Grains as Pinwheels'. In: *The Dusty Universe* (eds. G. B. Field and A. G. W. Cameron). Neale Watson Academic Publications, New York.
Purcell, E. M.: 1979, 'Suprathermal Rotation of Interstellar Grains', *Astrophys. J.* **231**, 404–416.
Purse, J.: 1974, *The Mystic Spiral. The Journey of the Soul*, Thames and Hudson.
Quinn, P. J. and Sussman, G. J.: 1985, 'A Model of the Radio Continuum Filaments in the Galactic Center', *Astrophys. J.* **288**, 377–384.
Quiroga, R. J.: 1980, 'Local Galactic Field of Forces and the Interstellar Matter', *Astrophys. Space Sci.* **68**, 393–422.

Quiroga, R. J.: 1983, 'Hydrodynamic and Turbulent Motions in the Galactic Disk', *Astrophys. Space Sci.* **93**, 37–54.
Rädler, K.-H.: 1980, 'Mean-field Approach to Spherical Dynamo Models', *Astron. Nachr.* **301**, 101–129.
Rädler, K.-H.: 1986a, 'Investigations of Spherical Kinematic Mean-field Dynamo Models', *Astron. Nachr.* **307**, 89–113.
Rädler, K.-H.: 1986b, 'On the Effect of Differential Rotation on Axisymmetric and Non-axisymmetric Magnetic Fields of Cosmical Bodies'. In: *Plasma Astrophysics. Proc. Joint Varenna–Abastumani School and Workshop, Sukhumi-USSR, 18–28 May 1986*, pp. 569–574. ESA Publ. SP-251.
Rao, A. P. and Ananthakrishnan, S.: 1984, 'Interstellar Scattering in the Inner Parts of the Galaxy', *Nature* **312**, 707–711.
Razin, V. A.: 1958, 'Polarization of Cosmic Radio Radiation at 1.45 m and 3.3 m', *Astron. J. (SSSR)* **35**, 241–252 (*Sov. Astron.* **2**, 216–225, 1958).
Razin, V. A.: 1960, 'On the Theory of Radiospectra of Discrete Sources at Frequency Range Below 30 MHz', *Izv. Vyssh. Uchebn. Zaved. Radiofizika* **3**, 584–594 (*Sov. Radiophys.*).
Readhead, A. C. S. and Duffet-Smith, P. J.: 1975, 'The Scale Height of Interstellar Ionized Hydrogen', *Astron. Astrophys.* **42**, 151–153.
Rees, M. J.: 1984, 'Black Hole Models for Active Galactic Nuclei', *Ann. Rev. Astron. Astrophys.* **22**, 471–506.
Rees, M. J., Begelman, M. C., Blandford, R. D., and Phinney, E. S.: 1982, 'Ion-supported Tori and the Origin of Radio Jets', *Nature* **295**, 17–21.
Rees, M. J. and Reinhardt, M.: 1972, 'Some Remarks on Intergalactic Magnetic Fields', *Astron. Astrophys.* **19**, 189–192.
Reinhardt, M.: 1971, 'Orientation of Galaxies and a Magnetic "Urfield"', *Astrophys. Space Sci.* **10**, 363–382.
Reinhardt, M.: 1972, 'Interpretation of Rotation Measures of Radio Sources', *Astron. Astrophys.* **19**, 104–108.
Reinhardt, M. and Roberts, M. S.: 1972, 'Orientation of Galaxies and the Local Supercluster', *Astrophys. Lett.* **12**, 201–206.
Reinhardt, M. and Thiel, M. A. F.: 1970, 'Does a Primaeval Magnetic Field Exist?', *Astrophys. Lett.* **7**, 101–106.
Richtmeier, R. D. and Teller, E.: 1949, 'On the Origin of Cosmic Rays', *Phys. Rev.* **75**, 1729–1731.
Rickett, B. J.: 1977, 'Interstellar Scattering and Scintillation of Radio Waves', *Ann. Rev. Astron. Astrophys.* **15**, 479–504.
Roberts, M. S. and Whitehurst, R. N.: 1975, 'The Rotation Curve and Geometry of M31 at Large Galactocentric Distances', *Astrophys. J.* **201**, 327–346.
Roberts, P. H.: 1972, 'Kinematic Dynamo Models', *Phil. Trans. Roy. Soc.* **A272**, 663–698.
Roberts, P. H. and Stix, M.: 1971, 'The Turbulent Dynamo: A Translation of a Series of Papers by F. Krause, K.-H. Rädler and M. Steenbeck', *Tech. Note 60*, NCAR, Colorado.
Roberts, W. W., Jr.: 1969, 'Large-scale Shock Formation in Spiral Galaxies and its Implications on Star Formation', *Astrophys. J.* **158**, 123–144.
Roberts, W. W., Jr. and Shu, F. H.: 1972, 'The Role of Gaseous Dissipation in Density Waves of Finite Amplitude', *Astrophys. Lett.* **12**, 49–52.
Roberts, W. W., Jr. and Yuan, C.: 1970, 'Application of the Density-wave Theory to the Spiral Structure of the Milky Way System. III. Magnetic Field: Large-scale Hydromagnetic Shock Formation', *Astrophys. J.* **161**, 877–902.
Rogstad, D. H., Shostak, G. S., and Rots, A. H.: 1973, 'Aperture Synthesis Study of Neutral Hydrogen in the Galaxies NGC 6946 and IC 342', *Astron. Astrophys.* **22**, 111–119.
Rohlfs, K. and Kreitschmann, J.: 1980, 'A Two Component Mass Model for M81 (NGC 3031)', *Astron. Astrophys.* **87**, 175–182.
Rohlfs, K. and Kreitschmann, J.: 1981, 'A Realistic Model of the Galaxy', *Astrophys. Space Sci.* **79**, 289–320.
Rohlfs, K., Chini, R., Wink, J. E., and Böhme, R.: 1986, 'The Rotation Curve of the Galaxy', *Astron. Astrophys.* **158**, 181–190.

Rose, W. K.: 1987, 'Magnetic Fields in Extragalactic Jets', *Astrophys. J.* **313**, 146—151.
Roy, J.-R. and Joncas, G.: 1985, 'Structure and Origin of Velocity Fluctuations in the HII Region Sharpless 142', *Astrophys. J.* **288**, 142—147.
Rubin, V. C.: 1983, 'The Rotation of Spiral Galaxies', *Science* **220**, 1339—1344.
Rubin, V. C., Burstein, D., Ford, W. K., Jr., and Thonnard, N.: 1985, 'Rotation Velocities of 16 Sa Galaxies and a Comparison of Sa, Sb, and Sc Rotation Properties', *Astrophys. J.* **289**, 81—104.
Rubin, V. C. and Ford, W. K., Jr.: 1970, 'Rotation of the Andromeda Nebula from a Spectroscopic Survey of Emission Regions', *Astrophys. J.* **159**, 379—404.
Rubin, V. C., Ford, W. K., Jr., and Thonnard, N.: 1978, 'Extended Rotation Curves of High-luminosity Spiral Galaxies. IV. Systematic Dynamical Properties, Sa → Sc', *Astrophys. J. Lett.* **225**, L107—L111.
Rubin, V. C., Ford, W.K., Jr., and Thonnard, N.: 1980, 'Rotational Properties of 21 Sc Galaxies with a Large Range of Luminosities and Radii, from NGC 4605 (R = 4 kpc) to UGC 2885 (R = 122 kpc)', *Astrophys. J.* **238**, 471—487.
Ruzmaikin, A. A.: 1976, 'On the Rôle of Spiral Arms in Generation of Galactic Magnetic Field', in *Plasma Astrophysics*, (ed. V. E. Stepanov), Irkutsk (in Russian).
Ruzmaikin, A. A.: 1987, 'Magnetic Fields of Galaxies'. In: *Proc. Workshop Interstellar Magnetic Fields, 8—12 Sept. 1986, Schloss Ringberg/Tegernsee*, pp. 16—22. Springer Verlag, Berlin.
Ruzmaikin, A. A. and Shukurov, A. M.: 1981, 'Magnetic Field Generation in the Galactic Disc', *Astron. J.* (*SSSR*) **58**, 969—978. (*Sov. Astron.* **25**, 553—558, 1981).
Ruzmaikin, A. A. and Shukurov, A. M.: 1982, 'Spectrum of the Galactic Magnetic Fields', *Astrophys. Space Sci.* **82**, 397—407.
Ruzmaikin, A. A. and Shukurov, A. M.: 1983, 'Magnetic Fields of Spiral Galaxies', *Proc. XV All-Union Conf. Galactic and Extragalactic Radioastron., Kharkov, 11—13 Oct. 1983*, pp. 42—43. Int. Radiophys. Electron. AS Ukrainian SSR, Kharkov.
Ruzmaikin, A. A. and Sokoloff, D. D.: 1977a, 'The Nonhomogeneous Large Scale Magnetic Field and the Global Structure of the Universe', *Astrofizika* **13**, 95—102.
Ruzmaikin, A. A. and Sokoloff, D. D.: 1977b, 'The Scale and Strength of the Galactic Magnetic Field According to the Pulsar Data', *Astrophys. Space Sci.* **52**, 365—376.
Ruzmaikin, A. A. and Sokoloff, D. D.: 1977c, 'The Interpretation of Rotation Measures of Extragalactic Radio Sources', *Astron. Astrophys.* **58**, 247—253.
Ruzmaikin, A. A. and Sokoloff, D. D.: 1979, 'The Calculation of Faraday Rotation Measures of Cosmic Radio Sources', *Astron. Astrophys.* **78**, 1—6.
Ruzmaikin, A. A., Sokoloff, D. D., and Kovalenko, A. V.: 1977, 'Determination of the Galactic Magnetic Field Parameters from Faraday Rotations for Radiosources', Preprint *Inst. Appl. Math.*, No. 20, Moscow.
Ruzmaikin, A. A., Sokoloff, D. D., and Kovalenko, A. V.: 1978, 'Galactic Magnetic Field Parameters Determined from Faraday Rotation of Radio Sources', *Astron. J.* (*SSSR*) **55**, 692—701 (*Sov. Astron.* **22**, 395—401, 1978).
Ruzmaikin, A. A., Sokoloff, D. D., and Shukurov, A. M.: 1980a, 'Disk Dynamo with Concentrated Helicity', *Magn. Gidrodin.* No. 1, 20—26 (*Magnetohydrodynamics* **16**, 15—20, 1980).
Ruzmaikin, A. A., Sokoloff, D. D., and Turchaninoff, V. I.: 1980b, 'The Turbulent Dynamo in a Disk', *Astron. J.* (*SSSR*) **57**, 311–320 (*Sov. Astron.* **24**, 182–187, 1980).
Ruzmaikin, A. A., Sokoloff, D. D. and Shukurov, A. M.: 1985, 'Magnetic Field Distribution in Spiral Galaxies', *Astron. Astrophys.* **148**, 335—343.
Ruzmaikin, A. A., Sokoloff, D. D., and Shukurov, A. M.: 1986, 'Magnetic Fields of Spiral Galaxies'. In: *Plasma Astrophysics. Proc. Joint Varenna—Abastumani Intern. School and Workshop, Sukhumi-USSR, 19—28 May 1986*, pp. 539—544. ESA Publ. SP–251.
Ruzmaikin, A. A., Sokoloff, D. D., and Shukurov, A. M.: 1987, 'Hydromagnetic Screw Dynamo', *Preprint Pr-1249*, Inst. Space Res. As USSR, Moscow, USSR.
Ruzmaikin, A. A., Sokoloff, D. D. Turchaninoff, V. I., and Zeldovich, Ya. B.: 1979, 'The Disk Dynamo', *Astrophys. Space Sci.* **66**, 369—384.
Ruzmaikina, T. V.: 1983, 'The Role of Magnetic Fields in Star Formation', *Chapter 14* in Zeldovich *et al.* (1983).

Sagdeev, R. Z. and Galeev, A. A.: 1969, *Nonlinear Plasma Theory* (eds. T. O'Neil and D. Book). Benjamin Inc., New York-Amsterdam.
Sancisi, R. and Allen, R. J.: 1979, 'Neutral Hydrogen Observations of the Edge-on Disk Galaxy NGC 891', *Astron. Astrophys.* **74**, 73–84.
Sanders, D. B., Solomon, P. M., and Scoville, N. Z.: 1984, 'Giant Molecular Clouds in the Galaxy. I. The Axisymmetric Distribution of H_2', *Astrophys. J.* **276**, 182–203.
Sanders, D. B., Scoville, N. Z., and Solomon, P. M.: 1985, 'Giant Molecular Clouds in the Galaxy. II. Characteristics of Discrete Features', *Astrophys. J.* **289**, 373–387.
Sanders, R. H.: 1984a, 'Anti-gravity and Galaxy Rotation Curves', *Astron. Astrophys.* **136**, L21–L23.
Sanders, R. H.: 1984b, 'Alignment of Distant Radio Sources', *Nature* **309**, 35–37.
Sanders, R. H. and Wrixon, G. T.: 1973, 'Observations of Neutral Hydrogen Near the Galactic Center. II. The Nuclear Disc', *Astron. Astrophys.* **26**, 365–377.
Sarazin, C. L.: 1986, 'X-ray Emission from Clusters of Galaxies', *Rev. Mod. Phys.* **58**, 1–115.
Savage, B. D. and de Boer, K. S.: 1981, 'Ultraviolet Absorption by Interstellar Gas at Large Distances from the Galactic Plane', *Astrophys. J.* **243**, 460–484.
Savage, B. D. and Massa, D.: 1987, 'Highly Ionized Interstellar Gas Located in the Galactic Disk and Halo', *Astrophys. J.* **314**, 380–396.
Sawa, T. and Fujimoto, M.: 1980, 'Bisymmetric Open-spiral Configuration of Magnetic Fields in Disk Galaxies in Differential Rotation', *Publ. Astron. Soc. Japan* **32**, 551–566.
Sawa, T. and Fujimoto, M.: 1986, 'Bisymmetric Spiral Configuration of Magnetic Fields in Spiral Galaxies. I. Local Theory', *Publ. Astron. Soc. Japan* **38**, 133–150.
Sazonov, V. N.: 1969a, 'Polarization of Normal Waves and Synchrotron Radiation Transfer in a Relativistic Plasma', *JETP* **56**, 1075–1086 (*Sov. Phys. JETP* **29**, 578).
Sazonov, V. N.: 1969b, 'The Generation and Transfer of Polarized Synchrotron Emission', *Astron. J. (SSSR)* **46**, 502–511.
Scalo, J. M.: 1984, 'Turbulent Velocity Structure in Interstellar Clouds', *Astrophys. J.* **277**, 556–561.
Scalo, J. M. and Pumphrey, W. A.: 1982, 'Dissipation of Supersonic Turbulence in Interstellar Clouds', *Astrophys. J. Lett.* **258**, L29–L33.
Scarrott, S. M., Ward-Thompson, D., and Warren-Smith, R. F.: 1987, 'Evidence for a Spiral Magnetic Field Configuration in the Galaxy M51', *Monthly Notices Roy. Astron. Soc.* **224**, 299–305.
Schlüter, A. and Biermann, L.: 1950, 'Interstellare Magnetfelder', *Z. Naturforsch.* **5a**, 237–251.
Schmidt, G. D., Peterson, B. M., and Beaver, E. A.: 1978, 'Imaging Polarimetry of the Jets of M87 and 3C 273', *Astrophys. J. Lett.* **220**, L31–L36.
Schneider, S. E. and Terzian, Y.: 1983, 'Planetary Nebulae and the Galactic Rotation Curve', *Astrophys. J. Lett.* **274**, L61–L64.
Schreier, E. J., Burns, J. O., and Feigelson, E. D.: 1981, 'Detection of Radio Emission from the Jet in Centaurus A', *Astrophys. J.* **251**, 523–529.
Schwarz, U. J., Troland, T. H., Albinson, J. C., Bregman, J. D., Goss, W. M., and Heiles, C.: 1986, 'Aperture Synthesis Observations of the 21 Centimeter Zeeman Effect Toward Cassiopeia A', *Astrophys. J.* **301**, 320–330.
Scott, E. H. and Black, D. C.: 1980, 'Numerical Calculations of the Collapse of Nonrotating Magnetic Gas Clouds', *Astrophys. J.* **239**, 166–172.
Scoville, N. Z. and Solomon, P. M.: 1975, 'Molecular Clouds in the Galaxy', *Astrophys. J. Lett.* **199**, L105–L109.
Segalovitz, A., Shane, W. W., and Bruyn, A. G. de: 1976, 'Polarisation Detection at Radio Wavelengths in Three Spiral Galaxies', *Nature* **264**, 222–226.
Seiradakis, J. H., Lasenby, A. N., Yusef-Zadeh, F., Wielebinski, R., and Klein, U.: 1985, 'A New Symmetrical Polarization Structure Near the Galactic Centre', *Nature* **317**, 697–699.
Serabyn, E. and Lacy, J. H.: 1985, '[NeII] Observations of the Galactic Center: Evidence for a Massive Black Hole', *Astrophys. J.* **293**, 445–458.
Shajn, G.: 1955, 'Diffuse Nebulae and the Interstellar Magnetic Field', *Astron. J. (SSSR)* **32**, 110–117.

Shakura, N. I. and Sunyaev, R. A.: 1973, 'Black Holes in Binary Systems. Observational Appearance', *Astron. Astrophys.* **24**, 337—355.

Shakura, N. I., Sunyaev, R. A., and Zilitinkevich, S. S.: 1978, 'On the Turbulent Energy Transport in Accretion Discs', *Astron. Astrophys.* **62**, 179—187.

Shapiro, P. R. and Field, G. B.: 1976, 'Consequences of a New Hot Component of the Interstellar Medium', *Astrophys. J.* **205**, 762—765.

Shapiro, S. L. and Wasserman, I.: 1981, 'Massive Neutrinos, Helium Production and the Primordial Magnetic Field', *Nature* **289**, 657—658.

Shapirovskaya, N. Ya. and Bocharov, A. A.: 1986a, 'The Two Phases of Interstellar Medium. The Distribution of Scattering Matter in the Galaxy', *Astron. J. (SSSR)* **63**, 666—673.

Shapirovskaya, N. Ya. and Bocharov, A. A.: 1986b, 'The Two Phases of Interstellar Medium. The Distribution of Electron Density in the Galactic Disk', *Astron. J. (SSSR)* **63**, 1111—1117.

Sharov, A. S.: 1982, *The Andromeda Nebula*, Nauka, Moscow (in Russian).

Shklovsky, I. S.: 1956, *The Cosmic Radioemission*, Gostekhizdat, Moscow (in Russian).

Shklovsky, I. S.: 1976, *Supernovae*, Nauka, Moscow (in Russian).

Shukurov, A. M., Sokoloff, D. D., and Ruzmaikin, A. A.: 1983, 'Explosive Growth of Magnetic Energy in a Turbulent Medium', *Magn. Gidrodin.*, No. 3, 53—58.

Shukurov, A. M., Sokoloff, D. D., and Ruzmaikin, A. A.: 1985, 'Oscillatory α^2-dynamo', *Magn. Gidrodin.*, No. 1, 9—13.

Shvartsman, V. F.: 1971, 'Halos Around Black Holes', *Astron. J. (SSSR)* **48**, 479—488 (*Sov. Astron.* **15**, 377—384, 1971).

Silk, J.: 1985, 'Macroscopic Turbulence in Molecular Clouds', *Astrophys. J. Lett.* **292**, L71—L74.

Silverglate, P. R.: 1984, 'Upper Limits to Magnetic Fields in CII Regions', *Astrophys. J.* **279**, 694—697.

Simard-Normandin, M. and Kronberg, P. P.: 1979, 'New Large-scale Magnetic Features of the Milky Way', *Nature* **279**, 115—118.

Simonetti, J. H. and Cordes, J. M.: 1986, 'Small-scale Variations in the Galactic Magnetic Field: Rotation Measure Variations Across Extragalactic Radio Sources', *Astrophys. J.* **310**, 160—171.

Simonetti, J. H., Cordes, J. M., and Spangler, S. R.: 1984, 'Small-scale Variations in the Galactic Magnetic Field: The Rotation Measure Structure Function and Birefringence in Interstellar Scintillations', *Astrophys. J.* **284**, 126—134.

Simonson, S. C.: 1976, 'A Density-wave Map of the Galactic Spiral Structure', *Astron. Astrophys.* **46**, 261—268.

Slysh, V. I.: 1963, 'Angular Size of Radio Sources', *Nature* **199**, 682.

Smith, B. W.: 1977, 'Hot Interstellar Tunnels. I. Simulation of Interacting Supernova Remnants', *Astrophys. J.* **211**, 404—420.

Sofue, Y.: 1987, 'Vertical Dust Lanes and Magnetic Field in Spiral Galaxies', *Publ. Astron. Soc. Japan* **39**, 547–558.

Sofue, Y. and Fujimoto, M.: 1983, 'A Bisymmetric Spiral Magnetic Field and the Spiral Arms in Our Galaxy', *Astrophys. J.* **265**, 722—729.

Sofue, Y., Fujimoto, M., and Kawabata, K.: 1968, 'Faraday Rotation by Metagalactic Magnetic Field', *Publ. Astron. Soc. Japan* **20**, 388—394.

Sofue, Y., Fujimoto, M., and Wielebinski, R.: 1986, 'Global Structure of Magnetic Field in Spiral Galaxies', *Ann. Rev. Astron. Astrophys.* **24**, 459—498.

Sofue, Y. and Handa, T.: 1984, 'A Radio Lobe Over the Galactic Centre', *Nature* **310**, 568—569.

Sofue, Y., Klein, U., Beck, R., and Wielebinski, R.: 1985, 'Large-scale Configuration of Magnetic Fields in Spiral Galaxies', *Astron. Astrophys.* **144**, 257—260.

Sofue, Y., Reich, W., Inoue, M., and Seiradakis, J. H.: 1987, 'The Magnetic Field Near the Galactic Center', *Publ. Astron. Soc. Japan* **39**, 95—108.

Sofue, Y. and Takano, T.: 1981, 'The Configuration of Magnetic Fields in the Spiral Galaxies M31 and M33', *Publ. Astron. Soc. Japan* **33**, 47—55.

Sofue, Y., Takano, T., and Fujimoto, M.: 1980, 'Bisymmetric Open-spiral Configuration of Magnetic Fields in the Galaxies M51 and M81', *Astron. Astrophys.* **91**, 335—340.

Sokoloff, D. D.: 1975, 'Large-scale Magnetic Field and Global Structure of the Universe', *Pisma Astron. J.* (SSSR), **1**, No. 6, 3—5.

Sokoloff, D. D., Shukurov, A. M., and Ruzmaikin, A. A.: 1983, 'Asymptotic Solution of the α^2-dynamo Problem', *Geophys. Astrophys. Fluid Dyn.* **25**, 293–307.
Solomon, P. M., Scoville, N. Z., and Sanders, D. B.: 1979, Giant Molecular Clouds in the Galaxy: The Distribution of ^{13}CO Emission in the Galactic Plane', *Astrophys. J. Lett.* **232**, L89–L93.
Soward, A.: 1978, 'A Thin Disc Model of the Galactic Dynamo', *Astron. Nachr.* **299**, 25–33.
Spangler, S. R.: 1982, 'The Transport of Polarized Synchrotron Radiation in a Turbulent Medium', *Astrophys. J.* **261**, 310–320.
Spangler, S. R.: 1983, 'Determination of the Properties of Magnetic Turbulence in Radio Sources', *Astrophys. J. Lett.* **271**, L49–L53.
Spicker, J. and Feitzinger, J. V.: 1987a, 'Improved Methods for Turbulence in Astrophysics', *Astron. Astrophys.*, in press.
Spicker, J. and Feitzinger, J. V.: 1987b, 'Analysis of Turbulence in the Large Magellanic Cloud: The HI Radial Velocity Field', *Astron. Astrophys.*, in press.
Spitzer, L., Jr.: 1956, 'On a Possible Interstellar Galactic Corona', *Astrophys. J.* **124**, 20–34.
Spitzer, L., Jr.: 1978, *Physical Processes in the Interstellar Medium*, Wiley-Interscience Publ., New York.
Spoelstra, T. A. Th.: 1972, 'A Survey of Linear Polarization at 1415 MHz. IV. Discussion of the Results for the Galactic Spurs', *Astron. Astrophys.* **21**, 61–84.
Spoelstra, T. A. Th.: 1977, 'The Galactic Magnetic Field', *Usp. Fiz. Nauk* **121**, 679–694 (*Sov. Phys. Usp.* **20**, 336–342, 1977).
Steenbeck, M., Kirko, I. M., Gailitis, A., Klawina, A. P., Krause, F., Laumanis, I. J., and Lielausis, O. A.: 1968, 'An Experimental Verification of Electromotive Force along External Magnetic Field that is Produced by a Flow of Liquid Metal (α-effect)', *Dokl. Akad. Nauk SSSR* **180**, 326–329.
Steenbeck, M. and Krause, F.: 1966, 'The Generation of Stellar and Planetary Magnetic Fields by Turbulent Dynamo Action', *Z. Naturforsch.* **21a**, 1285–1296.
Steenbeck, M., Krause, F., and Rädler, K.-H.: 1966, 'Berechnung der mittleren Lorentz-Feldstärke $\mathbf{v} \times \mathbf{B}$ für ein elektrisch leitendes Medium in turbulenter, durch Coriolis-Kräfte beeinflusster Bewegung', *Z. Naturforsch.* **21a**, 369–376.
Stephens, S. A.: 1981, 'An Analysis of the Longitudinal Distribution of Gamma Rays from SAS-II Data', *Astrophys. Space Sci.* **79**, 419–434.
Stix, M.: 1971, 'A Non-Axisymmetric α-effect Dynamo', *Astron. Astrophys.* **13**, 203–208.
Stix, M.: 1975, 'The Galactic Dynamo', *Astron. Astrophys.* **42**, 85–89. Erratum, **68**, 459, 1978.
Strauss, H. R.: 1985, 'The Dynamo Effect in Fusion Plamas', *Phys. Fluids* **28**, 2786–2792.
Strukov, I. A. and Skulachev, I. D.: 1986, 'Investigation of the Background Radiation Anisotropy from Satellite', *Sov. Sci. Rev. E: Astrophys. Space Phys.* **5** (in press).
Sukumar, S. and Velusamy, T.: 1985, 'Radio Continuum Observation of NGC 4631 at 327 MHz', *Monthly Notices Roy. Astron. Soc.* **212**, 367–374.
Syrovatskii, S. I.: 1970, in: *Interstellar Gas Dynamics. Proc. IAU Symp. No. 39* (ed. H. J. Habing), pp. 192–193. D. Reidel, Dordrecht.
Tabara, H., and Inoue, M.: 1980, 'A Catalogue of Linear Polarization of Radio Sources', *Astron. Astrophys. Suppl.* **39**, 379–394.
Takahara, F.: 1979, 'Magnetic Flare Model of Quasars and Active Galactic Nuclei — Magnetized Accretion Disk Around a Massive Black Hole', *Progr. Theor. Phys.* **62**, 629–643.
Tarrab, I.: 1983, 'Estimated Energy and Momentum Input to the Interstellar Medium for Several External Galaxies', *Astron. Astrophys.* **125**, 308–312.
Taylor, J. H. and Manchester, R. N.: 1977, 'Galactic Distribution and Evolution of Pulsars', *Astrophys. J.* **215**, 885–896.
Tchen, C. M.: 1965, 'Spectrum of Stationary Homogeneous Magnetohydrodynamic Turbulence', *Phys. Fluids* **8**, 1659–1667.
Terzian, Y. and Davidson, K.: 1976, 'Pulsars: Observational Parameters and a Discussion on Dispersion Measures', *Astrophys. Space Sci.* **44**, 479–500.
Thorne, K. S.: 1967, 'Primordial Element Formation, Primordial Magnetic Fields, and the Isotropy of the Universe', *Astrophys. J.* **148**, 51–68.

Tikhonov, A. N. and Arsenin, V. Ya.: 1979, *Methods of Solution of Improperly Posed Problems*, Nauka, Moscow (in Russian).

Tosa, M. and Fujimoto, M.: 1978, 'The Configuration of Magnetic Fields in the Spiral Galaxy M51', *Publ. Astron. Soc. Japan* **30**, 315—325.

Troland, T. H. and Heiles, C.: 1977, 'The Zeeman Effect in Radio Frequency Recombination Lines', *Astrophys. J.* **214**, 703—705.

Troland, T. H. and Heiles, C.: 1986, 'Interstellar Magnetic Field Strengths and Gas Densities: Observational and Theoretical Perspectives', *Astrophys. J.* **301**, 339—345.

Tsuboi, M., Inoue, M., Handa, T., Tabara, H., and Kato, T.: 1985, 'Two Highly Polarized Lobes Near the Galactic Center', *Publ. Astron. Soc., Japan* **37**, 359—368.

Tsuboi, M., Inoue, M., Handa, T., Tabara, H., Kato, T., Sofue, Y., and Kaifu, N.: 1986, 'Prominent Polarized Plumes in the Galactic Center Region and Their Magnetic Field', *Astron. J.* **92**, 818—824.

Tully, R. B.: 1974, 'The Kinematics and Dynamics of M51. II. Axisymmetric Properties', *Astrophys. J. Suppl.* **27**, 437—448.

Turner, B. E. and Verschuur, G. L.: 1970, 'An Attempt to Measure the Zeeman Splitting of "Normal" Interstellar OH Lines', *Astrophys. J.* **162**, 341—343.

Uchida, Y., Shibata, K., and Sofue, Y.: 1985, 'Origin of the Galactic Centre Lobes', *Nature* **317**, 699—701.

Unwin, S. C.: 1980a, 'Neutral Hydrogen in the Andromeda Nebula — I. HI Emission in the South-west Region of the Nebula', *Monthly Notices Roy. Astron. Soc.* **190**, 551—574.

Unwin, S. C.: 1980b, 'Neutral Hydrogen in the Andromeda Nebula — II. HI Emission in the North-east Region of the Nebula', *Monthly Notices Roy. Astron. Soc.* **192**, 243—262.

Vainshtein, S. I. and Ruzmaikin, A. A.: 1971, 'Generation of the Large-scale Galactic Magnetic Field. I', *Astron. J. (SSSR)* **48**, 902—909 (*Sov. Astron.* **15**, 714—719, 1972).

Vainshtein, S. I. and Ruzmaikin, A. A.: 1972, 'Generation of the Large-scale Galactic Magnetic Field. II', *Astron. J. (SSSR)* **49**, 449—452. (*Sov. Astron.* **16**, 365—367).

Vainshtein, S. I., Zeldovich, Ya. B., and Ruzmaikin, A. A.: 1980, *Turbulent Dynamo in Astrophysics*, Nauka, Moscow (in Russian).

Vallée, J. P.: 1975, 'Magnetic Field in the Intergalactic Region', *Nature* **254**, 23—26.

Vallée, J. P.: 1980, 'The Rotation Measures of Radio Sources and Their Data Processing', *Astron. Astrophys.* **86**, 251—253.

Vallée, J. P.: 1983a, 'The Rotation Measures of Radio Sources in Selected Celestial Zones. — The Perseus Arm Window', *Astron. Astrophys. Suppl.* **51**, 127—129.

Vallée, J. P.: 1983b, 'Large-scale Magnetic Field in the Perseus Spiral Arm', *Astron. Astrophys.* **124**, 147—150.

Vallée, J. P.: 1983c, 'Intergalactic and Galactic Magnetic Fields. — An Updated Test', *Astrophys. Lett.* **23**, 85—94.

Vallée, J. P.: 1984, 'Effects of Interstellar Magnetic Bubbles and of Galactic Tides on Galactic Magnetic Fields', *Astron. Astrophys.* **136**, 373—377.

Vallée, J. P.: 1986, 'A Possible Relationship between Ring Magnetism and HI Gas in Spiral Galaxies', *Astron. J.* **91**, 541—545.

Vallée, J. P. and Bignell, R. C.: 1983, 'Magnetic Field Structure in the Gum Nebula Area', *Astrophys. J.* **272**, 131—136.

Vallée, J. P., Broten, N. W., and MacLeod, J. M.: 1984, 'Magnetic Field Structure Associated with the Large "Monogem" X-ray Ring', *Astron. Astrophys.* **134**, 199—203.

Vallée, J. P. and Kronberg, P. P.: 1973, 'Magnetic Field in the Galactic Spiral Arm', *Nature Phys. Sci.* **246**, 49—51.

Vallée, J. P. and Kronberg, P. P.: 1975, 'The Rotation Measures of Radio Sources and Their Interpretation', *Astron. Astrophys.* **43**, 233—242.

Vallée, J. P., MacLeod, J. M., and Broten, N. W.: 1986, 'A Large-Scale Magnetic Feature in the Galaxy Cluster A2319', *Astron. Astrophys.* **156**, 386—390.

Varshalovich, D. A. and Burdjuzha, V. V.: 1975, 'Polarization of Cosmic Maser OH Emission', *Astron. J. (SSSR)* **52**, 1178—1186. (*Sov. Astron.* **19**, 705—710, 1975).

Vaucouleurs, A. de, Vaucouleurs, G. de, and Corvin, H. G.: 1976, *Second Reference Catalogue of Bright Galaxies*, Austin and London.

Vaucouleurs, G. de: 1974, 'Structure, Dynamics and Statistical Properties of Galaxies'. In: *The Formation and Dynamics of Galaxies* (ed. J. R. Shakeshaft), pp. 1—54. D. Reidel, Dordrecht.

Verschuur, G. L.: 1968, 'Positive Determination of an Interstellar Magnetic Field by Measurement of the Zeeman Splitting of the 21-cm Hydrogen Line', *Phys. Rev. Lett.* **21**, 775—778.

Verschuur, G. L.: 1970, 'Observational Aspects of Galactic Magnetic Fields'. In: *Interstellar Gas Dynamics. Proc. IAU Symp. No. 39* (ed. H. J. Habing), pp. 150—167. D. Reidel, Dordrecht.

Verschuur, G. L.: 1971, 'Recent Measurements of the Zeeman Effect at 21-centimeter Wave-length', *Astrophys. J.* **165**, 651—653.

Verschuur, G. L.: 1979, 'Observations of the Galactic Magnetic Field', *Fund. Cosm. Phys.* **5**, 113—191.

Vilenkin, A. and Leahy, D. A.: 1982, 'Parity Nonconservation and the Origin of Cosmic Magnetic Fields', *Astrophys. J.* **254**, 77—81.

Vorontsov-Velyaminov, B. A.: 1978, *Extragalactic Astronomy*, Nauka, Moscow (in Russian).

Voschinnikov, N. V.: 1986, 'The Interstellar Dust', *Space Research*, vol. 25 pp. 98—202 (*Itogi Nauki i Tekhniki, VINITI AS SSSR*), Moscow (in Russian).

Voshchinnikov, N., Il'in, A. E., and Il'in, V. B.: 1986, 'Absorption and Polarization of Light by Dust in the Interstellar Medium: The Interstellar Linear Polarization', *Astrofizika* **24**, 523—536.

Voshchinnikov, N. V. and Marchenko, P. E.: 1982, 'Polarization of the Starlight in *R*-associations. Observational Data', *Astron. J.* (*SSSR*) **59**, 1115—1123.

Wannier, P. G., Scoville, N. Z., and Barvainis, R.: 1983, 'The Polarization of Millimeter-wave Emission Lines in Dense Interstellar Clouds', *Astrophys. J.* **267**, 126—136.

Warner, P. J., Wright, M. C. H., and Baldwin, J. E.: 1973, 'High Resolution Observations of Neutral Hydrogen in M33-II. The Velocity Field', *Monthly Notices Roy. Astron. Soc.* **163**, 182.

Wasserman, I.: 1978, 'On the Origin of Galaxies, Galactic Angular Momenta, and Galactic Magnetic Fields', *Astrophys. J.* **224**, 337—343.

Webber, W. R.: 1983, 'Cosmic Ray Electrons and Positrons — A Review of Current Measurements and Some Interpretations'. In: *Composition and Origin of Cosmic Rays* (ed. M. M. Shapiro), pp. 83—100. D. Reidel, Dordrecht.

Webber, W. R., Simpson, G. A., and Cane, H. V.: 1980, 'Radio Emission, Cosmic Ray Electrons, and the Production of γ-rays in the Galaxy', *Astrophys. J.* **236**, 448—459.

Webster, A.: 1978, 'The Shape and Size of the Radio Halo of the Galaxy', *Monthly Notices Roy. Astron. Soc.* **185**, 507—520.

Weiler, K. W. and de Pater, I.: 1983, 'A Catalog of High Accuracy Circular Polarization Measurements', *Astrophys. J. Suppl.* **52**, 293—327.

Weinberg, S.: 1972, *Gravitation and Cosmology: Principles and Applications of the General Theory of Relativity*, John Wiley and Sons, New York-London-Syndey-Toronto.

Weisheit, J. S.: 1978, 'On the Use of Line Shapes in the Analysis of QSO Absorption Spectra', *Astrophys. J.* **219**, 829—835.

Weiss, N. O.: 1966, 'The Expulsion of Magnetic Flux by Eddies', *Proc. Roy. Soc.* **A293**, 310—328.

Weizsäcker, C. F. von: 1951, 'The Evolution of Galaxies and Stars', *Astrophys. J.* **114**, 165—186.

Welter, G. L., Perry, J. J., and Kronberg, P. P.: 1984, 'The Rotation Measure Distribution of QSOs and of Intervening Clouds: Magnetic Fields and Column Densities', *Astrophys. J.* **279**, 19—39.

Werner, W.: 1984, Ph.D. Thesis, Rheinische Friedrich-Wilhelms-Universität, Bonn.

Werner, W.: 1985, 'A Dynamical Halo Around the Edge-on Galaxy NGC 4631', *Astron. Astrophys.* **141**, 502—505.

Westerhout, G.: 1957, 'The Distribution of Atomic Hydrogen in the Outer Parts of the Galactic System', *Bull. Astron. Inst. Netherlands* **13**, No. 475, 201—246.

Westerhout, G., Seeger, Ch. L., Brouw, W. N., and Tinbergen, J.: 1962, 'Polarization of the Galactic 75-cm Radiation', *Bull. Astron. Inst. Netherlands* **16**, 187—212.

Western, L. R. and Watson, W. D.: 1984, 'Linear Polarization of Astronomical Masers and Magnetic Fields', *Astrophys. J.* **285**, 158—173.

White, M. P.: 1978, 'Numerical Models of the Galactic Dynamo', *Astron. Nachr.* **299**, 209—216.

Wielebinski, R. and von Kap-herr, A.: 1977, 'The Radiocontinuum of NGC4631 at High Radiofrequencies', *Astron. Astrophys.* **59**, L17—L18.
Wielebinski, R., Shakeshaft, J. R., and Pauliny-Toth, I. I. K.: 1962, 'A Search for a Linearly Polarized Component of the Galactic Radio Emission at 408 Mc/s', *Observatory* **82**, 158—164.
Wilkinson, J. H. and Reinsch, C.: 1971, *Handbook for Automatic Computation. Linear Algebra*, Springer Verlag, Heidelberg-New York-Berlin.
Woerden, H. van: 1979, 'Warping and Thickness of Galactic Gas Layer'. In: *The Large-Scale Characteristics of the Galaxy, Proc. IAU Symp. No. 84* (ed. W. B. Burton), pp. 501–502. D. Reidel Dordrecht.
Woltjer, L.: 1972, 'Supernova Remnants', *Ann. Rev. Astron. Astrophys.* **10**, 129—158.
York, D. G.: 1977, 'On the Temperature and the Interstellar Nature of Coronal Gas Observed by "Copernicus"', *Astrophys. J.* **213**, 43—51.
York, D. G.: 1982, 'Gas in the Galactic Halo', *Ann. Rev. Astron. Astrophys.* **20**, 221—248.
York, D. G., Blades, J. C., Cowie, L. L., Morton, D. G., Songaila, A., and Wu, C.-C.: 1982, 'The Gaseous Galactic Halo as Interred from the Line Spectra of the Galaxies Markarian 509 and Fairall 9', *Astrophys. J.* **255**, 467—474.
Young, D. S. de: 1980, 'Turbulent Generation of Magnetic Fields in Extended Extragalactic Radio Sources', *Astrophys. J.* **241**, 81—97.
Young, D. S. de: 1984, 'The Physics of Extragalactic Radio Sources', *Phys. Rep.* **111**, 373—408.
Yuan, C.: 1969, 'Application of the Density-wave Theory to the Spiral Structure of the Milky Way System. II. Migration of Stars', *Astrophys. J.* **158**, 889—898.
Yusef-Zadeh, F., Morris, M., and Chance, D.: 1984, 'Large, Highly Organized Radiostructure Near the Galactic Centre', *Nature* **310**, 557—561.
Zakharov, V. E. and Sagdeev, R. Z.: 1970, 'On the Spectrum of Acoustic Turbulence', *Dokl. Akad. Nauk SSSR* **192**, 297—300.
Zasov, A. V. and Kyazumov, G. A.: 1983, 'Rotation Curves of Normal Galaxies', *Astron. J.* (*SSSR*) **60**, 656—665.
Zeldovich, Ya. B.: 1956, 'The Magnetic Field in the Two-dimensional Motion of a Conducting Turbulent Liquid', *JETP* **31**, 154—156 (*Sov. Phys. JETP* **4**, 460—462, 1957).
Zeldovich, Ya. B.: 1965, 'A Magnetic Model of the Universe', *JETP* **48**, 986—988 (*Sov. Phys. JETP* **21**, 656—659, 1965).
Zeldovich, Ya. B.: 1969, 'A Hypothesis of the Magnetic Cosmological Inhomogeneity', *Astron. J.* (*SSSR*) **46**, 775—778.
Zeldovich, Ya. B.: 1981, 'On the Friction of Fluids between Rotating Cylinders', *Proc. Roy. Soc. London* **A374**, 299—312.
Zeldovich, Ya. B., Molchanov, S. A., Ruzmaikin, A. A., and Sokoloff, D. D.: 1985a, 'Transport Intermittency in Random Media', *JETP* **89**, 2061—2072.
Zeldovich, Ya. B., Ruzmaikin, A. A., and Sokoloff, D. D.: 1985b, 'On the Representation of a Three-dimensional Vector Field by Scalar Potentials', *Dokl. Akad. Nauk. SSSR* **284**, 103—106.
Zeldovich, Ya. B., Molchanov, S. A., Ruzmaikin, A. A., and Sokoloff, D. D.: 1987, 'Intermittency in a Random Medium', *Usp. Fiz. Nauk* **152**, 3—32.
Zeldovich, Ya. B. and Novikov, I. D.: 1975, *The Structure and Evolution of the Universe*, Nauka, Moscow (English translation: Univ. Press, Chicago, 1982).
Zeldovich, Ya. B. and Ruzmaikin, A. A.: 1980, 'The Magnetic Field in Conducting Fluid in Two-dimensional Motion', *JETP* **78**, 980—986 (*Sov. Phys. JETP* **51**, 493—497, 1980).
Zeldovich, Ya. B., Ruzmaikin, A. A., and Sokoloff, D. D.: 1983, *Magnetic Fields in Astrophysics*, Gordon and Breach, New York.
Zeldovich, Ya. B. and Sokoloff, D. D.: 1985, 'Fractals, Self-similarity, Intermediate Asymptotics', *Usp. Fiz. Nauk* **146**, 493—506.

INDEX

GPSR Compliance

The European Union's (EU) General Product Safety Regulation (GPSR) is a set of rules that requires consumer products to be safe and our obligations to ensure this.

If you have any concerns about our products, you can contact us on ProductSafety@springernature.com

In case Publisher is established outside the EU, the EU authorized representative is:

Springer Nature Customer Service Center GmbH
Europaplatz 3
69115 Heidelberg, Germany

Batch number: 10815974

Printed by Printforce, the Netherlands